# High Performance Computing Symposium 2010



# (HPCS 2010)

Journal of Physics: Conference Series Volume 256

## Toronto, Ontario, Canada
## 5 – 9 June 2010

ISBN: 978-1-61782-246-9
ISSN: 1742-6588

**Printed from e-media with permission by:**

Curran Associates, Inc.
57 Morehouse Lane
Red Hook, NY 12571

**Some format issues inherent in the e-media version may also appear in this print version.**



Printed by Curran Associates, Inc. (2011)



Institute of Physics
Dirac House, Temple Back
Bristol BS1 6BE UK

Phone: 44 1 17 929 7481
Fax: 44 1 17 920 0979

techtracking@iop.org

**Additional copies of this publication are available from:**

Curran Associates, Inc.
57 Morehouse Lane
Red Hook, NY 12571 USA
Phone: 845-758-0400
Fax: 845-758-2634
Email: curran@proceedings.com
Web: www.proceedings.com

# TABLE OF CONTENTS

# PREFACE

Welcome to the Proceedings for HPCS2010!

HPCS (High Performance Computing Symposium) is Canada's foremost HPC conference – a multidisciplinary conference where computational researchers from all disciplines in industry and academia, computer scientists, and vendors exchange new tools, techniques and interesting results in and for HPC computational research. HPCS2010, the 24th in the series, took place at the University of Toronto, 5–9 June.

We are very pleased with the range and depth of excellent work that was presented at this year's High Performance Computing Symposium. The work that follows is a collection of papers presented by authors from institutions in five countries, covering topics ranging from squeezing every last cycle out of a single GPU to the design of an entire data centre, and almost everything in between.

We particularly encourage you to look through the 'Award Winners' section, where works are presented by junior and student researchers which were awarded prizes for being amongst the best work presented at HPCS2010.

Enjoy the excellent research that follows, and we look forward to seeing you in Montreal for HPCS2011!

**Jonathan Dursi**, for all of members of the Scientific Organizing Committee of HPCS2010.

Journal of Physics: Conference Series **256** (2010) 012001 doi:10.1088/1742-6596/256/1/012001

# Primal and dual-primal iterative substructuring methods of stochastic PDEs

**Waad Subber[1] and Abhijit Sarkar[2]**

Department of Civil and Environmental Engineering, Carleton University, Ottawa, Ontario K1S5B6, Canada

E-mail: [1]wsubber@connect.carleton.ca, [2]abhijit_sarkar@carleton.ca

**Abstract.** A novel non-overlapping domain decomposition method is proposed to solve the large-scale linear system arising from the finite element discretization of stochastic partial differential equations (SPDEs). The methodology is based on a Schur complement based geometric decomposition and an orthogonal decomposition and projection of the stochastic processes using Polynomial Chaos expansion. The algorithm offers a direct approach to formulate a two-level scalable preconditioner. The proposed preconditioner strictly enforces the continuity condition on the corner nodes of the interface boundary, while weakly satisfying the continuity condition over the remaining interface nodes. This approach relates to a primal version of an iterative substructuring method. Next, a Lagrange multiplier based dual-primal domain decomposition method is introduced in the context of SPDEs. In the dual-primal method the continuity condition on the corner nodes is strictly satisfied while Lagrange multipliers are used to enforce continuity on the remaining part of the interface boundary. For numerical illustrations, a two dimensional elliptic SPDE with non-Gaussian random coefficients is considered. The numerical results demonstrate the scalability of these algorithms with respect to the mesh size, subdomain size, fixed problem size per subdomain, order of Polynomial Chaos expansion and level of uncertainty in the input parameters. The numerical experiments are performed on a Linux cluster using MPI and PETSc libraries.

## 1. Introduction

A domain decomposition method of SPDEs is introduced in [1, 2] to quantify uncertainty in large-scale linear systems. The methodology is based on a Schur complement based geometric decomposition and an orthogonal decomposition and projection of the stochastic processes. A parallel preconditioned conjugate gradient method (PCGM) is adopted in [3] to solve the interface problem without explicitly constructing the Schur complement system. The parallel performance of the algorithms is demonstrated using a lumped preconditioner for non-Gaussian systems arising from a hydraulic problem having random soil permeability properties.

A one-level Neumann-Neumann domain decomposition preconditioner for SPDEs is introduced in [4] in order to enhance the performance of the parallel PCGM iterative solver in [3]. The implementation of the algorithm requires a local solve of a stochastic Dirichlet problem followed by a local solve of a stochastic Neumann problem in each iteration of the PCGM solver. The multilevel sparsity structure of the coefficient matrices of the stochastic system, namely (a) the sparsity structure due to the finite element discretization and (b) the block sparsity structure due to the Polynomial Chaos expansion is exploited for computational efficiency. The one-level

Neumann-Neumann preconditioner in [4] demonstrates a good (strong and weak) scalability for the moderate range of CPUs considered.

In this paper we first describe a primal version of iterative substructuring methods for the solution of the large-scale linear system arising from stochastic finite element method. The algorithm offers a straightforward approach to formulate a two-level scalable preconditioner. The continuity condition is strictly enforced on the corner nodes (nodes shared among more than two subdomains including the nodes at the ends of the interface edges). For the remaining part of the interface boundary, the continuity condition is satisfied weakly (in an average sense). Note that the continuity of the solution field across the entire interface boundary is eventually satisfied at the convergence of the iterations. This approach naturally leads to a *coarse grid* which connects the subdomains globally via the corner nodes. The coarse grid provides a mechanism to propagate information globally which makes the algorithm scalable with respect to subdomain size. In the second part of the paper, a dual-primal iterative substructuring method is introduced for SPDEs, which maybe viewed as an extension of the Dual-Primal Finite Element Tearing and Interconnecting method (FETI-DP) [5] in the context of SPDEs. In this approach, the continuity condition on the corner nodes is strictly satisfied by *partial assembly* while Lagrange multipliers are used to enforce continuity on the remaining part of the interface boundary. A system of Lagrange multiplier (also called the *dual variable*) is solved iteratively using PCGM method equipped with Dirichlet preconditioner. PETSc [6] and MPI [7] libraries are used for efficient parallel implementation of the primal and dual-primal algorithms. The graph partitioning tool METIS [8] is used for optimal decomposition of the finite element mesh for load balancing and minimum interprocessor communication. The parallel performance of the algorithms is studied for a two dimensional stochastic elliptic PDE with non-Gaussian random coefficients. The numerical experiments are performed using a Linux cluster.

## 2. Uncertainty representation by stochastic processes

This section provides a brief review of the theories of stochastic processes relevant to the subsequent developments of the paper [9, 1]. We assume the data induces a representation of the model parameters as random variables and processes which span the Hilbert space $\mathcal{H}_G$. A set of basis functions $\{\xi_i\}$ is identified to characterize this space using Karhunen-Loeve expansion. The state of the system resides in the Hilbert space $\mathcal{H}_L$ with basis functions $\{\Psi_i\}$ being identified with the Polynomial Chaos expansion (PC). The Karhunen-Loeve expansion of a stochastic process $\alpha(x,\theta)$ is based on the spectral expansion of its covariance function $R_{\alpha\alpha}(x,y)$. The expansion takes the following form

$$\alpha(x,\theta) = \bar{\alpha}(x) + \sum_{i=1}^{\infty} \sqrt{\lambda_i}\xi_i(\theta)\phi_i(x),$$

where $\bar{\alpha}(x)$ is the mean of the stochastic process, $\theta$ represents the random dimension, and $\xi_i(\theta)$ is a set of uncorrelated (but not generally independent for non-Gaussian processes) random variables, $\phi_i(x)$ are the eigenfunctions and $\lambda_i$ are the eigenvalues of the covariance kernel which can be obtained as the solution to the following integral equation

$$\int_{\mathcal{D}} R_{\alpha\alpha}(x,y)\phi_i(y)dy = \lambda_i\phi_i(x).$$

where $\mathcal{D}$ denotes the spatial dimension over which the process is defined. The covariance function of the solution process is not known a priori, and hence the Karhunen-Loeve expansion cannot be used to represent it. Therefore, a generic basis, that is complete in the space of all second-order random variables will be identified and used in the approximation process. Since the solution process is a function of the material properties, nodal solution variables, denoted by $u(\theta)$ can be formally expressed as some nonlinear functional of the set $\xi_i(\theta)$ used to represent the material stochasticity. It has been shown that this functional dependence can be expanded in terms of polynomials in Gaussian random variables, namely Polynomial Chaos [9] as

$$\mathbf{u}(\theta) = \sum_{j=0}^{N} \Psi_j(\theta)\mathbf{u}_j.$$

These polynomials are orthogonal in the sense that their inner product $\langle \Psi_j \Psi_k \rangle$, defined as the statistical average of their product, is equal to zero for $j \neq k$.

## 3. Review of Schur complement based domain decomposition method of SPDEs

Consider an elliptic stochastic PDE defined on a domain $\Omega$ with a given boundary conditions on $\partial\Omega$. Finite element discretization of the stochastic PDE leads to the following linear system

$$\mathbf{A}(\theta)\mathbf{u}(\theta) = \mathbf{f}, \tag{1}$$

where $\mathbf{A}(\theta)$ is the stiffness matrix with random coefficients, $\mathbf{u}(\theta)$ is the stochastic process representing the response vector and $\mathbf{f}$ is the applied force. For large-scale system, Eq.(1) can be solved efficiently using domain decomposition method [1, 2].

In domain decomposition method, the spatial domain $\Omega$ is partitioned into $n_s$ non-overlapping subdomains $\{\Omega_s, 1 \leq s \leq n_s\}$ such that

$$\Omega = \bigcup_{s=1}^{n_s} \Omega_s, \quad \Omega_s \bigcap \Omega_r = 0, \quad s \neq r$$

and

$$\Gamma = \bigcup_{s=1} \Gamma_s \text{ where } \Gamma_s = \partial\Omega_s \backslash \partial\Omega$$

For a typical subdomain $\Omega_s$ the nodal vector $\mathbf{u}^s(\theta)$ is partitioned into a set of interior unknowns $\mathbf{u}_I^s(\theta)$ associated with nodes in the interior of $\Omega_s$ and interface unknowns $\mathbf{u}_\Gamma^s(\theta)$ associated with nodes that are shared among two or more subdomains, as shown in Fig.(1).

Consequently, the subdomain equilibrium equation can be represented as

$$\begin{bmatrix} \mathbf{A}_{II}^s(\theta) & \mathbf{A}_{I\Gamma}^s(\theta) \\ \mathbf{A}_{\Gamma I}^s(\theta) & \mathbf{A}_{\Gamma\Gamma}^s(\theta) \end{bmatrix} \left\{ \begin{array}{c} \mathbf{u}_I^s(\theta) \\ \mathbf{u}_\Gamma^s(\theta) \end{array} \right\} = \left\{ \begin{array}{c} \mathbf{f}_I^s \\ \mathbf{f}_\Gamma^s \end{array} \right\}.$$

The Polynomial Chaos expansion can be used to represent the uncertainty in the model parameters as

$$\sum_{i=0}^{L} \Psi_i \begin{bmatrix} \mathbf{A}_{II,i}^s & \mathbf{A}_{I\Gamma,i}^s \\ \mathbf{A}_{\Gamma I,i}^s & \mathbf{A}_{\Gamma\Gamma,i}^s \end{bmatrix} \left\{ \begin{array}{c} \mathbf{u}_I^s(\theta) \\ \mathbf{u}_\Gamma^s(\theta) \end{array} \right\} = \left\{ \begin{array}{c} \mathbf{f}_I^s \\ \mathbf{f}_\Gamma^s \end{array} \right\}.$$

A Boolean restriction operator $\mathbf{R}_s$ of size $(n_\Gamma^s \times n_\Gamma)$ which maps the global interface vector $\mathbf{u}_\Gamma(\theta)$ to the local interface vector $\mathbf{u}_\Gamma^s(\theta)$ is defined as

$$\mathbf{u}_\Gamma^s(\theta) = \mathbf{R}_s \mathbf{u}_\Gamma(\theta).$$

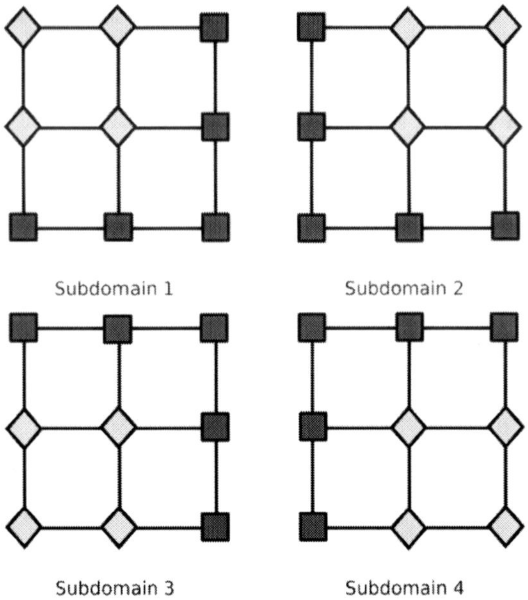


**Figure 1.** Partitioning domain nodes into: interior (◆) and interface (■)

Enforcing the transmission conditions (compatibility and equilibrium) along the interfaces, the global equilibrium equation of the stochastic system can be expressed in the following block linear systems of equations:

$$\sum_{i=0}^{L} \Psi_i \begin{bmatrix} \mathbf{A}_{II,i}^{1} & \dots & 0 & \mathbf{A}_{I\Gamma,i}^{1}\mathbf{R}_1 \\ \vdots & \ddots & \vdots & \vdots \\ 0 & \dots & \mathbf{A}_{II,i}^{n_s} & \mathbf{A}_{I\Gamma,i}^{n_s}\mathbf{R}_{n_s} \\ \mathbf{R}_1^T\mathbf{A}_{\Gamma I,i}^{1} & \dots & \mathbf{R}_{n_s}^T\mathbf{A}_{\Gamma I,i}^{n_s} & \sum_{s=1}^{n_s}\mathbf{R}_s^T\mathbf{A}_{\Gamma\Gamma,i}^{s}\mathbf{R}_s \end{bmatrix} \begin{Bmatrix} \mathbf{u}_I^1(\theta) \\ \vdots \\ \mathbf{u}_I^{n_s}(\theta) \\ \mathbf{u}_\Gamma(\theta) \end{Bmatrix} = \begin{Bmatrix} \mathbf{f}_I^1 \\ \vdots \\ \mathbf{f}_I^{n_s} \\ \sum_{s=1}^{n_s}\mathbf{R}_s^T\mathbf{f}_\Gamma^s \end{Bmatrix}. \quad (2)$$

The solution process can be expanded using the same Polynomial Chaos basis as

$$\begin{Bmatrix} \mathbf{u}_I^1(\theta) \\ \vdots \\ \mathbf{u}_I^{n_s}(\theta) \\ \mathbf{u}_\Gamma(\theta) \end{Bmatrix} = \sum_{j=0}^{N} \Psi_j(\theta) \begin{Bmatrix} \mathbf{u}_{I,j}^1 \\ \vdots \\ \mathbf{u}_{I,j}^{n_s} \\ \mathbf{u}_{\Gamma,j} \end{Bmatrix}. \quad (3)$$

Substituting Eq.(3) into Eq.(2) and performing Galerkin projection to minimize the error over the space spanned by the Polynomial Chaos basis [1], the following coupled deterministic systems of equations is obtained

$$\begin{bmatrix} \mathcal{A}_{II}^{1} & \dots & 0 & \mathcal{A}_{I\Gamma}^{1}\mathcal{R}_1 \\ \vdots & \ddots & \vdots & \vdots \\ 0 & \dots & \mathcal{A}_{II}^{n_s} & \mathcal{A}_{I\Gamma}^{n_s}\mathcal{R}_{n_s} \\ \mathcal{R}_1^T\mathcal{A}_{\Gamma I}^{1} & \dots & \mathcal{R}_{n_s}^T\mathcal{A}_{\Gamma I}^{n_s} & \sum_{s=1}^{n_s}\mathcal{R}_s^T\mathcal{A}_{\Gamma\Gamma}^{s}\mathcal{R}_s \end{bmatrix} \begin{Bmatrix} \mathcal{U}_I^1 \\ \vdots \\ \mathcal{U}_I^{n_s} \\ \mathcal{U}_\Gamma \end{Bmatrix} = \begin{Bmatrix} \mathcal{F}_I^1 \\ \vdots \\ \mathcal{F}_I^{n_s} \\ \sum_{s=1}^{n_s}\mathcal{R}_s^T\mathcal{F}_\Gamma^s \end{Bmatrix}, \quad (4)$$

where

$$[\mathcal{A}^s_{\alpha\beta}]_{jk} = \sum_{i=0}^{L} \langle \Psi_i \Psi_j \Psi_k \rangle \mathbf{A}^s_{\alpha\beta,i} \quad , \quad \mathcal{F}^s_{\alpha,k} = \langle \Psi_k \mathbf{f}^s_\alpha \rangle,$$
$$\mathcal{U}^m_I = (\mathbf{u}^m_{I,0}, \ldots, \mathbf{u}^m_{I,N})^T, \;\; \mathcal{U}_\Gamma = (\mathbf{u}_{\Gamma,0}, \ldots, \mathbf{u}_{\Gamma,N})^T,$$

the subscripts $\alpha$ and $\beta$ represent the index $I$ and $\Gamma$. The coefficient matrix in Eq.(4) is of order $n(N+1) \times n(N+1)$ where $n$ and $(N+1)$ denote the total number of the degrees of freedom and chaos coefficients respectively. The stochastic counterpart of the restriction operator in Eq.(4) takes the following form

$$\mathcal{R}_s = blockdiag(\mathbf{R}^0_s, \ldots, \mathbf{R}^N_s),$$

where $(\mathbf{R}^0_s, \ldots, \mathbf{R}^N_s)$ are the deterministic restriction operators. In parallel implementation $\mathcal{R}_s$ acts as a *scatter* operator while $\mathcal{R}^T_s$ acts as a *gather* operator and are not constructed explicitly.

A block Gaussian elimination reduces the system in Eq.(4) to the following *extended Schur complement system* for the interface variable $\mathcal{U}_\Gamma$:

$$\mathcal{S}\, \mathcal{U}_\Gamma = \mathcal{G}_\Gamma, \tag{5}$$

where the global extended Schur complement matrix $\mathcal{S}$ is given by

$$\mathcal{S} = \sum_{s=1}^{n_s} \mathcal{R}^T_s [\mathcal{A}^s_{\Gamma\Gamma} - \mathcal{A}^s_{\Gamma I} (\mathcal{A}^s_{II})^{-1} \mathcal{A}^s_{I\Gamma}] \mathcal{R}_s,$$

and the corresponding right hand vector $\mathcal{G}_\Gamma$ is

$$\mathcal{G}_\Gamma = \sum_{s=1}^{n_s} \mathcal{R}^T_s [\mathcal{F}^s_\Gamma - \mathcal{A}^s_{\Gamma I} (\mathcal{A}^s_{II})^{-1} \mathcal{F}^s_I].$$

Once the interface unknowns $\mathcal{U}_\Gamma$ is available, the interior unknowns can be obtained concurrently by solving the interior problem on each subdomain as

$$\mathcal{A}^s_{II}\, \mathcal{U}^s_I = \mathcal{F}^s_I - \mathcal{A}^s_{\Gamma I} \mathcal{R}_s \mathcal{U}_\Gamma.$$

## 4. Solution methods for the extended Schur complement system

Solution methods for linear systems are broadly categorized into direct methods and iterative methods. The direct methods, generally, are based on sparse Gaussian elimination technique and are popular for their robustness. However, they are expensive in computation time and memory requirements and therefore cannot be applied to the solution of large-scale linear systems [10]. On the other hand, the iterative methods generate a sequences of approximate solutions which converge to the true solutions. In the iterative methods the main arithmetic operation is the matrix-vector multiplication. Therefore the linear system itself need not be constructed explicitly and only a procedure for matrix-vector product is required. This property makes iterative methods more suitable to parallel processing than direct methods.

### *4.1. Preconditioned Conjugate Gradient Method (PCGM)*

Non-overlapping domain decomposition method or *iterative substructuring* can be viewed as a preconditioned iterative method to solve the Schur complement system of the form [11]

$$\mathcal{S}\,\mathcal{U}_\Gamma = \mathcal{G}_\Gamma.$$

For symmetric positive-definite system such as Schur complement system, the Conjugate Gradient Method (CGM) is generally used. The performance of CGM mainly depends on the spectrum of the coefficient matrix. However, the rate of convergence of the iterative method can generally be improved by transforming the original system into an equivalent system that has better spectral properties (i.e. lower condition number $\kappa(\mathcal{S})$) of the coefficient matrix. This transformation is called preconditioning and the matrix used in the transformation is called the preconditioner. In other words, the transformed linear system becomes:

$$\mathcal{M}^{-1}\mathcal{S}\,\mathcal{U}_\Gamma = \mathcal{M}^{-1}\mathcal{G}_\Gamma,$$

In general, $\kappa(\mathcal{M}^{-1}\mathcal{S})$ is much smaller than $\kappa(\mathcal{S})$ and the eigenvalues of $\mathcal{M}^{-1}\mathcal{S}$ are clustered near one. This procedure known as Preconditioned Conjugate Gradient Method (PCGM). In practice, the explicit construction of $\mathcal{M}^{-1}$ is not needed. Instead, for a given vector $r_\Gamma$, a system of the the following form is solved

$$\mathcal{M}\mathcal{Z} = r_\Gamma.$$

The PCGM algorithm to solve the Schur complement system proceeds as follows [10]

**Algorithm** 1: The PCGM Algorithm

1. *Initialize* $\mathcal{U}_{\Gamma_0} := 0$
2. *Compute* $r_{\Gamma_0} := \mathcal{G}_\Gamma - \mathcal{S}\,\mathcal{U}_{\Gamma_0}$
3. *Precondition* $\mathcal{Z}_0 := \mathcal{M}^{-1} r_{\Gamma_0}$
4. *First search direction* $\mathcal{P}_0 := \mathcal{Z}_0$
5. *Initialize* $\rho_0 := (r_{\Gamma_0}, \mathcal{Z}_0)$
6. *For* $j = 0, 1, \cdots,$ *until convergence Do:*
7. $\quad \mathcal{Q}_j := \mathcal{S}\mathcal{P}_j$
8. $\quad \rho_{tmp_j} := (\mathcal{Q}_j, \mathcal{P}_j)$
9. $\quad \alpha_j := \rho_j / \rho_{tmp_j}$
10. $\quad \mathcal{U}_{\Gamma_{j+1}} := \mathcal{U}_{\Gamma_j} + \alpha_j \mathcal{P}_j$
11. $\quad r_{\Gamma_{j+1}} := r_{\Gamma_j} - \alpha_j \mathcal{Q}_j$
12. $\quad \mathcal{Z}_{j+1} = \mathcal{M}^{-1} r_{\Gamma_{j+1}}$
13. $\quad \rho_{j+1} := (r_{\Gamma_{j+1}}, \mathcal{Z}_{j+1})$
14. $\quad \beta_j := \rho_{j+1} / \rho_j$
15. $\quad \mathcal{P}_{j+1} = \mathcal{Z}_{j+1} + \beta_j \mathcal{P}_j$
16. *EndDo*

The PCGM algorithm indicates that the main arithmetic operations are calculating the product $\mathcal{Q} = \mathcal{S}\mathcal{P}$ in step 7 and the preconditioned residual $\mathcal{Z} = \mathcal{M}^{-1} r_\Gamma$ in step 12. These operations can be performed in parallel as outlined next.

Given the subdomain Schur complement matrices $\mathcal{S}_s$ and a global vector $\mathcal{P}$, the matrix-vector product $\mathcal{Q} = \mathcal{S}\mathcal{P}$ can be calculated in parallel as

$$\mathcal{Q} = \sum_{s=1}^{n_s} \mathcal{R}_s^T \mathcal{S}_s \mathcal{R}_s \mathcal{P},$$

where $n_s$ is the number of subdomains and $\mathcal{R}_s$ and $\mathcal{R}_s^T$ are scatter and gather operator respectively. The parallel implementation of this procedure is summarized in Algorithm (2).

**Algorithm** 2: Parallel Matrix-Vector Product Procedure

1. *Input* ($\mathcal{P}$)
2. *Scatter* : $\mathcal{P}^s = \mathcal{R}_s\mathcal{P}$
3. *Local operation* : $\mathcal{Q}^s = \mathcal{S}_s\mathcal{P}^s$
4. *Gather* : $\mathcal{Q} = \sum_{s=1}^{n_s} \mathcal{R}_s^T \mathcal{Q}^s$
5. *Output* ($\mathcal{Q}$)

The working vectors $\mathcal{P}^s$ and $\mathcal{Q}^s$ are defined on the subdomain level.

Similarly, the effect of a parallel preconditioner on a residual vector $\mathcal{Z} = \mathcal{M}^{-1} r_\Gamma$ can be computed as

$$\mathcal{Z} = \sum_{s=1}^{n_s} \mathcal{R}_s^T \mathcal{M}_s^{-1} \mathcal{R}_s r_\Gamma.$$

This procedure is outlined in following algorithm.

**Algorithm** 3: Parallel Preconditioner Effect Procedure

1. *Input* ($r_\Gamma$)
2. *Scatter* : $r_\Gamma^s = \mathcal{R}_s r_\Gamma$
3. *Local Solve* : $\mathcal{M}_s \mathcal{Z}^s = r_\Gamma^s$
4. *Gather* : $\mathcal{Z} = \sum_{s=1}^{n_s} \mathcal{R}_s^T \mathcal{Z}^s$
5. *Output* ($\mathcal{Z}$)

The local preconditioner $\mathcal{M}_s$ and the working vectors $r_\Gamma^s$ and $\mathcal{Z}^s$ are defined on the subdomain level.

## 5. Iterative substructuring methods of SPDEs

Next sections describe the primal and dual-primal substructuring methods in the context of SPDEs. In the primal method, the interface problem is solved iteratively using PCGM solver equipped with a scalable preconditioner. At each iteration of the iterative solver loop, local problems are solved on each subdomain in parallel. These local problems are used to construct a subdomain level precondtioner. Moreover, a coarse problem is required to propagate information globally across the subdomains. This global exchange of information leads to a scalable preconditioner. In the dual-primal method, a system of Lagrange multiplier that enforces continuity constraints across the interface boundary, is solved iteratively using PCGM solver. The global coarse problem is already embedded in the operator of the Lagrange multiplier system and therefore a one-level preconditioner such as lumped or Dirichlet is sufficient for scalability. A framework of the primal and dual-primal iterative substructuring methods for SPDEs is detailed next.

## 6. A primal iterative substructuring method of SPDEs

In order to define local problems over each of the subdomains, we partition the subdomain nodal vector $\mathbf{u}^s(\theta)$ into a set of interior unknowns $\mathbf{u}_i^s(\theta)$, corner unknowns $\mathbf{u}_c^s(\theta)$ and remaining unknowns $\mathbf{u}_r^s(\theta)$, as schematically shown in Fig.(2).

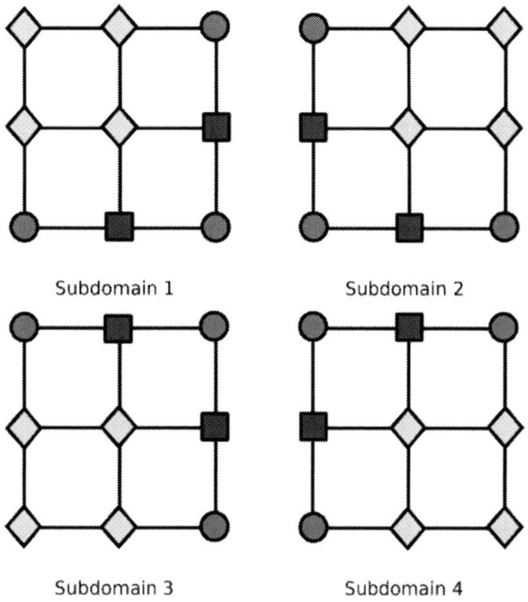


**Figure 2.** Partitioning domain nodes into: interior (◆), remaining (■) and corner(●) nodes

According to this partitioning scheme, the subdomain equilibrium equation can be represented as

$$\begin{bmatrix} \mathbf{A}_{ii}^s(\theta) & \mathbf{A}_{ir}^s(\theta) & \mathbf{A}_{ic}^s(\theta) \\ \mathbf{A}_{ri}^s(\theta) & \mathbf{A}_{rr}^s(\theta) & \mathbf{A}_{rc}^s(\theta) \\ \mathbf{A}_{ci}^s(\theta) & \mathbf{A}_{cr}^s(\theta) & \mathbf{A}_{cc}^s(\theta) \end{bmatrix} \begin{Bmatrix} \mathbf{u}_i^s(\theta) \\ \mathbf{u}_r^s(\theta) \\ \mathbf{u}_c^s(\theta) \end{Bmatrix} = \begin{Bmatrix} \mathbf{f}_i^s \\ \mathbf{f}_r^s \\ \mathbf{f}_c^s \end{Bmatrix}.$$

The Polynomial Chaos representation of uncertain model parameters leads to the following subdomain equilibrium equation

$$\sum_{l=0}^{L} \Psi_l \begin{bmatrix} \mathbf{A}_{ii,l}^s & \mathbf{A}_{ir,l}^s & \mathbf{A}_{ic,l}^s \\ \mathbf{A}_{ri,l}^s & \mathbf{A}_{rr,l}^s & \mathbf{A}_{rc,l}^s \\ \mathbf{A}_{ci,l}^s & \mathbf{A}_{cr,l}^s & \mathbf{A}_{cc,l}^s \end{bmatrix} \begin{Bmatrix} \mathbf{u}_i^s(\theta) \\ \mathbf{u}_r^s(\theta) \\ \mathbf{u}_c^s(\theta) \end{Bmatrix} = \begin{Bmatrix} \mathbf{f}_i^s \\ \mathbf{f}_r^s \\ \mathbf{f}_c^s \end{Bmatrix}. \tag{6}$$

The solution process is expressed using the same Polynomial Chaos basis as

$$\left\{\begin{array}{c}\mathbf{u}_i^s(\theta)\\ \mathbf{u}_r^s(\theta)\\ \mathbf{u}_c^s(\theta)\end{array}\right\} = \sum_{j=0}^{N}\Psi_j(\theta)\left\{\begin{array}{c}\mathbf{u}_{i,j}^s\\ \mathbf{u}_{r,j}^s\\ \mathbf{u}_{c,j}^s\end{array}\right\}. \tag{7}$$

Substituting Eq.(7) into Eq.(6) and performing Galerkin projection leads to the following coupled deterministic systems of equations

$$\begin{bmatrix}\mathcal{A}_{ii}^s & \mathcal{A}_{ir}^s & \mathcal{A}_{ic}^s\\ \mathcal{A}_{ri}^s & \mathcal{A}_{rr}^s & \mathcal{A}_{rc}^s\\ \mathcal{A}_{ci}^s & \mathcal{A}_{cr}^s & \mathcal{A}_{cc}^s\end{bmatrix}\left\{\begin{array}{c}\mathcal{U}_i^s\\ \mathcal{U}_r^s\\ \mathcal{U}_c^s\end{array}\right\} = \left\{\begin{array}{c}\mathcal{F}_i^s\\ \mathcal{F}_r^s\\ \mathcal{F}_c^s\end{array}\right\}, \tag{8}$$

where

$$\begin{aligned}[\mathcal{A}_{\alpha\beta}^s]_{jk} &= \sum_{l=0}^{L}\langle\Psi_l\Psi_j\Psi_k\rangle\mathbf{A}_{\alpha\beta,l}^s,\\ \mathcal{F}_{\alpha,k}^s &= \langle\Psi_k\mathbf{f}_\alpha^s\rangle,\\ \mathcal{U}_\alpha^s &= (\mathbf{u}_{\alpha,0}^s,\cdots,\mathbf{u}_{\alpha,N}^s)^T,\end{aligned}$$

the subscripts $\alpha$ and $\beta$ represent the index $i$, $r$ and $c$.

Enforcing the transmission conditions along the boundary interfaces, the subdomain equilibrium equation can be expressed as

$$\begin{bmatrix}\mathcal{A}_{ii}^s & \mathcal{A}_{ir}^s\mathcal{B}_r^s & \mathcal{A}_{ic}^s\mathcal{B}_c^s\\ \sum_{s=1}^{n_s}\mathcal{B}_r^{sT}\mathcal{A}_{ri}^s & \sum_{s=1}^{n_s}\mathcal{B}_r^{sT}\mathcal{A}_{rr}^s\mathcal{B}_r^s & \sum_{s=1}^{n_s}\mathcal{B}_r^{sT}\mathcal{A}_{rc}^s\mathcal{B}_c^s\\ \sum_{s=1}^{n_s}\mathcal{B}_c^{sT}\mathcal{A}_{ci}^s & \sum_{s=1}^{n_s}\mathcal{B}_c^{sT}\mathcal{A}_{cr}^s\mathcal{B}_r^s & \sum_{s=1}^{n_s}\mathcal{B}_c^{sT}\mathcal{A}_{cc}^s\mathcal{B}_c^s\end{bmatrix}\left\{\begin{array}{c}\mathcal{U}_i^s\\ \mathcal{U}_r\\ \mathcal{U}_c\end{array}\right\} = \left\{\begin{array}{c}\mathcal{F}_i^s\\ \sum_{s=1}^{n_s}\mathcal{B}_r^{sT}\mathcal{F}_r^s\\ \sum_{s=1}^{n_s}\mathcal{B}_c^{sT}\mathcal{F}_c^s\end{array}\right\}, \tag{9}$$

where $\mathcal{B}_r^s$ is a Boolean rectangular matrix that maps the global remaining vector $\mathcal{U}_r$ to the local remaining vector $\mathcal{U}_r^s$ as

$$\mathcal{U}_r^s = \mathcal{B}_r^s\mathcal{U}_r. \tag{10}$$

Similarly, the restriction operator $\mathcal{B}_c^s$ is a Boolean rectangular matrix that maps the global corner vector $\mathcal{U}_c$ to the local corner vector $\mathcal{U}_c^s$ as

$$\mathcal{U}_c^s = \mathcal{B}_c^s\mathcal{U}_c. \tag{11}$$

In parallel implementation both $\mathcal{B}_r^s$ and $\mathcal{B}_c^s$ act as *scatter* operators, while $\mathcal{B}_r^{sT}$ and $\mathcal{B}_c^{sT}$ act as *gather* operators.

The first block equation in Eq.(9) can be solved for $\mathcal{U}_i^s$ in parallel as

$$\mathcal{U}_i^s = [\mathcal{A}_{ii}^s]^{-1}(\mathcal{F}_i^s - \mathcal{A}_{ir}^s\mathcal{B}_r^s\mathcal{U}_r - \mathcal{A}_{ic}^s\mathcal{B}_c^s\mathcal{U}_c). \tag{12}$$

Substituting Eq.(12) into Eq.(9) leads to the following condensed system which represents the Schur compliment system in Eq.(5), but now the interface boundary nodes are split into remaining and corner nodes, as shown schematically in Fig.(3).

$$\begin{bmatrix} \sum_{s=1}^{n_s} \mathcal{B}_r^{s\,T} \mathcal{S}_{rr}^s \mathcal{B}_r^s & \sum_{s=1}^{n_s} \mathcal{B}_r^{s\,T} \mathcal{S}_{rc}^s \mathcal{B}_c^s \\ \sum_{s=1}^{n_s} \mathcal{B}_c^{s\,T} \mathcal{S}_{cr}^s \mathcal{B}_r^s & \sum_{s=1}^{n_s} \mathcal{B}_c^{s\,T} \mathcal{S}_{cc}^s \mathcal{B}_c^s \end{bmatrix} \begin{Bmatrix} \mathcal{U}_r \\ \mathcal{U}_c \end{Bmatrix} = \begin{Bmatrix} \sum_{s=1}^{n_s} \mathcal{B}_r^{s\,T} \mathcal{G}_r^s \\ \sum_{s=1}^{n_s} \mathcal{B}_c^{s\,T} \mathcal{G}_c^s \end{Bmatrix}, \tag{13}$$

where

$$\mathcal{S}_{\alpha\beta}^s = \mathcal{A}_{\alpha\beta}^s - \mathcal{A}_{\alpha i}^s [\mathcal{A}_{ii}^s]^{-1} \mathcal{A}_{i\beta}^s,$$
$$\mathcal{G}_\alpha^s = \mathcal{F}_\alpha^s - \mathcal{A}_{\alpha i}^s [\mathcal{A}_{ii}^s]^{-1} \mathcal{F}_i^s.$$

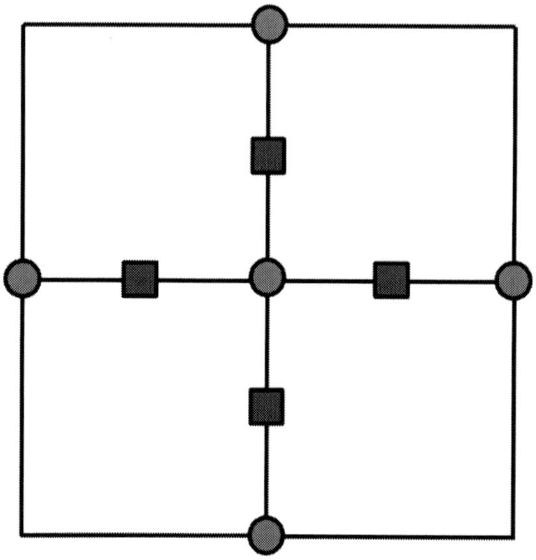

**Figure 3.** The interface boundary nodes split into: remainder (■) and corner(●) nodes

The corner nodal vector $\mathcal{U}_c$ in Eq.(13) is eliminated next to obtain the following (symmetric positive definite) reduced interface problem:

$$(F_{rr} - F_{rc}[F_{cc}]^{-1} F_{cr})\mathcal{U}_r = d_r - F_{rc}[F_{cc}]^{-1} d_c \tag{14}$$

where

$$F_{\alpha\beta} = \sum_{s=1}^{n_s} \mathcal{B}_\alpha^{s\,T} \mathcal{S}_{\alpha\beta}^s \mathcal{B}_\beta^s,$$
$$d_\alpha = \sum_{s=1}^{n_s} \mathcal{B}_\alpha^{s\,T} \mathcal{G}_\alpha^s.$$

and $\alpha$ and $\beta$ denotes subscripts $r$ and $c$. The above system can be solved using PCGM with an appropriate preconditioner $\mathcal{M}^{-1}$ defined in the next section.

*6.1. A two-level preconditioner*
As mentioned previously, the continuity condition at the corner nodes is enforced strictly while those for the remaining interface boundary nodes is satisfied in a weak sense. This fact is schematically illustrated in Fig.(4) [12]. As the iterations converge, the continuity condition at all interface nodes (both corner and remaining boundary nodes) is satisfied strictly. The assembly of the unknown vector at the corner nodes leads to the following partially assembled Schur complement system:

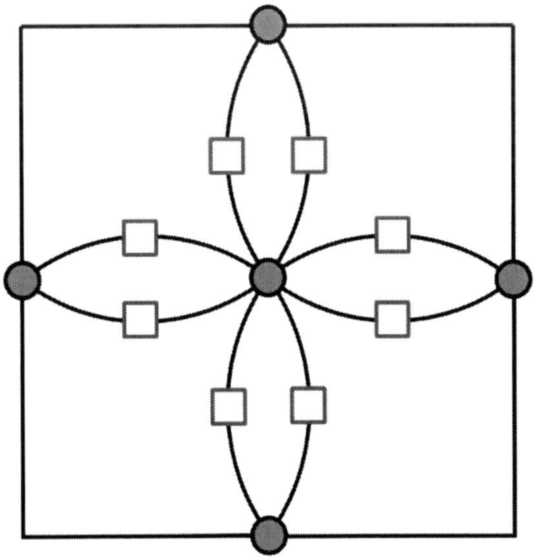

**Figure 4.** Partial assembly of corner nodes (●)

$$\begin{bmatrix} \mathcal{S}_{rr}^s & \mathcal{S}_{rc}^s \mathcal{B}_c^s \\ \sum_{s=1}^{n_s} \mathcal{B}_c^{sT} \mathcal{S}_{cr}^s \mathcal{B}_r^s & \sum_{s=1}^{n_s} \mathcal{B}_c^{sT} \mathcal{S}_{cc}^s \mathcal{B}_c^s \end{bmatrix} \begin{Bmatrix} \mathcal{U}_r^s \\ \mathcal{U}_c \end{Bmatrix} = \begin{Bmatrix} \mathcal{F}_r^s \\ 0 \end{Bmatrix}, \tag{15}$$

where

$$\mathcal{F}_r^s = \mathcal{D}_r^s \mathcal{B}_r^s r_j,$$

and $r_j$ is the residual at the $j^{th}$ iteration of the PCGM and $\mathcal{D}_r^s$ represents a block diagonal weighting matrix which satisfies the following property:

$$\sum_{s=1}^{n_s} \mathcal{B}_r^{sT} \mathcal{D}_r^s \mathcal{B}_r^s = \mathcal{I}.$$

The diagonal entries of each block of $\mathcal{D}_r^s$ are the reciprocal of the number of subdomains that share the interface boundary nodes.

The subdomain level remaining unknown vector $\mathcal{U}_r^s$ can be eliminated in parallel from Eq.(15) as

$$\mathcal{U}_r^s = [\mathcal{S}_{rr}^s]^{-1} (\mathcal{F}_r^s - \mathcal{S}_{rc}^s \mathcal{B}_c^s \mathcal{U}_c) \tag{16}$$

Substituting $\mathcal{U}_r^s$ into the second block of Eq.(15) leads to the following coarse problem

$$F_{cc}^* \mathcal{U}_c = d_c^*, \tag{17}$$

where

$$F_{cc}^* = \sum_{s=1}^{n_s} \mathcal{B}_c^{sT} (\mathcal{S}_{cc}^s - \mathcal{S}_{cr}^s [\mathcal{S}_{rr}^s]^{-1} \mathcal{S}_{rc}^s) \mathcal{B}_c^s,$$

$$d_c^* = -\sum_{s=1}^{n_s} \mathcal{B}_c^{sT} \mathcal{S}_{cr}^s [\mathcal{S}_{rr}^s]^{-1} \mathcal{F}_r^s.$$

The continuity of the solution field over the remaining interface nodes is satisfied next by averaging the local results as

$$\mathcal{U}_r = \sum_{s=1}^{n_s} \mathcal{B}_r^{sT} \mathcal{D}_r^s \mathcal{U}_r^s.$$

After some algebraic manipulations, the preconditioner can be expressed as

$$\mathcal{M}^{-1} = \sum_{s=1}^{n_s} \mathcal{B}_r^{sT} \mathcal{D}_r^s [\mathcal{S}_{rr}^s]^{-1} \mathcal{D}_r^s \mathcal{B}_r^s + R_0^T [F_{cc}^*]^{-1} R_0, \tag{18}$$

where

$$R_0 = \sum_{s=1}^{n_s} \mathcal{B}_c^{sT} \mathcal{S}_{cr} [\mathcal{S}_{rr}^s]^{-1} \mathcal{D}_r^s \mathcal{B}_r^s.$$

## 7. A dual-primal iterative substructuring method of SPDEs

In this section, the dual-primal domain decomposition method is introduced in the context of stochastic PDEs. This approach is an extension of FETI-DP [5] for SPDEs. In this approach, the continuity condition at the corner nodes is enforced strictly, and Lagrange multipliers are used to enforce the continuity condition weakly over the remaining interface nodes.

Partial assembly of Eq.(8) leads to the following subdomain equilibrium equation

$$\begin{bmatrix} \mathcal{A}_{ii}^s & \mathcal{A}_{ir}^s & \mathcal{A}_{ic}^s \mathcal{B}_c^s \\ \mathcal{A}_{ri}^s & \mathcal{A}_{rr}^s & \mathcal{A}_{rc}^s \mathcal{B}_c^s \\ \sum_{s=1}^{n_s} \mathcal{B}_c^{sT} \mathcal{A}_{ci}^s & \sum_{s=1}^{n_s} \mathcal{B}_c^{sT} \mathcal{A}_{cr}^s & \sum_{s=1}^{n_s} \mathcal{B}_c^{sT} \mathcal{A}_{cc}^s \mathcal{B}_c^s \end{bmatrix} \begin{Bmatrix} \mathcal{U}_i^s \\ \mathcal{U}_r^s \\ \mathcal{U}_c \end{Bmatrix} = \begin{Bmatrix} \mathcal{F}_i^s \\ \mathcal{F}_r^s \\ \sum_{s=1}^{n_s} \mathcal{B}_c^{sT} \mathcal{F}_c^s \end{Bmatrix}, \tag{19}$$

where $\mathcal{B}_c^s$ is a Boolean restriction operator that maps the global corner vector $\mathcal{U}_c$ to the local corner vector $\mathcal{U}_c^s$ as

$$\mathcal{U}_c^s = \mathcal{B}_c^s \mathcal{U}_c,$$

Eq.(19) can be rewritten in compact form as

$$\tilde{\mathcal{A}}^s \tilde{\mathcal{U}}^s = \tilde{\mathcal{F}}^s. \tag{20}$$

Let $\mathcal{B}_r^s$ be a block diagonal signed Boolean continuity matrix defined as

$$\sum_{s=1}^{s=n_s} \mathcal{B}_r^s \mathcal{U}_r^s = 0.$$

Next, the original finite element problem can be reformulated as an equivalent constrained minimization problem as

$$\frac{1}{2}\tilde{\mathcal{U}}^T \tilde{\mathcal{A}}\tilde{\mathcal{U}} - \tilde{\mathcal{U}}^T \tilde{\mathcal{F}} \rightarrow min \tag{21}$$
$$\text{subject to } \tilde{\mathcal{B}}\mathcal{U} = 0,$$

where

$$\tilde{\mathcal{A}} = \begin{bmatrix} \tilde{\mathcal{A}}^1 & & & & \\ & \ddots & & & \\ & & \tilde{\mathcal{A}}^s & & \\ & & & \ddots & \\ & & & & \tilde{\mathcal{A}}^{n_s} \end{bmatrix}, \quad \tilde{\mathcal{U}} = \left\{ \begin{array}{c} \tilde{\mathcal{U}}^1 \\ \vdots \\ \tilde{\mathcal{U}}^s \\ \vdots \\ \tilde{\mathcal{U}}^{n_s} \end{array} \right\}, \quad \tilde{\mathcal{F}} = \left\{ \begin{array}{c} \tilde{\mathcal{F}}^1 \\ \vdots \\ \tilde{\mathcal{F}}^s \\ \vdots \\ \tilde{\mathcal{F}}^{n_s} \end{array} \right\},$$

$$\tilde{\mathcal{B}} = \left[ (0 \;\; \mathcal{B}_r^1 \;\; 0), \cdots, (0 \;\; \mathcal{B}_r^s \;\; 0), \cdots, (0 \;\; \mathcal{B}_r^{n_s} \;\; 0) \right].$$

By introducing a vector of Lagrange multipliers to enforce the weak compatibility constraint, the saddle point formulation of Eq.(21) can be expressed as:

$$\mathcal{L}(\tilde{\mathcal{U}}, \Lambda) = \frac{1}{2}\tilde{\mathcal{U}}^T \tilde{\mathcal{A}}\tilde{\mathcal{U}} - \tilde{\mathcal{U}}^T \tilde{\mathcal{F}} + \tilde{\mathcal{U}}^T \tilde{\mathcal{B}}^T \Lambda. \tag{22}$$

Minimizing Eq.(22) with respect to $\tilde{\mathcal{U}}$ and $\Lambda$ leads to the following equilibrium system

$$\begin{bmatrix} \mathcal{A}_{ii}^s & \mathcal{A}_{ir}^s & \mathcal{A}_{ic}^s \mathcal{B}_c^s & 0 \\ \mathcal{A}_{ri}^s & \mathcal{A}_{rr}^s & \mathcal{A}_{rc}^s \mathcal{B}_c^s & \mathcal{B}_r^{sT} \\ \sum_{s=1}^{n_s} \mathcal{B}_c^{sT} \mathcal{A}_{ci}^s & \sum_{s=1}^{n_s} \mathcal{B}_c^{sT} \mathcal{A}_{cr}^s & \sum_{s=1}^{n_s} \mathcal{B}_c^{sT} \mathcal{A}_{cc}^s \mathcal{B}_c^s & 0 \\ 0 & \sum_{s=1}^{n_s} \mathcal{B}_r^s & 0 & 0 \end{bmatrix} \left\{ \begin{array}{c} \mathcal{U}_i^s \\ \mathcal{U}_r^s \\ \mathcal{U}_c \\ \Lambda \end{array} \right\} = \left\{ \begin{array}{c} \mathcal{F}_i^s \\ \mathcal{F}_r^s \\ \sum_{s=1}^{n_s} \mathcal{B}_c^{sT} \mathcal{F}_c^s \\ 0 \end{array} \right\}, \tag{23}$$

where

$$\Lambda = \left\{ \begin{array}{c} \lambda_0 \\ \vdots \\ \lambda_N \end{array} \right\},$$

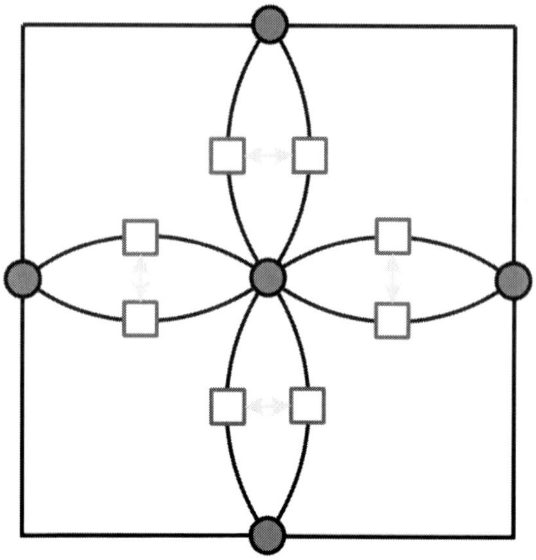

**Figure 5.** Lagrange multipliers are the forces required to connect the tore interface boundary

and $\lambda_j$ is the nodal force required to satisfy compatibility at the remaining interface nodes as shown schematically in Fig.(5).

Eliminating the interior unknowns $\mathcal{U}_i^s$ from Eq.(23) as

$$\mathcal{U}_i^s = [\mathcal{A}_{ii}^s]^{-1}(\mathcal{F}_i^s - \mathcal{A}_{ir}^s\mathcal{U}_r^s - \mathcal{A}_{ic}^s\mathcal{B}_c^s\mathcal{U}_c) \tag{24}$$

Substituting Eq.(24) into Eq.(23), leads to

$$\begin{bmatrix} \mathcal{S}_{rr}^s & \mathcal{S}_{rc}^s\mathcal{B}_c^s & \mathcal{B}_r^{sT} \\ \sum_{s=1}^{n_s}\mathcal{B}_c^{sT}\mathcal{S}_{cr}^s & \sum_{s=1}^{n_s}\mathcal{B}_c^{sT}\mathcal{S}_{cc}^s\mathcal{B}_c^s & 0 \\ \sum_{s=1}^{n_s}\mathcal{B}_r^s & 0 & 0 \end{bmatrix} \begin{Bmatrix} \mathcal{U}_r^s \\ \mathcal{U}_c \\ \Lambda \end{Bmatrix} = \begin{Bmatrix} \mathcal{G}_r^s \\ \sum_{s=1}^{n_s}\mathcal{B}_c^{sT}\mathcal{G}_c^s \\ 0 \end{Bmatrix}, \tag{25}$$

where

$$\mathcal{S}_{\alpha\beta}^s = \mathcal{A}_{\alpha\beta}^s - \mathcal{A}_{\alpha i}^s[\mathcal{A}_{ii}^s]^{-1}\mathcal{A}_{i\beta}^s,$$
$$\mathcal{G}_\alpha^s = \mathcal{F}_\alpha^s - \mathcal{A}_{\alpha i}^s[\mathcal{A}_{ii}^s]^{-1}\mathcal{F}_i^s.$$

The subdomain level remaining unknown vector $\mathcal{U}_r^s$ can be obtained in parallel from Eq.(25) as

$$\mathcal{U}_r^s = [\mathcal{S}_{rr}^s]^{-1}(\mathcal{G}_r^s - \mathcal{S}_{rc}^s\mathcal{B}_c^s\mathcal{U}_c - \mathcal{B}_r^{sT}\Lambda) \tag{26}$$

Substituting Eq.(26) into Eq.(25) leads to

$$\begin{bmatrix} \tilde{F}_{cc} & -\tilde{F}_{cr} \\ \tilde{F}_{rc} & \tilde{F}_{rr} \end{bmatrix} \begin{Bmatrix} \mathcal{U}_c \\ \Lambda \end{Bmatrix} = \begin{Bmatrix} \tilde{d}_c \\ \tilde{d}_r \end{Bmatrix}, \tag{27}$$

where

$$\bar{F}_{cc} = \sum_{s=1}^{n_s} \mathcal{B}_c^{sT}(\mathcal{S}_{cc}^s - \mathcal{S}_{cr}^s[\mathcal{S}_{rr}^s]^{-1}\mathcal{S}_{rc}^s)\mathcal{B}_c^s$$

$$\bar{F}_{cr} = \sum_{s=1}^{n_s} \mathcal{B}_c^{sT}\mathcal{S}_{cr}^s[\mathcal{S}_{rr}^s]^{-1}\mathcal{B}_r^{sT}$$

$$\bar{F}_{rc} = \sum_{s=1}^{n_s} \mathcal{B}_r^s[\mathcal{S}_{rr}^s]^{-1}\mathcal{S}_{rc}^s\mathcal{B}_c^s$$

$$\bar{F}_{rr} = \sum_{s=1}^{n_s} \mathcal{B}_r^s[\mathcal{S}_{rr}^s]^{-1}\mathcal{B}_r^{sT}$$

$$\bar{d}_c = \sum_{s=1}^{n_s} \mathcal{B}_c^{sT}(\mathcal{G}_c^s - \mathcal{S}_{cr}^s[\mathcal{S}_{rr}^s]^{-1}\mathcal{G}_r^s)$$

$$\bar{d}_r = \sum_{s=1}^{n_s} \mathcal{B}_r^s[\mathcal{S}_{rr}^s]^{-1}\mathcal{G}_r^s$$

Solving for $\mathcal{U}_c$ from Eq.(27) gives the following coarse problem

$$\bar{F}_{cc}\mathcal{U}_c = (\bar{d}_c + \bar{F}_{cr}\Lambda) \tag{28}$$

Substituting $\mathcal{U}_c$ into Eq.(27) leads to the following symmetric positive-definite Lagrange multiplier system

$$(\bar{F}_{rr} + \bar{F}_{rc}[\bar{F}_{cc}]^{-1}\bar{F}_{cr})\Lambda = \bar{d}_r - \bar{F}_{rc}[\bar{F}_{cc}]^{-1}\bar{d}_c \tag{29}$$

Eq.(29) is solved using PCGM with a Dirichlet precondtioner defined as

$$\bar{\mathcal{M}} = \sum_{s=1}^{n_s} \mathcal{B}_r^s\mathcal{D}_r^s\mathcal{S}_{rr}^s\mathcal{D}_r^s\mathcal{B}_r^{sT} \tag{30}$$

## 8. Connection between the methods

The explicit forms of the coarse problem operators for the primal preconditioner in Eq.(17) and for the dual-primal operator in Eq.(28) are the same, and can be expressed as

$$F_{cc}^* = \bar{F}_{cc} = \sum_{s=1}^{n_s} \mathcal{B}_c^{sT}(\mathcal{S}_{cc}^s - \mathcal{S}_{cr}^s[\mathcal{S}_{rr}^s]^{-1}\mathcal{S}_{rc}^s)\mathcal{B}_c^s.$$

Furthermore, the algebraic form of the primal preconditioner in Eq.(18) can be re-casted as

$$\begin{aligned}\mathcal{M}^{-1} &= \sum_{s=1}^{n_s} \mathcal{B}_r^{sT}\mathcal{D}_r^s[\mathcal{S}_{rr}^s]^{-1}\mathcal{D}_r^s\mathcal{B}_r^s \\ &+ \sum_{s=1}^{n_s} \mathcal{B}_r^{sT}\mathcal{D}_r^s[\mathcal{S}_{rr}^s]^{-1}\mathcal{S}_{rc}\mathcal{B}_c^s\left[\sum_{s=1}^{n_s} \mathcal{B}_c^{sT}(\mathcal{S}_{cc}^s - \mathcal{S}_{cr}^s[\mathcal{S}_{rr}^s]^{-1}\mathcal{S}_{rc}^s)\mathcal{B}_c^s\right]^{-1} \\ &\sum_{s=1}^{n_s} \mathcal{B}_c^{sT}\mathcal{S}_{cr}[\mathcal{S}_{rr}^s]^{-1}\mathcal{D}_r^s\mathcal{B}_r^s,\end{aligned}$$

which has the same form of the dual-primal operator in Eq.(29).

## 9. Parallel implementation

In this section we give an outline on parallel implementation of PCGM to solve the primal Eq.(14) and dual-primal Eq.(29) interface problems. As mentioned previously, in PCGM, the coefficient matrix need not be constructed explicitly as only its effect on a vector is required. This matrix-vector product can be obtained concurrently by solving subdomain level problems (Dirichlet and Neumann) and a global level coarse problem.

### *9.1. Primal method*

In this subsection we give a brief description of parallel implementation of Algorithm (1) to solve the primal interface problem in Eq.(14).

For the $j^{th}$ iteration of Algorithm (1), the matrix-vector product in step 7 defined as

$$\mathcal{Q}_j = (F_{rr} - F_{rc}[F_{cc}]^{-1}F_{cr})\mathcal{P}_j,$$

can be computed using the following algorithm:

**Algorithm** 4: Parallel Matrix-Vector Product for Primal Method

1. *Input* $(\mathcal{P})$
2. *Scatter* : $\mathcal{P}^s = \mathcal{B}_r^s\mathcal{P}$
3. *Compute* : $v_1^s = \mathcal{S}_{cr}^s\mathcal{P}^s$
4. *Gather* : $V_1 = \sum_{s=1}^{n_s} \mathcal{B}_r^{sT} v_1^s$
5. *Global Solve* : $F_{cc}V_2 = V_1$
6. *Scatter* : $v_2^s = \mathcal{B}_c^s V_2$
7. *Compute* : $v_3^s = \mathcal{S}_{rc}^s v_2^s$
8. *Update* : $\mathcal{Q}^s = \mathcal{S}_{rr}^s\mathcal{P}^s - v_3^s$
9. *Gather* : $\mathcal{Q} = \sum_{s=1}^{n_s} \mathcal{B}_r^{sT}\mathcal{Q}^s$
10. *Output* $(\mathcal{Q})$

Multiplication of Schur complement matrix by a vector in step 3, step 7 and step 8 in Algorithm (4) is computed by solving a corresponding Dirichlet problem as

$$v_\alpha^s = \mathcal{S}_{\alpha\beta}^s v_\beta^s,$$
$$v_\alpha^s = (\mathcal{A}_{\alpha\beta}^s - \mathcal{A}_{\alpha i}^s[\mathcal{A}_{ii}^s]^{-1}\mathcal{A}_{i\beta}^s)v_\beta^s$$

This procedure is outlined in the following algorithm:

**Algorithm** 5: Dirichlet Solver Procedure

1. *Input* $(v_\beta^s)$
2. *Compute* : $v_1^s = \mathcal{A}_{i\beta}^s v_\beta^s$
3. *Solve* : $\mathcal{A}_{ii}^s v_2^s = v_1^s$
4. *Compute* : $v_3^s = \mathcal{A}_{\alpha i}^s v_2^s$
5. *Compute* : $v_4^s = \mathcal{A}_{\alpha\beta}^s v_\beta^s$

6. *Compute* : $v_\alpha^s = v_4^s - v_3^s$
7. *Output* $(v_\alpha^s)$

The global problem in step 5 of Algorithm (4) is solved iteratively using PCGM equipped with lumped preconditioner as

$$\mathcal{M}_{cc}^{-1} F_{cc} V_2 = \mathcal{M}_{cc}^{-1} V_1,$$

where

$$\mathcal{M}_{cc}^{-1} = \sum_{s=1}^{n_s} \mathcal{B}_c^{sT} \mathcal{A}_{cc}^s \mathcal{B}_c^s$$

Next the effect of the two-level preconditioner in step 12 of Algorithm (1) is computed by solving a subdomain level Neumann problem and a global coarse problem. The procedure is outlined in the following algorithm:

**Algorithm** 6: Two-Level Preconditioner Effect Procedure

1. *Input* $(r_\Gamma)$
2. *Scatter* : $\mathcal{F}_r^s = \mathcal{D}_r^s \mathcal{B}_r^s r_\Gamma$
3. *Local Solve* : $\mathcal{S}_{rr}^s v_1^s = \mathcal{F}_r^s$
4. *Compute* : $d_c^s = \mathcal{S}_{cr}^s v_1^s$
5. *Gather* : $d_c = \sum_{s=1}^{n_s} \mathcal{B}_c^{sT} d_c^s$
6. *Global Solve* : $F_{cc}^* \mathcal{Z}_c = -d_c$
7. *Scatter* : $\mathcal{Z}_c^s = \mathcal{B}_s \mathcal{Z}_c$
8. *Update* : $v_2^s = \mathcal{F}_r^s + \mathcal{S}_{rc}^s \mathcal{Z}_c^s$
9. *Local Solve* : $\mathcal{S}_{rr}^s \mathcal{Z}_f^s = v_2^s$
10. *Gather* : $\mathcal{Z} = \sum_{s=1}^{n_s} \mathcal{B}_r^{sT} \mathcal{D}_r^s \mathcal{Z}_f^s$
11. *Output* $(\mathcal{Z})$

The local solve in step 3 and step 9 of Algorithm (6) constitute a subdomain level Neumann problem of the form: $\mathcal{S}_{rr}^s \mathcal{U}_r^s = r_r^s$ which can be solved using the following algorithm:

**Algorithm** 7: Neumann-Solver Procedure

1. *Input* $(r_r^s)$
2. *Solve*:
$$\begin{bmatrix} \mathcal{A}_{ii}^s & \mathcal{A}_{ir}^s \\ \mathcal{A}_{ri}^s & \mathcal{A}_{rr}^s \end{bmatrix} \begin{Bmatrix} \mathcal{X}^s \\ \mathcal{U}_r^s \end{Bmatrix} = \begin{Bmatrix} 0 \\ r_r^s \end{Bmatrix}$$
3. *Output* $(\mathcal{U}_r^s)$

The global solve of the coarse problem in step 6 of Algorithm (6) is conducted in parallel using PCGM equipped with lumped preconditioner as

$$\mathcal{M}_{cc}^{-1} F_{cc}^{*} \mathcal{Z}_c = -\mathcal{M}_{cc}^{-1} d_c,$$

where

$$\mathcal{M}_{cc}^{-1} = \sum_{s=1}^{n_s} \mathcal{B}_c^{sT} \mathcal{A}_{cc}^{s} \mathcal{B}_c^{s}.$$

Finally, we summarize the parallel implementation of the PCGM to solve the primal interface problem in the following flow chart.

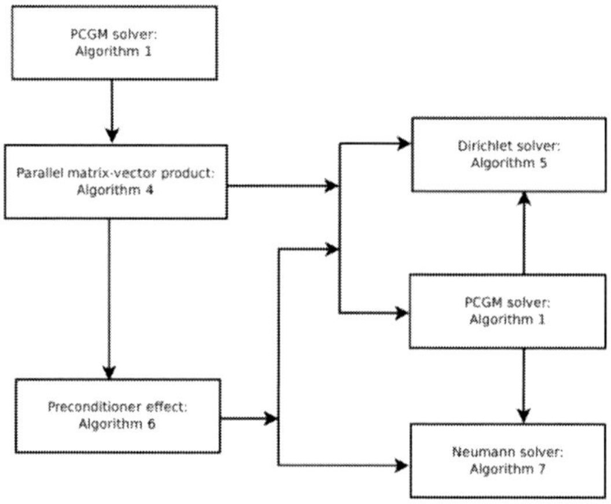


**Figure 6.** Flowchart of Parallel PCGM to solve the primal interface problem

*9.2. Dual-primal method*

In this subsection, we outline the parallel implementation of the Algorithm (1) to solve the dual-primal interface problem Eq.(29).

For the $j^{th}$ iteration of Algorithm (1), the matrix-vector product in step 7 defined as

$$\mathcal{Q}_j = (\bar{F}_{rr} + \bar{F}_{rc}[\bar{F}_{cc}]^{-1}\bar{F}_{cr})\mathcal{P}_j$$

can be computed using the following algorithm:

**Algorithm** 8 : Parallel Matrix-Vector Product for Dual-Primal Method

1. *Input* $(\mathcal{P})$
2. *Scatter* : $\mathcal{P}^s = \mathcal{B}_r^{sT}\mathcal{P}$
3. *Local Solve* : $\mathcal{S}_{rr}^s v_1^s = \mathcal{P}^s$
4. *Compute* : $v_2^s = \mathcal{S}_{cr}^s v_1^s$
5. *Gather* : $V_2 = \sum_{s=1}^{n_s} \mathcal{B}_c^{sT} v_2^s$

6. *Global Solve* : $\tilde{F}_{cc}V_3 = V_2$
7. *Scatter* : $v_3^s = \mathcal{B}_c^s V_3$
8. *Compute* : $v_4^s = \mathcal{S}_{rc}^s v_3^s$
9. *Update* : $v_5^s = \mathcal{P}^s + v_4^s$
10. *Local Solve* : $\mathcal{S}_{rr}^s \mathcal{Q}^s = v_5^s$
11. *Gather* : $\mathcal{Q} = \sum_{s=1}^{n_s} \mathcal{B}_r^s \mathcal{Q}^s$
12. *Output* $(\mathcal{Q})$

The local solve in step 3 and step 10 of Algorithm (8) is calculated by solving a subdomain level Neumann problem as outlined in Algorithm (7). The global coarse problem in step 6 of Algorithm (8) is solved in parallel using PCGM with lumped preconditioner similar to the procedure of solving the coarse problem in the primal preconditioner.

Next the effect of the Dirichlet Preconditioner in step 12 of Algorithm (1) is obtained using the following algorithm:

**Algorithm** 9 : Dirichlet Preconditioner Effect Procedure

1. *Input* $(r_\Gamma)$
2. *Scatter* : $r_\Gamma^s = \mathcal{D}_r^s \mathcal{B}_r^{s\,T} r_\Gamma$
3. *Compute* : $\mathcal{Z}^s = \mathcal{S}_{rr}^s r_\Gamma^s$
4. *Gather* : $\mathcal{Z} = \sum_{s=1}^{n_s} \mathcal{B}_r^s \mathcal{D}_r^s \mathcal{Z}^s$
5. *Output* $(\mathcal{Z})$

We summarize the parallel implementation of the PCGM to solve the dual-primal interface problem in the following flow chart.

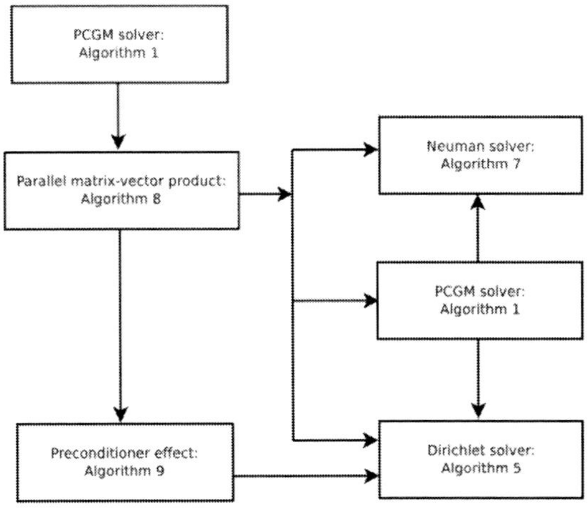


**Figure 7.** Flowchart of Parallel PCGM to solve the dual-primal interface problem

## 10. Numerical results

For numerical illustrations to the aforementioned mathematical framework, we consider a stationary stochastic Poisson's equation with randomly heterogeneous coefficients given as

$$\frac{\partial}{\partial x}[c_x(x,y,\theta)\frac{\partial u(x,y,\theta)}{\partial x}] + \frac{\partial}{\partial y}[c_y(x,y,\theta)\frac{\partial u(x,y,\theta)}{\partial y}] = f(x,y) \quad in\ \Omega,$$

where the forcing term is

$$f(x,y) = 1.0.$$

For simplicity, a homogeneous Dirichlet boundary condition is imposed as

$$u(x,y,\theta) = 0 \quad on\ \partial\Omega.$$

The random coefficients $c_x(x,y,\theta)$ and $c_y(x,y,\theta)$ are modeled as independent lognormal random variables. The underlying Gaussian random variable has a mean 1.0 and standard deviation 0.25.

In PCGM implementation, the forcing term is taken to be the initial residual, and the iterations are terminated when the ratio of $L_2$ norms of the current and the initial residual is less than $10^{-5}$

$$\frac{\|\mathcal{G}_\Gamma^k - \mathcal{S}\mathcal{U}_\Gamma^k\|_2}{\|\mathcal{G}_\Gamma^0\|_2} \leqslant 10^{-5}.$$

Numerical experiments are performed in a Linux cluster with InfiniBand interconnect (2 Quad-Core 3.0 GHz Intel Xeon processors and 32 GB of memory per node) using MPI [7] and PETSc [6] parallel libraries. The graph partitioning tool METIS [8] is used to decompose the finite element mesh.

### *10.1. Stochastic features*

Finite element discretization with linear triangular elements results in 202,242 elements and 101,851 nodes. The random coefficients and the response are represented by third order polynomial chaos expansion ($L = 7$, $N = 9$) leading to a linear system of order 1,018,510. Fig.(8) shows a typical finite element mesh while Fig.(9) shows a typical mesh decomposition. The mean and the associated standard deviation of the solution process are shown in Fig.(10) and Fig.(11), respectively. Clearly the maximum value of the coefficient of variation of the solution field is 0.20. Details of the stochastic features of the solution field are shown in Figs.(12-17) through there Polynomial Chaos coefficients. The mean and the standard deviation of the solution field computed using the dual-primal method (not shown here) exactly match the results from the primal method. In Figs.(18-23) the Polynomial Chaos coefficient of Lagrange multipliers are shown.

**Figure 8.** A typical FEM mesh

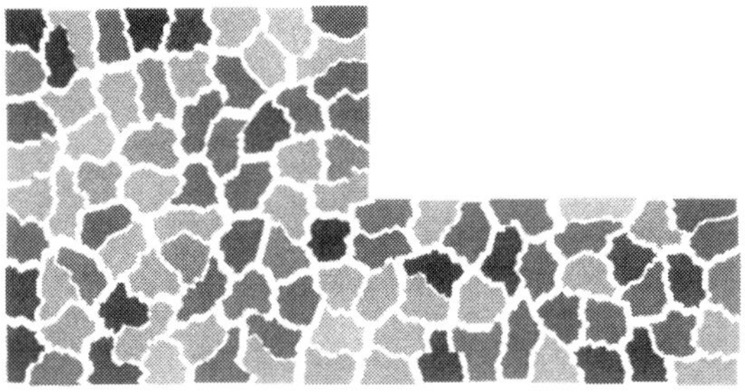

**Figure 9.** Mesh Partitioning using METIS

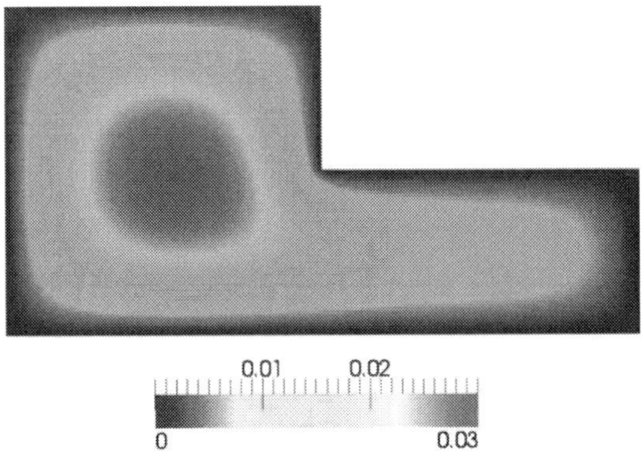


**Figure 10.** The mean of the solution filed.

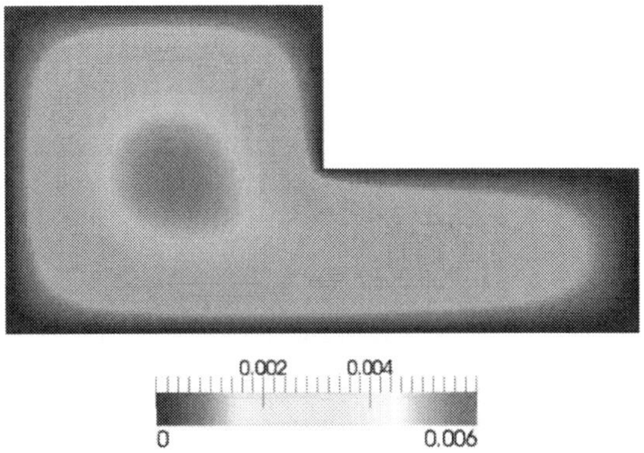


**Figure 11.** The standard deviation of the solution filed.

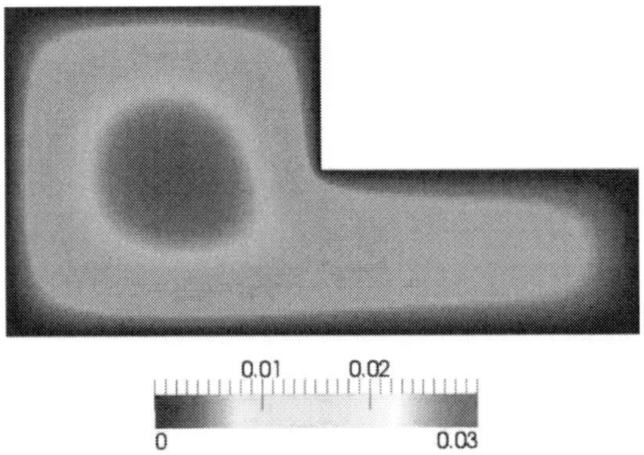


**Figure 12.** Chaos coefficients: $\mathbf{u}_0$

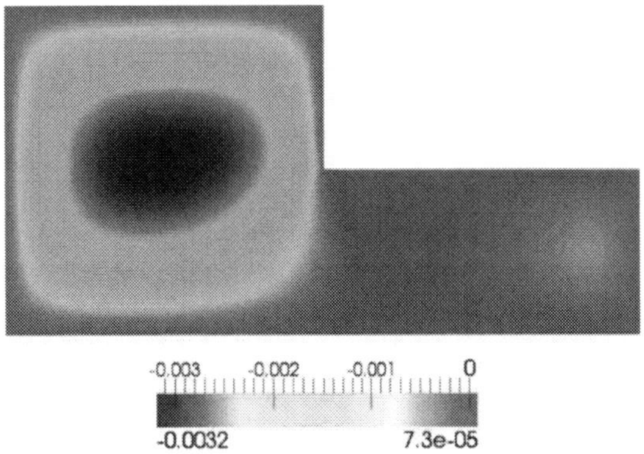


**Figure 13.** Chaos coefficients: $\mathbf{u}_1$

*10.2. Scalability study*

Firstly, we study the scalability of the algorithms with respect to the problem size where we fix the number of subdomains used to solve the problem to 100 while increasing both mesh resolution in the spatial dimension and the Polynomial Chaos order as reported in Table(1). Evidently increasing mesh resolution by factor $(10\times)$ does not deteriorate the performance of the primal and the dual-primal algorithms. Simultaneously, increasing Polynomial Chaos order from the first order to third order does not effect the performance of the methods. Note that for a given spatial problem size $(n)$, using the first order Polynomial Chaos expansion leads to a total problem size of $(3 \times n)$, and using the third order Polynomial Chaos expansion leads to

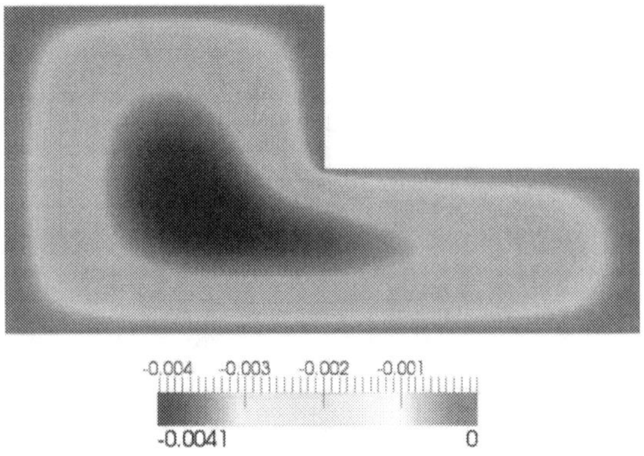


**Figure 14.** Chaos coefficients: $\mathbf{u}_2$

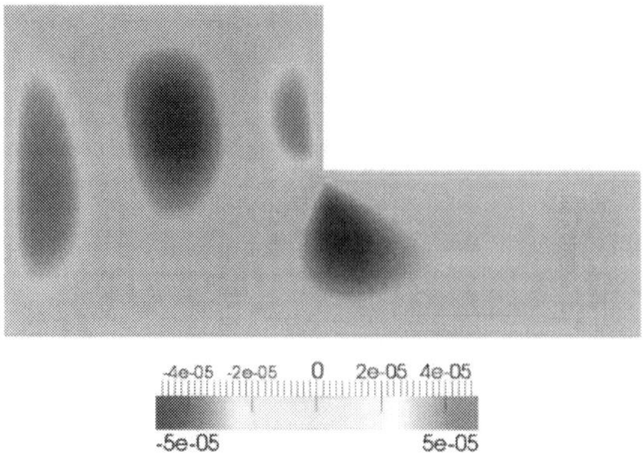


**Figure 15.** Chaos coefficients: $\mathbf{u}_3$

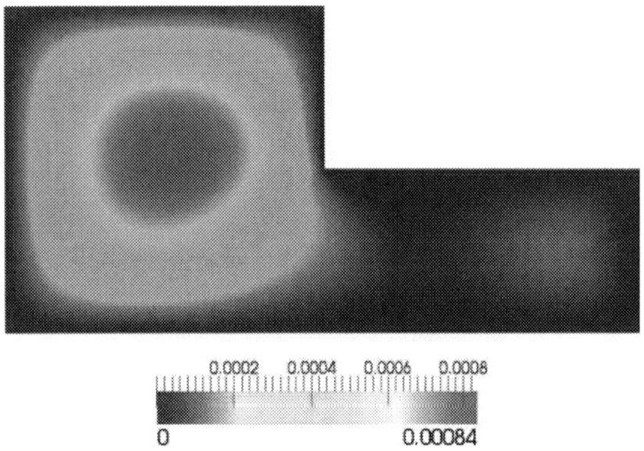


**Figure 16.** Chaos coefficients: $\mathbf{u}_4$

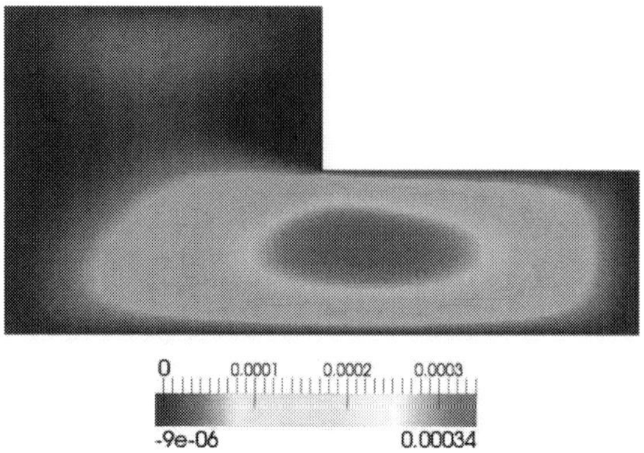


**Figure 17.** Chaos coefficients: $\mathbf{u}_5$

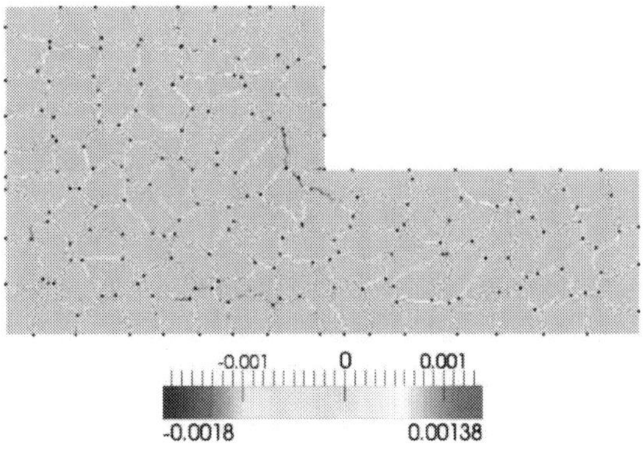


**Figure 18.** Lagrange multipliers: $\lambda_0$

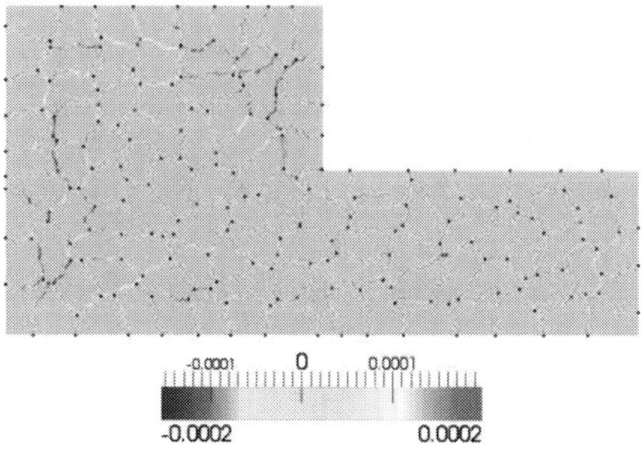


**Figure 19.** Lagrange multipliers: $\lambda_1$

a total problem size of $(10 \times n)$.

Secondly, we fix the problem size in the spatial domain to (7,1389 dofs) and increase the number of subdomains used to solve the problem the results are presented in Table(2). The results reported for first, second and third order Polynomial Chaos expansion. These performance results suggest that both the primal and the dual-primal methods are scalable with respect to number of subdomains. Clearly, the dual-primal method requires slightly less number of iter-

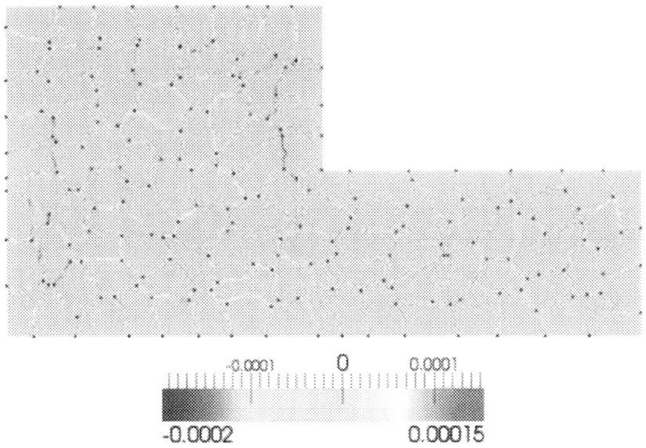


**Figure 20.** Lagrange multipliers: $\lambda_2$

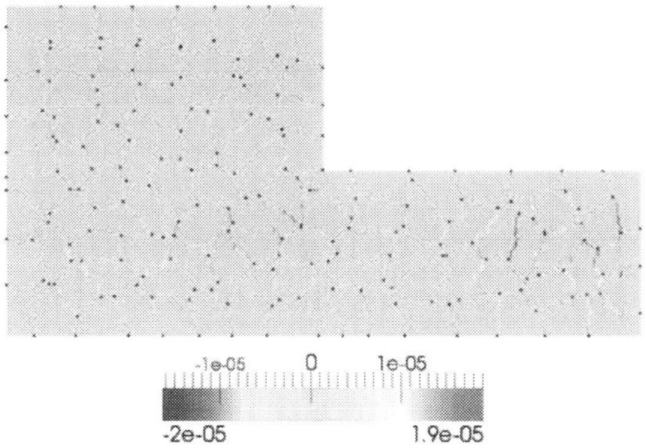


**Figure 21.** Lagrange multipliers: $\lambda_3$

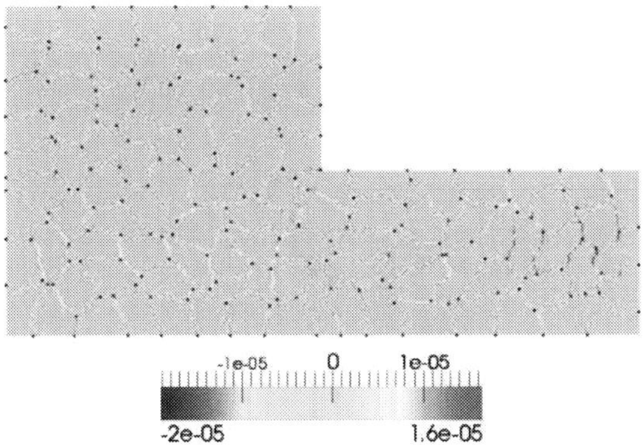


**Figure 22.** Lagrange multipliers: $\lambda_4$

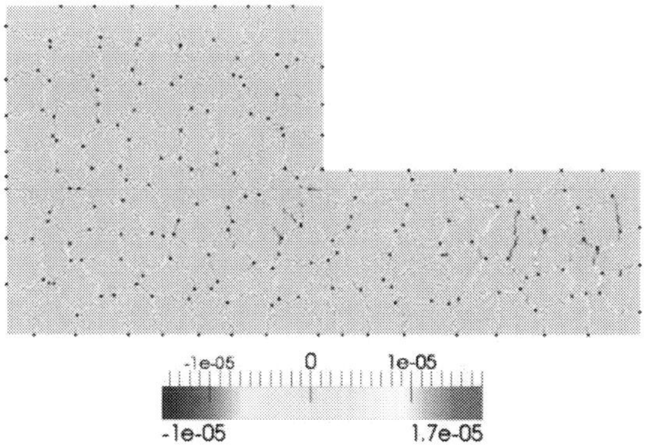


**Figure 23.** Lagrange multipliers: $\lambda_5$

**Table 1.** Iteration counts of the primal and dual-primal methods for fixed number of subdomain (100).

| Problem size | PDDM | | | DP-DDM | | |
|---|---|---|---|---|---|---|
| | *1st* | *2nd* | *3rd* | *1st* | *2nd* | *3rd* |
| 10,051 | 10 | 10 | 10 | 8 | 8 | 8 |
| 20,303 | 11 | 11 | 11 | 8 | 8 | 8 |
| 40,811 | 11 | 12 | 12 | 8 | 9 | 9 |
| 59,935 | 13 | 14 | 14 | 10 | 10 | 10 |
| 71,386 | 12 | 12 | 12 | 9 | 9 | 9 |
| 80,172 | 11 | 11 | 12 | 8 | 8 | 8 |
| 101,851 | 12 | 12 | 12 | 9 | 9 | 9 |

ations to converge than the primal method. This may be attributed to fact that the starting initial residual in the dual-primal method is smaller than the starting initial residual in the primal method. However, the rate of convergence of both the methods is almost the same as indicated in Figs.(24-26).

Thirdly, we fix problem size per subdomain while increase the overall problem size by adding

**Table 2.** Iteration counts of the primal and dual-primal methods for fixed problem size of 71,386 dof.

| CPUs | PDDM | | | DP-DDM | | |
|---|---|---|---|---|---|---|
| | $1^{st}$ | $2^{nd}$ | $3^{rd}$ | $1^{st}$ | $2^{nd}$ | $3^{rd}$ |
| 20 | 10 | 11 | 11 | 8 | 8 | 8 |
| 40 | 12 | 12 | 12 | 9 | 9 | 9 |
| 60 | 12 | 13 | 13 | 9 | 9 | 9 |
| 80 | 12 | 12 | 13 | 9 | 9 | 9 |
| 100 | 12 | 12 | 12 | 9 | 9 | 9 |
| 120 | 12 | 12 | 12 | 9 | 9 | 9 |
| 140 | 11 | 11 | 12 | 8 | 8 | 8 |
| 160 | 12 | 12 | 12 | 8 | 8 | 9 |

**Table 3.** Iteration counts of the primal and dual primal methods for fixed problem size per subdomain (101,851 dof).

| Subdomains | PDDM | | | DP-DDM | | |
|---|---|---|---|---|---|---|
| | $1^{st}$ | $2^{nd}$ | $3^{rd}$ | $1^{st}$ | $2^{nd}$ | $3^{rd}$ |
| 100 | 10 | 10 | 10 | 8 | 8 | 8 |
| 200 | 10 | 10 | 11 | 8 | 8 | 8 |
| 400 | 12 | 13 | 13 | 9 | 9 | 9 |
| 600 | 11 | 12 | 12 | 8 | 8 | 9 |
| 800 | 12 | 13 | 13 | 9 | 9 | 9 |

more subdomains. Table (3) shows the performance of the primal and the dual-primal methods for the first, second and third order Polynomial Chaos expansion. Again these results suggest that both the primal and the dual-primal methods are scalable with respect to fixed problem size per subdomain.

Fourthly, we study the performance of the primal and the dual-primal methods with respect the strength of randomness of the system parameters. Table (4) shows the performance of the algorithms when the Coefficient of variation ($CoV$) of the random parameters is varied from (5% to 50%). Clearly, the strength of the randomness does not degrade the performance of the algorithms as the number of PCGM iteration is nearly constant.

Finally, it worth mentioning that the performances of the primal method and the dual-primal method demonstrate similar trend and this fact points out the similarity (duality) between the two methods through numerical experiments.

## 11. Conclusion

Novel primal and dual-primal domain decomposition methods are proposed to solve the algebraic linear system arising from the stochastic finite element method. The primal method is equipped with a scalable two-level preconditioner. The numerical experiments illustrate that the proposed

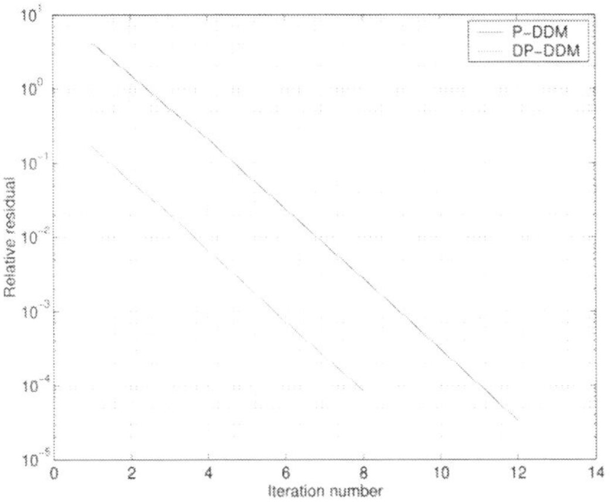


**Figure 24.** The relative PCGM residual history for the case of 160 subdomains and first PC order.

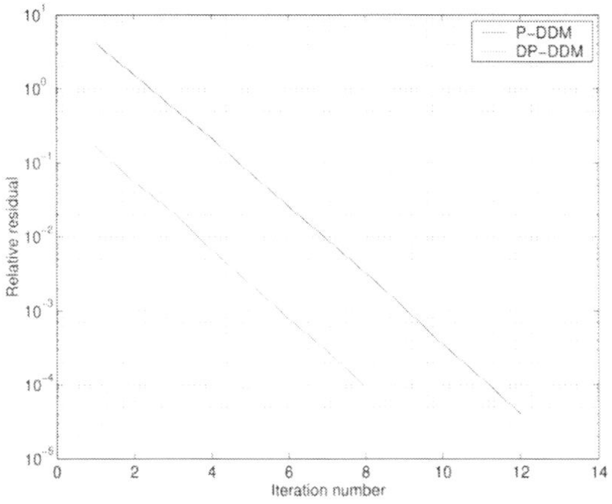


**Figure 25.** The relative PCGM residual history for the case of 160 subdomains and second PC order.

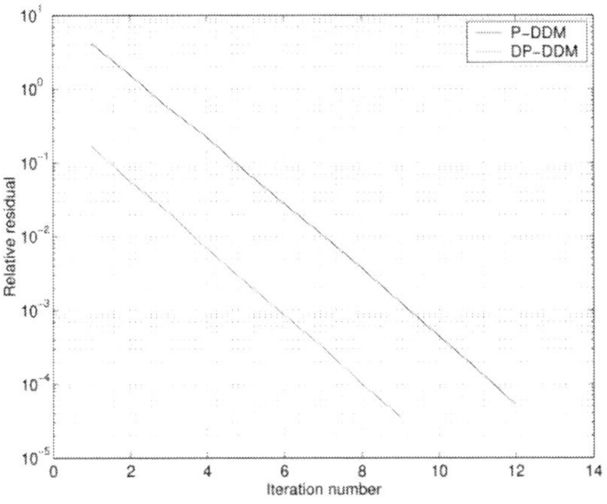


**Figure 26.** The relative PCGM residual history for the case of 160 subdomains and third PC order.

primal method and the dual-primal method are numerically scalable with respect to problem size, subdomain size and number of subdomains. Both algorithms exhibit similar convergence rates with respect to the coefficient of variation (i.e. the level of uncertainty) and Polynomial Chaos order. Both primal and dual-primal iterative substructuring methods exploit a coarse grid in the geometric space. At this point, it is worth mentioning that adding a coarse grid in the stochastic space would be beneficial in the cases where a large number of random variables are required to prescribe uncertainty in the input parameters. This aspect is currently being investigated by the authors.

## Acknowledgments

The authors gratefully acknowledge the financial support from the Natural Sciences and Engineering Research Council of Canada through a Discovery Grant, Canada Research Chair Program, Canada Foundation for Innovation and Ontario Innovation Trust. Dr. Ali Rebaine for his help with ParMETIS graph-partitioning software.

**Table 4.** Iteration counts of the primal and dual primal methods for different $CoV$, fixed problem size (10,1851 dofs) and fixed number of subdomains (100).

| $CoV$ | PDDM | | | DP-DDM | | |
|---|---|---|---|---|---|---|
| | $1^{st}$ | $2^{nd}$ | $3^{rd}$ | $1^{st}$ | $2^{nd}$ | $3^{rd}$ |
| 0.05 | 10 | 10 | 10 | 8 | 8 | 8 |
| 0.10 | 10 | 10 | 10 | 8 | 8 | 8 |
| 0.15 | 10 | 10 | 10 | 8 | 8 | 8 |
| 0.20 | 10 | 10 | 10 | 8 | 8 | 8 |
| 0.25 | 10 | 10 | 10 | 8 | 8 | 8 |
| 0.30 | 10 | 10 | 11 | 8 | 8 | 8 |
| 0.35 | 10 | 10 | 11 | 8 | 8 | 8 |
| 0.40 | 10 | 11 | 11 | 8 | 8 | 9 |
| 0.45 | 10 | 11 | 12 | 8 | 8 | 9 |
| 0.50 | 10 | 11 | 12 | 8 | 8 | 9 |

## References

[1] Sarkar A, Benabbou N and Ghanem R 2009 *International Journal for Numerical Methods in Engineering* **77** 689–701
[2] Sarkar A, Benabbou N and Ghanem R 2010 *International Journal of High Performance Computing Applications* Accepted
[3] Subber W, Monajemi H, Khalil M and Sarkar A 2008 *International Symposium on Uncertainties in Hydrologic and Hydraulic*
[4] Subber W and Sarkar A 2010 *High Performance Computing Systems and Applications* (*Lecture Notes in Computer Science* vol 5976) ed et al D M (Springer Berlin / Heidelberg) pp 251–268
[5] Farhat C, Lesoinne M and Pierson K 2000 *Numerical Linear Algebra with Application* **7** 687–714
[6] Balay S, Buschelman K, Gropp W D, Kaushik D, Knepley M G, McInnes L C, Smith B F and Zhang H 2009 PETSc Web page http://www.mcs.anl.gov/petsc
[7] Message passing interface forum http://www.mpi-forum.org
[8] Karypis G and Kumar V 1995 METIS unstructured graph partitioning and sparse matrix ordering system
[9] Ghanem R and Spanos P 1991 *Stochastic Finite Element: A Spectral Approach* (New York: Springer-Verlag)
[10] Saad Y 2003 *Iterative methods for sparse linear systems* 2nd ed (Philadelphia)
[11] Toselli A and Widlund O 2005 *Domain Decomposition Methods - Algorithms and Theory* (*Springer Series in Computational Mathematics* vol 34) (Berlin: Springer)
[12] Li J and Widlund O 2010 *International Journal for Numerical Methods in Engineering* **66**(2) 250–271

# CellPilot: Seamless communication within Cell BE and heterogeneous clusters

**N Girard, J Carter, W B Gardner, G Grewal**
School of Computer Science, University of Guelph, ON, Canada

E-mail: {ngirard, jcarter, gardnerw, ggrewal}@uoguelph.ca

**Abstract.** The Pilot library is targeted to novice scientific programmers within High Performance Computing. The CellPilot library extends the Pilot library to the Cell Broadband Engine processor and heterogeneous clusters. Using Pilot's process and channel abstractions, the CellPilot library can create a process on any of the processor types, both PPEs and SPEs, across the cluster. Communication is achieved by creating a channel between any two processes, and using the write/read channel functions in the participating processes. The CellPilot library uses MPI for the inter-node communication and the Cell SDK within a Cell node. All the architecture specific details of Cell communications are hidden from the user.

## 1. Introduction

The Cell BE processor by IBM, Sony and Toshiba offers tremendous speeds: the third fastest supercomputer is the Roadrunner, a Cell-based cluster located in Los Alamos, New Mexico, with a peak performance of over 1 petaflops [1]. The top 6 green supercomputers are Cell-based clusters, with 722.98 MFLOPS/W for first position [2].

However, Cell-based computing clusters available to the scientific research community can be underutilized due to the Cell's complex architecture and the difficulty programming for this environment. Programmers face small local storage on the accelerators, explicit address alignment for the Direct Memory Access (DMA) transfers and using several programming paradigms simultaneously in order to use the Cell cluster appropriately.

The Pilot library [3] is a new library for high-performance computing built on top of MPI, targeted to novice scientific programmers. The CellPilot library applies Pilot's concepts to the Cell BE processor and heterogeneous clusters. The goal of the CellPilot library is to have identical style communication, from a programmer's point-of-view, between any two types of processes within a Cell cluster. The CellPilot library hides the alignment of data for DMA transfers making it significantly easier to do communication on the cluster. Programmers will only need to work in a single paradigm when using the CellPilot library.

The following sections will give further details of the Cell BE processor, the Pilot library, the CellPilot library and the issues faced when communicating across a Cell cluster.

## 2. The Cell Broadband Engine processor

The Cell processor [4] is made up of 9 cores: one Power-PC Element (PPE) holds the operating system and is essentially the coordinator of the 8 Synergistic Processing Elements (SPEs), which are smaller processors used for accelerated computation and have a small local storage space of 256KB. These 9 cores are connected via a specialized high-bandwidth circular data bus called the Element

Interconnect Bus (EIB). Two Cell processors can be connected via their I/O elements to make a Cell blade. Typical high-performance Cell clusters are constructed using Cell blades and a few quad-core processors such as the Xeon processor. These types of clusters are hard to program due to their unconventional heterogeneous architectures.

### *2.1. The Cell Software Development Kit*

The Cell Software Development Kit (SDK) offers an API for creating, building, simulating and testing Cell applications. The SDK includes many libraries to optimize the performance of vector and linear mathematics computation, Fast Fourier Transforms, and data communication.

### *2.2. The Data Communication and Synchronization library*

The Data Communication and Synchronization library (DaCS), included in the Cell SDK, provides services to ease the development of applications on the Cell BE processor by providing a hierarchical topology of processing elements, resource and process management, and data communication services. However, there are limitations to the DaCS library. Communication between SPEs is not directly possible due to the hierarchical model of DaCS, and data transfers between nodes is not possible with the DaCS library without the use of an external library such as MPI.

### *2.3. Programming issues*

Many of the Cell's inconveniences are due to its heterogeneous multi-core design. All data transfers within the Cell blade is done via DMA calls. For optimal transfer performance, the data to be transferred should be aligned to a quad-word. The communication between two different Cell nodes must be done via the PPE. Therefore three steps are needed for the communication between two SPEs of different nodes.

Another constraint related to coding for the Cell processor is the local storage room on the SPEs. Each SPE has only 256KB of local storage for the application code and data. Programmers must pay special attention not to exceed this limitation while dividing their application.

## 3. The Pilot library

Pilot [5] is a new library for high-performance computing built on top of MPI, adding a layer of abstraction to the communication calls. Pilot's structure follows the Communicating Sequential Processes' (CSP, [6]) process-channel paradigm and it has a very small set of functions. The communication calls follow C's syntax for fprintf and fscanf, first specifying the channel, then the format and finally the values or variables.

Pilot involves two phases to help the novice programmer put together a working application. The first phase is the configuration phase where the application architecture, comprising processes and channels, is defined. The second phase, execution, is where all the calculations and communication are done. Pilot comes equipped with an integrated deadlock detection tool which can detect a variety of deadlock situations during the application's runtime.

## 4. CellPilot

The CellPilot library's aim is to extend the Pilot library for the global communication of heterogeneous clusters including the Cell BE processor. Cell Pilot uses a single communication abstraction for all communication types between two processors anywhere across the cluster. With the CellPilot library, data can be transferred from one end of a heterogeneous cluster to another with a single fprintf- or fscanf-style call by each of the sender and receiver processes, whether they be situated on different nodes or if the process is a PPE, an SPE or a non-Cell process.

With the Cell SDK and the DaCS library, the communication between two SPEs from different nodes involves code on both SPEs and on their corresponding parent PPEs. The CellPilot library follows the same conceptual model as Pilot, with minor changes to accommodate the Cell architecture. Only a single function is needed to communicate between any two processes across the cluster.

The Pilot paradigm helps programmers organize their parallel programs by first designing the process-channel diagram which should help reduce common communication errors.

*4.1. SPE process creation and deployment*
A new process type is introduced to the CellPilot library, the SPE process. The SPE process is similar to a regular process except that it will be run on a physical SPE processor, is created using a different function that specifies a "program handle" for the code that will be run, must be explicitly started, and has the SPE local store's memory size restriction. We require that the SPE processes be explicitly started to allow for SPEs to be reused by multiple processes under the control of their PPE process. SPE processes run to completion, after which an SPE is available for reuse. Starting more SPE processes than the currently available SPEs is an error. That is, CellPilot does not attempt to schedule or swap SPE processes.

Channels are still created using the same method as with Pilot and can be created between any two process types across the cluster whether they are an SPE, a PPE or a process of a non-Cell node. The channels are used for communication in the same way as is done using the Pilot library.

Thus there are effectively five types of channels, depending on the locations of their endpoints, as seen in Table 1. The local communication is done via DMA transfers and MPI is used for the communication between nodes.

| Channel Type | Processes involved | | |
|---|---|---|---|
| Type 1 | PPE/non-Cell | Remote | PPE/non-Cell |
| Type 2 | PPE | Local | SPE |
| Type 3 | PPE/non-Cell | Remote | SPE |
| Type 4 | SPE | Local | SPE |
| Type 5 | SPE | Remote | SPE |

**Table 1.** Channel types.

*4.2. Co-Pilot process*
The CellPilot library makes use of a second MPI process started on each Cell node. This process, called the Co-Pilot process, is a fully automated process that relays information to the SPEs and coordinates the intra-Cell and inter-node communication for the SPEs. The Co-Pilot process handles the start-up and completion of the SPE processes.

The Co-Pilot process coordinates the communication within the Cell node. The SPE processes signal the Co-Pilot process of a read or write on a channel involving a PPE or another SPE process of that node. The Co-Pilot process passes information such as the format, the location of the process and the buffer address to the other process.

With the CellPilot library, communication between a Cell node and a non-Cell node within a heterogeneous cluster is possible with a read or write over a channel. The Co-Pilot process formats the variables and calls the appropriate MPI functions to communicate with another Co-Pilot thread of another Cell node or a process of a non-Cell node. The Co-Pilot process allows the seamless communication across a heterogeneous cluster.

## 5. Communication scenario
The most complicated communication pattern on a Cell-based cluster is the communication between the SPEs of two different nodes. To do this, the SPEs communicate through their parent PPEs and use MPI for the communication between the two different nodes. Figure 1 illustrates this communication process.

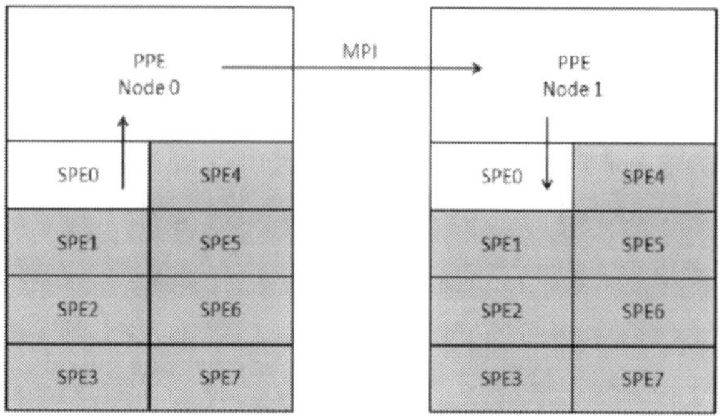


**Figure 1.** Communication across a Cell cluster.

### *5.1. Explanation*

In this communication scenario, the programmer wishes to send data from SPE0 of Node 0 to SPE0 of Node 1, as seen in Figure 1 above. Due to the hierarchical nature of the Cell, the data needs to be transferred through the PPEs of each corresponding node. Therefore, three separate transfer steps are coded: from SPE0 of node 0 to the PPE of node 0, from the PPE of node 0 to the PPE of node 1 using MPI, then from the PPE of node 1 to SPE0 of node 1.

### *5.2. Code comparison*

The appendix shows the code for the communication of the scenario in figure 1 using the Cell SDK, the DaCS library and the CellPilot library. The Cell SDK and the DaCS library have roughly the same amount of code across four processes for the communication between two SPEs of different nodes. The CellPilot library only requires one line of code on each of the SPEs for the communication, hiding the details of the three separate transfers.

The communication set-up (sharing of memory location between PPE and SPE) for the Cell SDK and DaCS library has to be done each time these two processes need to communicate. The CellPilot library's set-up is only done once (the creation of a channel object) at the beginning of the program during the configuration phase.

Table 2 shows the number of lines of code for a Cell SDK, DaCS and CellPilot implementation of a program where an array of 10 integers is sent from the PPE of node 0 to the SPE of node 0, then from SPE of node 0 to SPE of node 1, then from SPE of node 1 to PPE of node 1.

| Program Implementation | Lines of code |
|---|---|
| Cell SDK | 186 |
| DaCS | 114 |
| CellPilot | 80 |

**Table 2.** Number of lines of code for a simple program implemented with the Cell SDK, DaCS library and CellPilot library.

## 6. Conclusion and future work

It can be seen from the contrasting code samples that Cell communication using CellPilot is much simpler to program than either the Cell SDK or the DaCS library. The Pilot paradigm helps programmers organize their parallel programs and avoid common communication errors. A programmer that is familiar with the Pilot library should not have difficulty using the CellPilot library.

CellPilot is currently under development. Future work includes implementing Pilot's collective communication functions and deadlock detection on the Cell.

## 7. References

[1] http://www.top500.org, June 2010.

[2] http://www.green500.org, November 2009.

[3] Carter J, Gardner W B, Grewal G, 2010 The Pilot Approach to Cluster Programming in C *Proc. of the 24th IEEE International Parallel & Distributed Processing Symposium, Workshops and Phd Forum* (Atlanta) 8 pp.

[4] CBE programming tutorial ver 3.1, http://www.ibm.com/developerworks/power/cell/, 2008.

[5] http://carmel.cis.uoguelph.ca/pilot.

[6] Hoare C A R, 1978 Communicating Sequential Processes *Communications of the ACM* **21** (8) 666-677.

## Appendix A. Cell SDK code

```
--- Sender SPE ---
unsigned int add = (unsigned int) spu_read_in_mbox();
int Array[100] __attribute__ (__aligned__ (128));  //put data here

...

mfc_put(Array, add, sizeof(int)*100, 10, 0, 0);  // size in bits
mfc_write_tag_mask(1<<10);
mfc_read_tag_status_all();  //wait for transfer to complete
spu_write_out_mbox((unsigned int)1);  //notify PPE


--- Sender PPE ---
int Array[100] __attribute__ (__aligned__ (128));
int retVal;
unsigned int *add, data, temp;
spe_context_ptr_t ctx; //holds the SPE info

...

data = (unsigned int) &Array;
while (spe_in_mbox_status(ctx)==0);  //wait for room in mailbox

do {
     add = &data;
     retVal = spe_in_mbox_write(ctx, add, 1, SPE_MBOX_ANY_NONBLOCKING);
} while(retVal!=1); //retry until success
while(!spe_out_mbox_status(ctx));  //wait for message from SPE
spe_out_mbox_read(ctx, &temp, 1);

MPI_Send(&Array, 100, MPI_INT, 1, 0, MPI_COMM_WORLD);


--- Receiver PPE ---
//about 22 lines of code is needed to start the SPE process

int Array[100] __attribute__ (__aligned__ (128));
int retVal;
unsigned int *add, data, temp;
spe_context_ptr_t ctx; //holds the SPE info
MPI_Status status;

...

MPI_Recv(&Array, 100, MPI_INT, 0, 0, MPI_COMM_WORLD, &status);

data = (unsigned int)&Array;
while(spe_in_mbox_status(ctx)==0); //wait for room in mailbox

do {
```

```
        add = &data;
        retVal = spe_in_mbox_write(ctx, add, 1, SPE_MBOX_ANY_NONBLOCKING);
    } while(retVal!=1); //retry until success
    while(!spe_out_mbox_status(ctx)); //wait for a message from SPE
    spe_out_mbox_read(ctx, &temp, 1);


    --- Receiver SPE ---
    //about 22 lines of code is needed to start the SPE process

    unsigned int add = (unsigned int)spu_read_in_mbox();
    int Array[100] __attribute__ (__aligned__ (128));

    ...

    mfc_get(&Array, add, sizeof(int)*100, 31, 0, 0);
    mfc_write_tag_mask(1<<31);
    mfc_read_tag_status_all();
    spu_write_out_mbox((unsigned int)1); //notify PPE
```

## Appendix B. DaCS code

```
--- Sender SPE ---
uint64_t size;
int Array[100];
dacs_wid_t waitid;
dacs_remote_mem_t location;

...

//accept memory location
dacs_remote_mem_accept(DACS_DE_PARENT, DACS_PID_PARENT,  &location);
//get memory size
dacs_remote_mem_query(location, DACS_REMOTE_MEM_SIZE,  &size);

dacs_put(location, 0, &Array, size, waitid, DACS_ORDER_ATTR_NONE,
     DACS_BYTE_SWAP_DISABLE);
dacs_wait(waitid); //wait for transfer to complete
dacs_remote_mem_release(&location);


--- Sender PPE ---
de_id_t spe_de;
dacs_process_id_t spe_pid;

//2 more lines of code to start the SPE process

int Array[100];
dacs_remote_mem_t location;

...

dacs_remote_mem_create(&Array, sizeof(int)*100, DACS_WRITE_ONLY,
     &location); //create memory region
dacs_remote_mem_share(spe_de, spe_pid, location);
dacs_remote_mem_destroy(&location);

MPI_Send(&Array, 100, MPI_INT, 1, 0, MPI_COMM_WORLD);


--- Receiver PPE ---
de_id_t spe_de;
dacs_process_id_t spe_pid;

//2 more lines of code to start the SPE process
```

```

int Array[100];
dacs_remote_mem_t location;
MPI_Status status;

...

MPI_Recv(&Array,100,MPI_INT,0,0,MPI_COMM_WORLD,&status);

dacs_remote_mem_create(&Array, sizeof(int)*100, DACS_READ_ONLY,
    &location); //create memory region
dacs_remote_mem_share(spe_de, spe_pid, location);
dacs_remote_mem_destroy(&location);


--- Receiver SPE ---
uint64_t size;
int Array[100];
dacs_remote_mem_t location;
dacs_wid_t waitid;

...

//accept memory location
dacs_remote_mem_accept(DACS_DE_PARENT, DACS_PID_PARENT, &location);
dacs_remote_mem_query(location, DACS_REMOTE_MEM_SIZE, &size);

dacs_get(&Array, location, 0, size, waitid, DACS_ORDER_ATTR_NONE,
    DACS_BYTE_SWAP_DISABLE);
dacs_wait(waitid); //wait for transfer to complete
dacs_remote_mem_release(&location);
```

## Appendix C. CellPilot code

```
#include <stdio.h>
#include <pilot.h>

extern spe_program_handle_t spe_send, spe_recv;

PI_CHANNEL *betweenSPEs

--- Receiver PPE function ---
int recvFunc(int arg, void *ptr){
    PI_RunSPE(recvSPE, 0, NULL);

    return 0;
}

--- Sender PPE function ---
//(configuration phase also done by Receiver PPE)
int main(int argc, char *argv[]){
    int N = PI_Configure(&argc, &argv);

    PI_PROCESS *recvPPE = PI_CreateProcess(recvFunc, 0, NULL);
    PI_PROCESS *sendSPE = PI_CreateProcess(spe_send, PI_MAIN, 0);
    PI_PROCESS *recvSPE = PI_CreateProcess(spe_recv, recvPPE, 0);

    betweenSPEs = PI_CreateChannel(sendSPE, recvSPE);

    PI_StartAll();
    PI_RunSPE(sendSPE, 0, NULL);

    PI_StopMain(0);
    return 0;
}
```

```
--- Sender SPE ---
//contents of spe_send.c
PI_CHANNEL_SPE betweenSPEs

PI_SPE_PROCESS(int arg1, void *arg2)
     int Array[100], i;

     for(i=0; i<100; i++)
       Array[i]=i;

     PI_Write(betweenSPEs, "%100d", Array);
PI_SPE_END

--- Receiver SPE ---
//contents of spe_recv.c
PI_CHANNEL_SPE betweenSPEs

PI_SPE_PROCESS(int arg1, void *arg2)
     int Array[100], i;

     PI_Read(betweenSPEs, "%*d", 100, &Array);

     for(i=0; i<100; i++)
       printf("%d ", Array[i]);
     printf("\n");
PI_SPE_END
```

# Research computing in a distributed cloud environment

**K. Fransham[1], A. Agarwal[1], P. Armstrong[1], A. Bishop[1], A. Charbonneau[2], R. Desmarais[1], N. Hill[3], I. Gable[1], S. Gaudet[3], S. Goliath[3], R. Impey[2], C. Leavett-Brown[1], J. Ouellete[3], M. Paterson[1], C. Pritchet[1], D. Penfold-Brown[1], W. Podaima[2], D. Schade[3], R.J. Sobie[1,4]**

[1] Department of Physics and Astronomy, University of Victoria, Victoria, Canada V8W 3P6
[2] National Research Council Canada, 100 Sussex Drive, Ottawa, Canada
[3] National Research Council Herzberg Institute of Astrophysics, 5071 West Saanich Road, Victoria, BC, Canada, V9E 3E7
[4] Institute of Particle Physics of Canada.

E-mail: `fransham@uvic.ca`

**Abstract.** The recent increase in availability of Infrastructure-as-a-Service (IaaS) computing clouds provides a new way for researchers to run complex scientific applications. However, using cloud resources for a large number of research jobs requires significant effort and expertise. Furthermore, running jobs on many different clouds presents even more difficulty. In order to make it easy for researchers to deploy scientific applications across many cloud resources, we have developed a virtual machine resource manager (*Cloud Scheduler*) for distributed compute clouds. In response to a user's job submission to a batch system, the *Cloud Scheduler* manages the distribution and deployment of user-customized virtual machines across multiple clouds. We describe the motivation for and implementation of a distributed cloud using the *Cloud Scheduler* that is spread across both commercial and dedicated private sites, and present some early results of scientific data analysis using the system.

## 1. Introduction

Infrastructure as a Service (IaaS) cloud computing is emerging as a new and efficient way to provide computing to the research community. Clouds are considered to be a solution to some of the problems encountered with early adaptations of grid computing where the site retains control over the resources and the user must adapt their application to the local operating system, software and policies. This often leads to difficulties, especially when a single resource provider must meet the demands of multiple projects or when projects cannot conform to the configuration of the resource provider. IaaS clouds offer a solution to these challenges by delivering computing resources using virtualization technologies. Users lease the resources from the provider and install their application software within a virtual environment. This frees the providers from having to adapt their systems to specific application requirements and removes the software constraints on the user applications. In most cases, it is easy for a user or a small project to create their virtual machine (VM) images and run them on IaaS clouds. However, the complexity rapidly increases for projects with large user communities and significant computing requirements. In this paper we describe a system that simplifies and combines multiple IaaS clouds into a single distributed cloud for High Throughput Computing (HTC) workloads.

The growing interest in clouds can be attributed in part to the ease in encapsulating complex research applications in Virtual Machines (VMs), often with little or no performance degradation [1]. Studies have shown, for example, that particle physics application code runs equally well in a VM as on the native system [2]. Today, open source virtualization software such as Xen [3] and KVM [4] are incorporated into many Linux operating system distributions, resulting in the use of VMs for a wide variety of applications. Often, special purpose servers, particularly those requiring high availability or redundancy, are built inside a VM making them independent of the underlying hardware and allowing them to be easily moved or replicated. It is also not uncommon to find an old operating system lacking the drivers for new hardware, a problem which may be resolved by running the old software in a virtual environment.

The deployment and management of many VMs in an IaaS cloud is labour intensive. This can be simplified by attaching the VMs to a job scheduler and utilizing the VMs in a batch environment. The Nimbus Project [5] has developed the *one-click cluster* solution. This provides a batch system on multiple clouds using one type of VM [6].

We further simplify the management of VMs in an IaaS cloud and provide new functionality with the introduction of *Cloud Scheduler*. *Cloud Scheduler* provides a means of managing user-customized VMs on any number of science and commercial clouds[1].

In the following sections we present the architecture of our system and highlight the role of *Cloud Scheduler* to manage VMs on IaaS clouds in a batch HTC environment. We present early results on the operation, and highlight the successes of the system.

## 2. System architecture

The architecture of the HTC environment running user-customized VMs on distributed IaaS clouds is shown in fig. 1. A user creates their VM and stores it in the *VM image repository*. They write a job script that includes information about their VM and submits it to the *Condor Job Scheduler*. The *Cloud Scheduler* reads the queues of the *Condor Job Scheduler*, requests that one of the available cloud resources boot the user VM, the VM advertises itself to the *Condor Job Scheduler* which then dispatches the user job to that VM. In this section, we describe the components of the system and the method used to manage VMs.

### *2.1. System components*

In this section we describe the following components: *VM image repository*, the cloud resources, the *Condor Job Scheduler* and the *Cloud Scheduler*.

**VM image repository**

The user builds their VM by first retrieving a base image from a *VM image repository*. The base images are simple Linux VM images or images that include project or application based code. Once the user has modified their image, they will store it back in the *VM image repository*. The repository may reside at a single site or be distributed, however, it must be accessible to the cloud resources.

**Cloud resources**

The system currently supports Amazon EC2 and IaaS clouds using Nimbus [5]. Support for IaaS clouds using OpenNebula [7] and Eucalyptus [8] is under development.

**Job Scheduler**

The job scheduler used in the system is the Condor HTC job scheduler [9]. Condor was

[1] Commercial cloud providers include Amazon, RackSpace and IBM, to name a few. Science clouds use hardware resources funded by governments for research purposes and are located in universities or national laboratories.

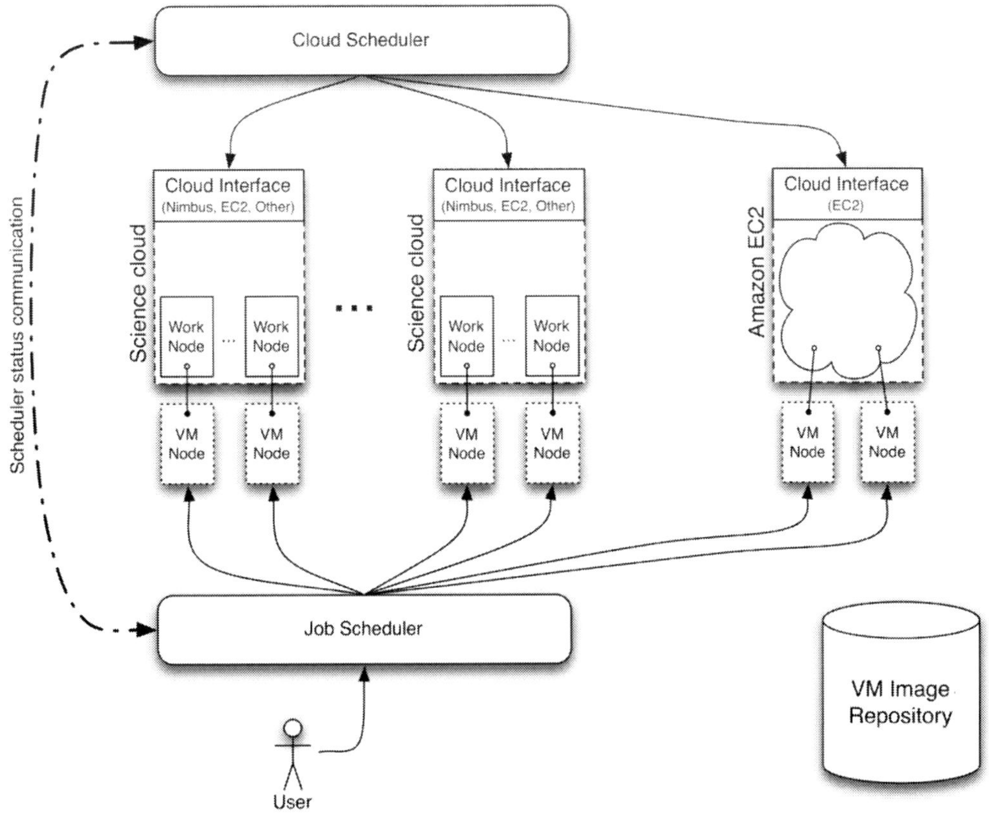


**Figure 1.** An overview of the architecture used for the system. A user prepares their VM image and a job script. The job script is submitted to the *Condor Job Scheduler*. The *Cloud Scheduler* reads the job queue and makes a request to boot the user VM on one of the available clouds. Once there are no more user jobs requiring that VM type, the *Cloud Scheduler* makes a request to the proper cloud to shutdown the user VM.

chosen because it was designed to utilize heterogeneous idle workstations which makes it ideal to use as a job scheduler for a dynamic VM environment. Condor has a central manager which matches user jobs to resources based on job and resource attributes. The Condor *startd daemon* must be installed and started when a VM image is booted. The VM then advertises its existence to the Condor central manager[2]. Users submit jobs by issuing the `condor_submit` command. The user must add a number of additional parameters specifying the location and properties of the VM. The description of the parameters is found in section 4.

**Cloud Scheduler**

The *Cloud Scheduler* is an object oriented python-based package designed to manage VMs for jobs based on the available cloud resources and job requirements. Users submit jobs to the *Condor Job Scheduler* after they have been authenticated using X.509 Proxy Certificates [10]. The certificates are also used to authenticate starting, shutting down, or polling VMs

[2] We use Condor Connection Brokering (CCB) to allow VM worker nodes that use Network Address Translation (NAT) to connect to the Condor Central Manager.

with Nimbus clusters. Authentication with EC2 is done by using a standard shared access key and secret key. The *Cloud Scheduler* then communicates with the IaaS clouds available to it, and manages the deployment and shutdown of VMs to run the jobs in the *Condor Job Scheduler* queue.

## 3. VM management

When *Cloud Scheduler* is started it reads configuration files that describe the available cloud resources. To schedule user jobs across the available clouds, the *Cloud Scheduler* starts the following threads that are run on a periodic basis.

(i) The *JobPoller* thread maintains the state and metadata of the jobs that are queued and running on the *Condor Job Scheduler*. It effectively maps the *Condor Job Scheduler* queue into the *JobPool*.

(ii) The *Scheduler* thread starts and stops VMs based on the information in the *JobPool*, satisfying the resource demands of the workload. The design goal of *Cloud Scheduler* is to leave prioritization and scheduling decisions in the domain of the *Condor Job Scheduler*. However, the order in which the *Scheduler* thread provides resources can impact the scheduling algorithms of the *Condor Job Scheduler*.
The *Scheduler* thread also monitors the VMs and, if necessary, updates the state of VMs in the *JobPool*. It will shut down VMs that are in an error state; if there are jobs that still require this VM, then the *Scheduler* will start a new instance of the VM to replace the one it has shut down.

(iii) The *CleanUp* thread stops VMs that are no longer required. It can correct the state of the job in the *JobPool*. If a VM is shut down due to an error, then the *CleanUp* thread changes the state of the job in the *JobPool* from "scheduled" to "new" so that a new VM can be created for that job.

When *Cloud Scheduler* is shut down, it can either shut down all the VMs that is has started, or it can persist its state. In the latter case, the VMs continue to run the jobs. *Cloud Scheduler* reloads the state when it is restarted and resumes managing the jobs and resources.

## 4. User Interaction

The users of the distributed cloud are typically researchers who have extensive knowledge of job submission to traditional HTC batch systems. The goal of the system developed around the *Cloud Scheduler* is to maintain as much of the user's existing workflow as possible. To that end, the user's job submission script follows the traditional Condor format, with some extra parameters required to configure the user's VM. A sample job script is shown below, and a description of the additional *Cloud Scheduler* attributes follows in table 1.

```
Regular Condor Attributes
Universe                = vanilla
Executable              = script.sh
Arguments               = one two three
Log                     = script.log
Output                  = script.out
Error                   = script.error
should_transfer_files   = YES
when_to_transfer_output = ON_EXIT
#
# Cloud Scheduler Attributes
Requirements =
+VMType                 = "vm-name"
+VMLoc                  = "http://repository.tld/your.vm.img.gz"
+VMAMI                  = "ami-dfasfds"
+VMCPUArch              = "x86"
+VMCPUCores             = "1"
+VMNetwork              = "private"
+VMMem                  = "512"
+VMStorage              = "20"
Queue
```

**Table 1.** Description of customizable attributes required by the *Cloud Scheduler*

| Attribute | Description |
|---|---|
| VMType | Unique name of required VM |
| VMLoc | URL (Nimbus) of the image |
| VMAMI | AMI (EC2-like) of the image |
| VMCPUArch (x86) | CPU architecture (x86 or x86_64) |
| VMCPUCores (1) | Number of CPU cores |
| VMStorage | Required storage space |
| VMMem | RAM required |
| VMNetwork | Network required (public/private) |

## 5. Results

Currently, a distributed cloud employing the architecture described above is being used to generate simulated data for the BaBar particle physics experiment. [3] This distributed cloud is managed by the *Cloud Scheduler*, and can boot up to 84 simultaneous virtual machines on three different cloud sites: the system has been configured to have access to 24 CPU cores on a Nimbus-based at the National Research Council (NRC) in Ottawa, Canada; 40 CPU cores on a Nimbus-based cloud at the University of Victoria (UVic) in Victoria, Canada; and 20 cores on Amazon EC2. The *Cloud Scheduler*, *Condor Job Scheduler*, and *VM Image Repository* are all located at the NRC.

[3] Based at the SLAC National Accelerator Laboratory in Stanford, California, the BaBar experiment collides high-energy electrons and positrons, and studies the results of the collision in order to gain insight into the properties of exotic sub-atomic particles.

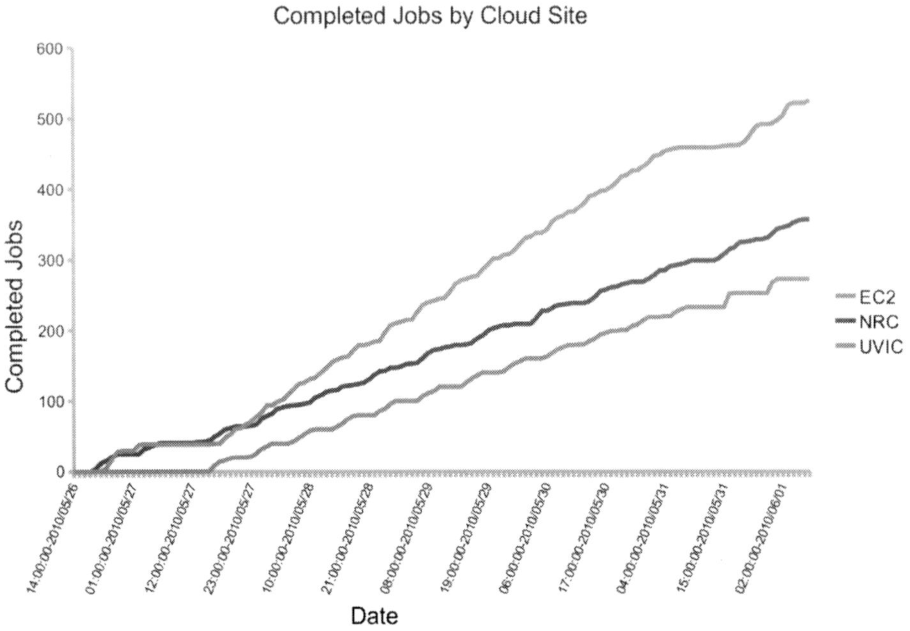


**Figure 2.** A comparison of the number of jobs that ran to completion at all three cloud sites in the distributed cloud. UVic, having the largest number of CPUs available, ran the most jobs over the course of the week.

A VM image containing the BaBar simulation application has been created, and is stored in the *VM Image Repository*. The application itself contains C++ and FORTRAN code, requires 1GB or more of RAM, and is completely contained in a VM approximately 16 GB in size. The BaBar application contains scripts to interface the production of simulated data with a local batch scheduler; in order to run on the distributed cloud, the only change to the software that was required was to point the simulation production manager to the cloud's *Condor Job Scheduler*. As described in section 4, a stated goal of the distributed cloud system was to avoid altering the cloud user's ordinary workflow as much as possible. In the case of Babar simulation production, this goal was certainly achieved, since any user familiar with running the simulation production application is able to use the same commands to run it on the distributed cloud.

When the system is functioning correctly, the *Cloud Scheduler* boots VMs and jobs are run on all of the available clouds sites available. Figure 2 shows the total number of completed jobs on the three different cloud sites over the course of a week. The increase in completed jobs from each of the three sites over the course of the week shows that the system is functioning as expected: the *Cloud Scheduler* has turned the three unique IaaS clouds into a single distributed cloud. The flat sections in the curves correspond to times when there were no jobs in the queue.

Furthermore, the BaBar collaboration places strict checks on the quality of any produced simulated data. The data that continues to be produced on the distributed cloud has passed all quality checks, further demonstrating the robustness and usefulness of the system.

The *Cloud Scheduler* is also a central component of a different distributed cloud used by CANFAR [14], an astronomy project led by researchers at the University of Victoria and the

National Research Council of Canada Herzberg Institute of Astrophysics. The goal is to provide researchers the ability to create custom environments for analyzing data from survey projects. Currently the system is available to early adopters, using two clusters: one at the WestGrid facility located at the University of Victoria (25 machines, 200 cores) and one at the Herzberg Institute of Astrophysics (6 machines, 32 cores). The system has been tested successfully with over 9,000 jobs utilizing more than 33,000 core hours. Work is currently underway to streamline the process of creating and sharing virtual machines with researchers, to allow them to easily make VM images that suit their needs and ready for deployment on cloud resources.

## 6. Summary

We have presented a new method for running large scale complex research applications across multiple IaaS computing clouds. The development of the *Cloud Scheduler* is central to the distributed cloud, as it simplifies the management of virtual machine resources across multiple science and commercial clouds [15] by hiding the complexity of VM management. We have demonstrated that the system works for both astronomy and particle physics applications, using multiple cloud resources.

The support of CANARIE, the Natural Sciences and Engineering Research Council, the National Research Council of Canada and Amazon are acknowledged.

## References

[1] M. Alef and I. Gable. HEP specific benchmarks of virtual machines on multi-core CPU architectures. J. Phys Conf. Ser. 219, 052015 (2008). doi: 10.1088/1742-6596/219/5/052015

[2] A. Agarwal, A. Charbonneau, R. Desmarais, R. Enge, I. Gable, D. Grundy, D. Penfold-Brown, R. Seuster, R.J. Sobie, and D.C. Vanderster. Deploying HEP Applications Using Xen and Globus Virtual Workspaces. J. Phys.: Conf. Ser. 119, 062002 (2008). doi: 10.1088/1742-6596/119/6/062002

[3] P. Barham, B. Dragovic, K. Fraser, S. Hand, T. Harris, A. Ho, R. Neugebauer, I. Pratt and A. Warfield. Xen and the art of virtualization. SOSP03: Proceedings of the Nineteenth ACM Symposium on Operating Systems Principles.

[4] KVM (Kernel Based Virtual Machine). http://www.linux-kvm.org/page/Main_Page.

[5] K.Keahey, I.Foster, T.Freeman, and X. Zhang. Virtual workspaces: Achieving quality of service and quality of life in the Grid. Sci. Program. Vol. 13 (2005) 265.

[6] K.Keahey and T.Freeman. Contextualization: Providing One-Click Virtual Clusters. SCIENCE08: Proceedings of the 2008 Fourth IEEE International Conference on eScience.

[7] Open Nebula (Open Source Toolkit for Cloud Computing). http://www.opennebula.org/.

[8] D. Nurmi, R. Wolski, C. Grzegorczyk, G. Obertelli, S. Soman, L. Youseff, and D. Zagorodnov. The Eucalyptus Open-Source Cloud-Computing System. CCGRID '09: Proceedings of the 2009 9th IEEE/ACM International Symposium on Cluster Computing and the Grid. doi:10.1109/CCGRID.2009.93

[9] D.Thani, T.Tannenbaum and M.Livny. Distributed computing in practice: the Condor experience. Concurrency and Computation: Practice and Experience Vol. 17 (2005) 323.

[10] S.Tuecke, V.Welch, D.Engert, L.Pearlman,and M.Thompson. Internet X.509 Public Key Infrastructure (PKI) Proxy Certificate Profile. http://www.ietf.org/rfc/rfc3820.txt.

[11] A. Agarwal, M. Ahmed, A. Berman, B.L. Caron, A. Charbonneau, D. Deatrich, R. Desmarais, A. Dimopoulos, I. Gable, L.S. Groer, R. Haria, R. Impey, L. Klektau, C. Lindsay, G. Mateescu, Q. Matthews, A. Norton, W. Podaima, D. Quesnel, R. Simmonds, R.J. Sobie, B.St. Arnaud, C. Usher, D.C. Vanderster, M. Vetterli, R. Walker, M. Yuen. GridX1: A Canadian computational grid. Future Generation Computer Systems **23** (2007) 680. doi:10.1016/j.future.2006.12.006

[12] A. Agarwal, P. Armstrong, R. Desmarais, I. Gable, S. Popov, S. Ramage, S. Schaffer, C. Sobie, R.J. Sobie, T. Sulivan, D.C. Vanderster, G. Mateescu, W. Podaima, A. Charbonneau, R. Impey, M. Viswanathan, D. Quesnel, BaBar MC Production on the Canadian Grid using a Web Services Approach. J. Phys.: Conf. Ser. 119 (2007) 072002. doi: 10.1088/1742-6596/119/7/072002

[13] CANARIE. Canada's Advanced Research and Innovation Network. http://www.canarie.ca.

[14] CANFAR. Canadian Advanced Network For Astronomical Research. http://astrowww.phys.uvic.ca/ canfar/

[15] The *Cloud Scheduler* code is available at http://cloudscheduler.org.

# GPU-accelerated 3-D model-based tracking

**J Anthony Brown and David W Capson**

Department of Electrical and Computer Engineering, McMaster University, Hamilton, Ontario, L8S 4K1 Canada

E-mail: {`brownja5, capson`}`@mcmaster.ca`

**Abstract.** Model-based approaches to tracking the pose of a 3-D object in video are effective but computationally demanding. While statistical estimation techniques, such as the particle filter, are often employed to minimize the search space, real-time performance remains unachievable on current generation CPUs. Recent advances in graphics processing units (GPUs) have brought massively parallel computational power to the desktop environment and powerful developer tools, such as NVIDIA Compute Unified Device Architecture (CUDA), have provided programmers with a mechanism to exploit it. NVIDIA GPUs' single-instruction multiple-thread (SIMT) programming model is well-suited to many computer vision tasks, particularly model-based tracking, which requires several hundred 3-D model poses to be dynamically configured, rendered, and evaluated against each frame in the video sequence. Using 6 degree-of-freedom (DOF) rigid hand tracking as an example application, this work harnesses consumer-grade GPUs to achieve real-time, 3-D model-based, markerless object tracking in monocular video.

## 1. Introduction and Background

Tracking the pose of a 3-D object in live video, without the use of position encoders, markers, or other measurement tools, has applications across many domains including augmented reality, visual servoing, surveillance, and human-computer interaction (HCI). Approaches to 3-D tracking can be coarsely divided into appearance- and model-based [1]. Appearance-based methods use cues from the video sequence (skin colour, contours, etc.) whereas model-based methods project a 3-D representation of the tracking target onto a plane for comparison with each video frame. In both cases, the goal is to estimate the set of parameters that best describes the tracking target's pose. Model-based approaches often generate superior results and are algorithmically simpler because the parameters being estimated are the same parameters being manipulated to control the model. However, because such approaches require many 3-D renderings of the tracking target to be produced and evaluated for each frame, they are computationally intensive and consequently less popular than appearance-based techniques.

In this work, it is demonstrated that recent advances in consumer-grade graphics cards and the maturation of tools that exploit their highly parallel computational power has made 3-D model-based approaches to object tracking possible in real-time, at a relatively low cost. Specifically, using hand tracking [2] as an example, NVIDIA'S Compute Unified Device Architecture (CUDA) [3] and Microsoft's Direct3D [4] are employed to generate several thousand 3-D model renderings, compare each rendering to the current frame of the video stream, and compute an estimate for the pose of the tracking target based on the quality-of-fit of the renderings. The *particle filter* [5], a sequential monte carlo approach to stochastic state estimation, is applied to ensure that all model renderings are statistically likely to be realistic predictions of the tracking target's pose based on tracking history. The massive parallel processing power of the graphics processing unit's (GPU) *streaming multiprocessors* (SMs), aided by the particle filter, facilitates robust 3-D model-based object tracking at 30+ frames per second on consumer-grade hardware.

In recent years, the GPU has emerged as a powerful, affordable, and adaptable parallel computing platform suitable for many high-performance computing (HPC) tasks. Programming models, such as ATI Stream, OpenCL, and NVIDIA CUDA, have allowed programmers to exploit hundreds of SMs for general purpose computing on the GPU (GPGPU) instead of the 3-D graphics rendering for which they were originally designed. Serving as a coprocessor to the central processing unit (CPU), the GPU has been proven capable of accelerating applications 10 to 1000 times over CPU-only implementations [6, 7]. NVIDIA CUDA is a highly scalable platform introduced in 2007 to provide GPU computing on modern NVIDIA GeForce and Quadro products, millions of which are already deployed in PCs and workstations around the world [8]. With a large, collaborative online community, an ever-expanding number of CUDA university courses and textbooks [9], and a massive user-base, NVIDIA GPUs are already well-established in the scientific community. CUDA has been a particularly popular target for object tracking and pose estimation applications, specifically face, body, and hand tracking [10–12].

Pose estimation and tracking using a particle filter is very well-suited to GPU-acceleration, yet there is relatively little literature available describing the integration of the two. Real-time, appearance-based particle filtering approaches to face tracking are described in [13] and [14], with the former using a sparse template matching algorithm and the latter integrating several visual cues, such as color and edges. Both report speedups of at least 10 times over CPU implementations. A hybrid model-based and appearance-based tracker is described by Cabido *et al.* [15] for 2D articulated object pose estimation. GPU filtering, projection and weighting are also used in [16] for a divide-and-conquer, machine learning approach to 3-D model-based tracking of articulated objects in a surgical setting with both mono and stereo image sequences. Finally, Lenz *et al.* [12] describe an approach similar to ours, reporting a speedup of up to 10 times over a CPU implementation in face and hand tracking experiments.

The remainder of this paper focuses on the GPU-acceleration aspects of the tracking system, beginning with a review of the NVIDIA GT200 GPU architecture and how it can be exploited with CUDA and Direct3D in Section 2. The 3-D model based tracking system is detailed in Section 3, Section 4 contains the experimental results, and Section 5 concludes the paper with a discussion and summary of future work.

## 2. GPU Computing

A complete overview of the broad field of GPGPU is beyond the scope of this paper; instead, the focus is limited to material that will clarify and justify this approach to model-based tracking. Specifically, the hardware architecture of the NVIDIA GT200-series GPU and the manner through which it can be exploited using CUDA and Direct3D is explored.

### *2.1. NVIDIA GT200-Series GPU Architecture*

NVIDIA's highly-scalable GT200-series GPU was released in 2008 and built on the company's previous G80 architecture. The GT200 is structured as a hierarchy (Fig. 1) [17]. At the lowest level is the *streaming processor* (SP), a processing element that contains two arithmetic logic units (ALUs) and an instruction pointer, but no hardware for fetching or scheduling instructions. Next is the SM, essentially an independent multiprocessor able to fetch, schedule, and execute instructions. Each SM contains eight SPs, two special function units for advanced operations (sine, square root, etc.), shared memory, a register file (divided equally between SPs), and a read-only cache for constant memory. SMs are organized in groups of three called *texture processing clusters* (TPCs), each of which contains a read-only L1 texture cache. At the top level of the hierarchy, a GT200 GPU consists of up to ten TPCs, a read-only L2 texture cache, up to 2

GB of GDDR3 device memory (constant, global, local, and texture), and a peripheral control interface express (PCI-E) bus to communicate with the CPU. To give some perspective, high-end consumer-grade GPUs built on this impressive architecture are comprised of approximately 1.4 billion transistors, have a bandwidth up to 159 GB/s, and are theoretically capable of executing over 1 trillion parallel 32-bit floating point instructions per second (1 teraflop) on their 240 SPs, whereas a similarly priced multicore CPU is theoretically capable of approximately 48 gigaflops.

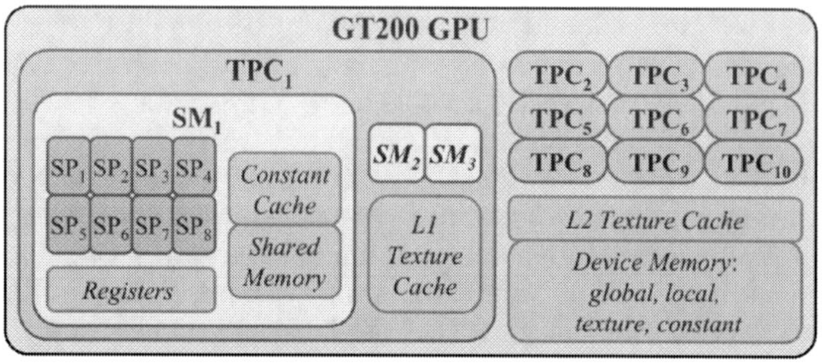


**Figure 1.** The GT200 processing element hierarchy consists of a GT200 GPU, 10 TPCs, 30 SMs, and 240 SPs.

This degree of parallel processing power can only be exploited by highly parallel tasks, such as the high-resolution 3-D graphics rendering for which it is intended. In fact, 1000's of concurrent threads are recommended to maximize hardware utilization in most applications. Unlike multicore CPU threads, a GPU thread is extremely lightweight and can be created and launched almost instantly. Additionally, general purpose CPUs are designed to maximize single-thread execution through branch prediction and other flow control techniques, whereas the majority of transistors on a GPU are dedicated to achieving maximum arithmetic intensity through massively parallel processing, while sacrificing individual thread performance.

Threads are partitioned by each SM into groups of 32 called *warps*, a paradigm referred to as *single-instruction multiple-thread* (SIMT), a variation on the single-instruction multiple-data (SIMD) class of parallel computers. Both SIMD and SIMT utilize parallel processing elements to simultaneously execute a common instruction on a vector of data; however, SIMD requires all threads to take the same execution path, whereas SIMT relaxes this constraint by allowing threads within a warp to diverge. *Warp divergence* forces all threads in a warp to execute every instruction branch taken by any thread in the warp, but unnecessary branch instructions are suppressed. As a result, computation time increases for highly-divergent code and is minimized when all warp threads agree on a common path. Additionally, SIMD threads have access to all vector elements whereas each SIMT thread maintains its own registers and must rely on shared memory (Section 2.2) to share data. Depending on resource utilization, up to 32 warps can be active on each SM, totaling 30,720 threads on a GPU. This allows a new warp to be swapped in by the SM scheduler while another is waiting on a high-latency instruction, such as a memory transfer. Because warp packaging and scheduling is handled by low-level hardware, the programmer could choose to completely ignore the concept of warps; however, significant performance gains can be realized if problems, such as warp divergence and other considerations (Section 2.2), are respected.

### *2.2. NVIDIA CUDA*

CUDA provides a set of C-language extensions that allow the programmer to create applications that execute serial code on the *host* (CPU) and can asynchronously launch functions called *kernels* to execute on the *device* (GPU), which serves as a coprocessor for data-intensive, parallelizable code [3, 18]. Scalability is a key tenet of CUDA. A well-written CUDA application is entirely independent of the underlying GPU and will scale to maximally utilize whatever hardware is available.

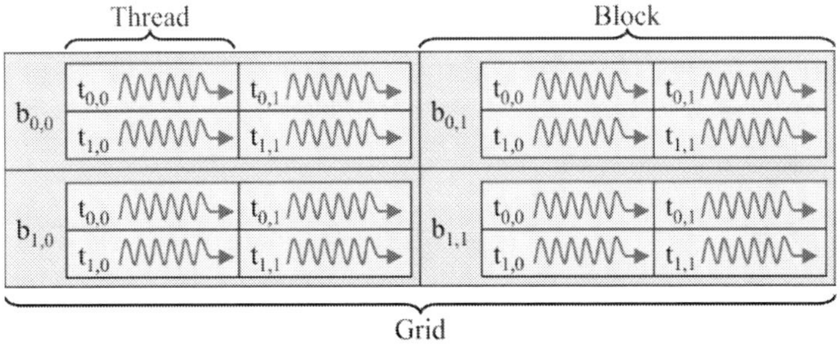


**Figure 2.** The GT200 thread hierarchy: a grid of blocks of threads.

This is achieved through a thread hierarchy as illustrated in Fig. 2. The kernel code is executed by fine-grained parallel threads, grouped into *blocks*. Blocks are arranged in a *grid*, providing second, coarser level of parallelism. A closer look at the GPU thread scheduling paradigm demonstrates the purpose of this hierarchy. One or more blocks are assigned to each SM and consecutive threads within those blocks are partitioned into warps and executed by SPs according to the SM's internal scheduler. However, there is no guarantee of concurrency for blocks within the grid. All blocks may be executed in parallel if enough SMs are available, but generally some will have to wait until other blocks are ready to be swapped out. Consequently, only threads within a block are able to share data and synchronize execution, meaning the programmer has no control over the coordination of warp-level or grid-level scheduling and interaction. The obvious benefit is scalability; as long as an arithmetically intense task can be partitioned into computationally independent blocks that do not need to communicate with other blocks, that task can be efficiently scheduled across all SMs of any capable NVIDIA GPU. Threads within each block can use the on-chip *shared memory* to efficiently exchange data and can synchronize execution through *barrier synchronization*, but the only way to synchronize all threads in the grid is to wait for a kernel to finish execution and launch another.

The manner in which a task is partitioned can significantly affect performance. If blocks are too large, there may not be enough to keep all available SMs busy; however, if they are too small, there may be too many and they will end up being serialized. The ratio of the number of active warps to total warp capacity is termed *multiprocessor occupancy*, which should be kept above 50% (16 active warps) for most applications. As a strong recommendation, block sizes on the GT200 architecture should be multiples of 32 to ensure even warp partitioning and there should be enough blocks active on an SM for the scheduler to hide high-latency instructions through block swapping.

CUDA also exposes various GPU memory spaces, which are described below and summarized in Table 1.

**Global Memory** is the largest, highest-latency memory (400 to 800 clock cycles) available on the GPU and acts as the primary data transfer channel between the host and device. The host and all threads in the grid can read and write global memory; however, it should be accessed sparingly to avoid the access latencies. Global memory is most efficient when consecutive threads access consecutive memory locations. Coalescing memory access in this manner facilitates batch transactions and ultimately increases throughput. A small amount of host memory can be *page-locked* and mapped into the global memory space to increase bandwidth and support memory transfers that are concurrent with kernel execution.

**Table 1.** Summary of the GT200 Memory Model

| Memory | Access | Scope | Cached | Latency |
|---|---|---|---|---|
| Global | R/W | Grid+host | No | High |
| Shared | R/W | Block | No | Low[1] |
| Register | R/W | Thread | No | Low |
| Constant | R | Grid+host | Yes | Low[2] |
| Texture | R | Grid+host | Yes | Low[2] |

[1]Assuming no bank conflicts
[2]Cached reads only

**Shared Memory** is an extremely fast, low-latency memory space found on each SM that has two main functions:

- Provide an intermediate storage area to avoid unnecessary global memory usage.
- Facilitate low-latency inter-thread data sharing.

Shared memory can be accessed by all threads within a block, but never by the host. The coalescing requirements of global memory do not apply to shared memory; however, if multiple threads of the same half-warp access the same 32-bit shared memory *bank* simultaneously, a *bank conflict* will occur and accesses will be serialized.

**Registers** are a fast, low-latency (typically 22 clock cycles), on-chip resource but, unlike shared memory, are only visible to their resident threads. Registers are generally used for intermediate results not shared with other threads.

**Constant Memory** also resides in global memory but data is cached on-chip to greatly reduce latency when multiple threads in a block access the same location. Constant memory is ideal for storing constants, such as filter kernel parameters, and can only be written to by the host and read by the device.

**Texture Memory** is another cached, read-only space that resides in global memory; however, unlike constant memory, it can span the entire global memory space (with some limitations) and has a less-restrictive access pattern. Texture memory requires an area of device memory (linear region or region allocated as a CUDA array) to be bound to a *texture reference* and can be accessed from kernels through *texture fetches.* Texture fetches are optimized for 2-D spatial locality within the memory space, providing a powerful alternative to global memory access patterns in some applications.

Both register and shared memory usage can have a significant effect on multiprocessor occupancy. Because each SM has a finite amount of register and shared memory space that is partitioned across all active warps on the SM, it is possible that there will not be enough to satisfy the memory requirements of all threads in a block. To compensate, the SM will lower the number of active warps to reduce SM occupancy and, consequently, performance. This is one of the many implementation guidelines that must be considered when writing effective code for the GPU. It is trivial to rework CPU code for CUDA but, to receive a tangible speedup, the programmer must be aware of nuances of the architecture and fully utilize the many tools CUDA provides.

### *2.3. Microsoft Direct3D*

Direct3D, an API available as part of Microsoft's DirectX SDK used primarily in video game development, facilitates optimized 3-D graphics rendering on the GPU. For a detailed look, the reader is directed to the programming guide [19] or one of the many textbooks [4, 20] or online tutorials on the topic. Here, a small subset of its features and terminology relevant to model-based tracking are briefly introduced.

There are a number of Direct3D resources available, such as *vertex buffers*, *index buffers*, *textures* and *surfaces*, all of which can be registered in CUDA's memory space and subsequently mapped there to be accessed by kernels. Here, the focus is on surfaces, a linear area of GPU memory that can be customized with a desired size (resolution) and pixel format (32-bit ARGB in this work). Surfaces have a number of useful features, such as the ability to be locked by the host to access and modify individual pixel data and efficient resizing and stretching options. They can also serve as render targets and are easily swapped into the video card's *back buffer*, which displays them to the screen. Textures are similar to surfaces, but can be applied to geometry during rendering. Rendering an image to a texture then applying it to a simple *quad* (square) is an efficient way of creating a modular 2-D projection of a model on a larger surface.

The Direct3D rendering pipeline is highly streamlined to produce realistic 2-D projections of a model with simulated lighting and shading. The projective properties of the virtual camera can be configured in a *projection transform matrix* and the position and direction of the camera are configured in a *view transform matrix*. A variety of virtual lights can be placed around the object to simulate real world lighting and shading; however, this increases computational complexity. Once the camera, lights, render target, transformation matrices, and texture properties are configured, a 3-D model rendering can be produced extremely quickly on the GPU by projecting mesh geometry onto a 2-D plane. Instead of displaying the renderings on the screen as they are intended, they are preserved in GPU memory for further processing using CUDA's Direct3D interoperability.

## 3. GPU-Accelerated 3-D Model-Based Tracking

The operation of the GPU-accelerated tracking system is broadly divided into the host (CPU) and device (GPU) subsystems as shown in Fig. 3. The remainder of this section examines these two subsystems in detail.

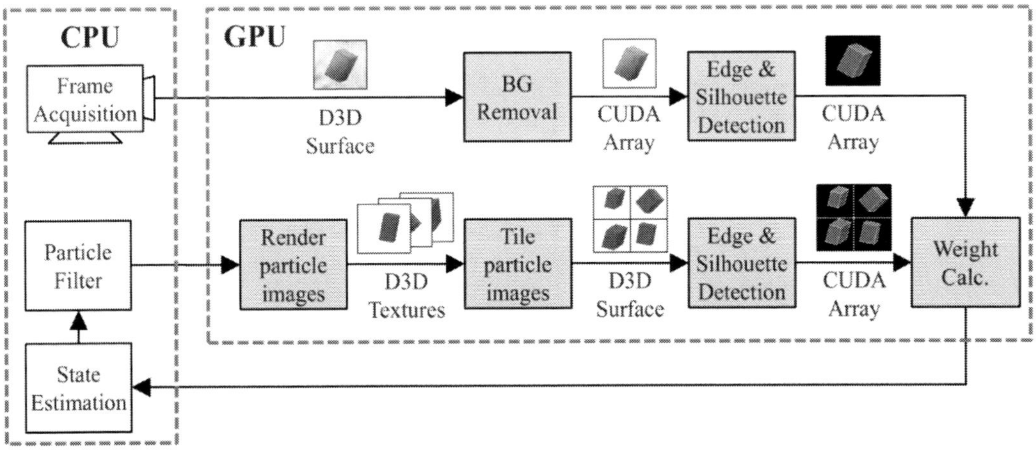


**Figure 3.** GPU-accelerated model-based tracking system block diagram

### *3.1. Host Subsystem: Particle Filtering and State Estimation*

Within the particle filter framework, the position and orientation of the tracking target is represented by a state vector:

$$\mathbf{x}_t = (O_x, O_y, O_z, r_x, r_y, r_z) \tag{1}$$

where $(O_x, O_y, O_z)$ are the coordinates of the tracking target's origin and $(r_x, r_y, r_z)$ are the rotations around the x-, y-, and z-axis, respectively. The goal of tracking is to estimate the parameters of this vector at discrete time points (e.g., once per frame). To obtain this estimate, the system maintains a set of $N$ possible states, known as *particles*, each of which is a statistically likely guess at the true tracking target location. Each particle has an associated *weight* that provides a measure of how well that particle represents the true state based on the most recent frame of video. Estimating the state that represents the tracking target's true pose can be obtained based on the:

- state of highest-weighted particle;
- average of $M$ highest-weighted particle states; or,
- weighted average of $M$ highest-weighted particle states

where $M \leq N$.

The value of the particle filter (also known as bootstrap filtering, the condensation algorithm [21], and survival of the fittest) is its ability to intelligently select statistically likely particles for simulation and evaluation based on particle weights and system dynamics. This work uses the *sampling importance resampling* (SIR) particle filter, which is described in detail by Arulampalam [5] and will only be outlined here. The motion tracking task is framed as a *Bayesian state estimation* problem wherein the weight of a particle describes the probability that its state represents the true pose of the tracking target. The weights are normalized and used to construct a cumulative density function (CDF). $N$ particles are then randomly selected from the CDF and the parameters of each are updated based on a system dynamics model, which monitors the velocity and acceleration of every DOF in the state estimates (*particle drift*), as well as some Gaussian noise (*particle diffusion*). By selecting particles from a CDF in this manner, the highly-weighted particles can be selected many times, whereas the lower-weighted particles may not get selected at all. The result is a set of particles that approximates the current posterior density function of the state sequence. Of course, reliable particle selection is contingent on meaningful particle weights generated by the GPU subsystem, which is the computationally demanding task described in the proceeding section.

### *3.2. Device Subsystem: Model Rendering and Weighting*

The goal of the device subsystem is to simulate and evaluate the particles selected by the host subsystem and generate a weight for each. In model-based tracking, *simulation* involves configuring a 3-D model of the tracking target according to the state of each particle, then projecting each configuration onto a 2-D plane, referred to as a *particle image*. *Evaluation* involves extracting features from the current video frame and comparing them to corresponding features extracted from each particle image.

To initialize the system, a 3-D model of a tracking target is imported (tracking frame rate is proportional to model polygon count) and Direct3D resources are created based on various system parameters, such as the number of particles and particle image resolution. Additionally, several values are precalculated and stored in CUDA constant memory to minimize unnecessary high-latency instructions within CUDA kernels (e.g., division, modulus). Finally, the image background data is identified based on a short training sequence and transferred into CUDA global memory for use in the background removal stage.

To generate the particle images, the model is oriented with respect to the virtual camera according to each of the $N$ particles' parameters and projected onto a texture. The $N$ textures are applied to quads and tiled onto a single, large Direct3D surface, referred to as the *particle grid* (Fig. 4). The particle grid must have the same x- and y-dimension, meaning the number of particles $N$ must be a perfect square. The particle grid is then mapped to a CUDA array in the device's texture memory along with the current frame from the video stream.

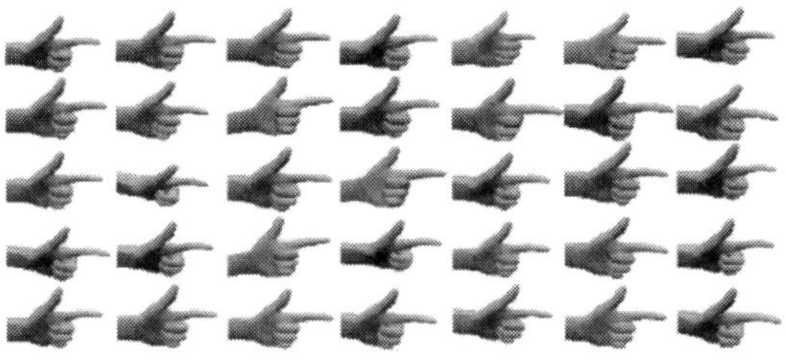

**Figure 4.** A 7x5 sample of the particle grid using a 3-D hand model

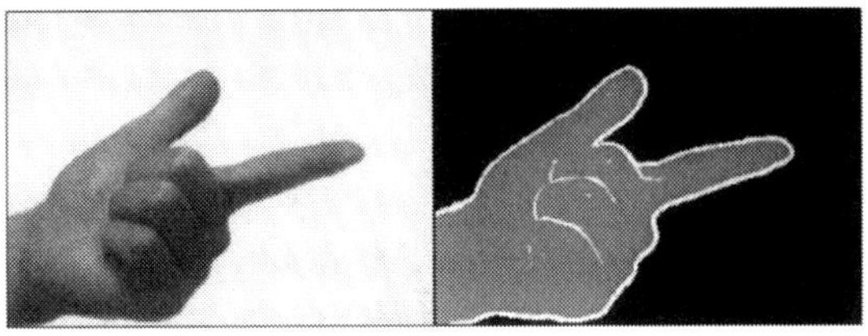

**Figure 5.** Silhouette and Sobel edge detection

CUDA's Direct3D interoperability allows individual pixels within the Direct3D surfaces to be accessed through texture fetches and read into the device's shared memory as floating point data. Because each SM's shared memory is on-chip, it is extremely low latency and facilitates the fast, scalable, synchronizable, parallel image processing required to efficiently generate a meaningful weight for all particles in the grid. The stages involved in weight generation are:

- Background subtraction
- Edge and silhouette detection
- Weight calculation

The remainder of this section describes how the weight generation task and is efficiently partitioned across SM's to minimize processing time and maximize GPU utilization.

A background subtraction CUDA kernel is executed on each frame in the video sequence to segment the tracking target from the background. Pixel data for each frame is compared to the background data acquired during initialization. If a pixel falls within a configurable threshold of the corresponding pixel in the background data, it is assumed that pixel represents image background and it is suppressed; otherwise, it is assumed the pixel is part of the tracking target and its value is preserved. While this approach requires a static camera position be maintained, this is not a significant restriction for the majority of target applications.

Edge and silhouette detection (Fig. 5) are performed by CUDA kernels on both the current frame and the particle grid, which reside in texture memory. The kernel processing the frame stores results as floating point data in global memory, whereas the kernel processing the particle grid maintains its results in shared memory for the weight calculation stage. Because the process is essentially identical for both, the algorithm is described in terms of the particle grid only.

A simple thresholding approach to silhouette detection is employed: white pixels are considered background and are set to black, whereas non-white pixels are identified as part of the tracking target and are set to gray. Edge detection is achieved using the Sobel edge detector, which involves the convolution of the pixel data with two 3x3 filters - a highly parallelizable task that is extremely well-suited to the SIMT parallel programming paradigm.

An efficient CUDA implementation partitions the algorithm in a manner that maximizes GPU utilization. In this work, a 32x8 thread block is used, regardless of particle image resolution or number of particles. The number of blocks in the CUDA grid, however, varies with these parameters. Texture fetches in consecutive threads are used to access consecutive pixels and store them as floating point data in shared memory. This adheres to the optimal access patterns for both shared and texture memory described in Section 2.2. To ensure blocks evenly tile the particle grid, the width of particle images are multiples of 32 and the height preserves the aspect ratio of the video source. The number of texture fetches is minimized by loading as much pixel data shared memory as possible, without excessively impacting multiprocessor occupancy. The CUDA occupancy calculator [22] was used to determine that each thread can

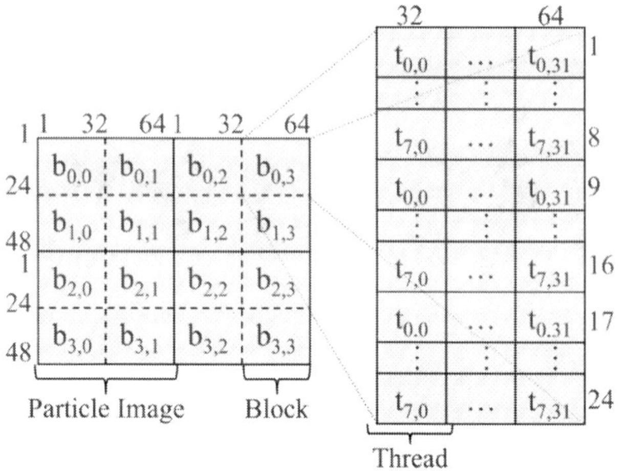


**Figure 6.** Partitioning of a 2x2 particle grid with particle image resolution of 64x48, block size of 32x8, each thread fetching 3 pixels along the y-direction. Pixel indices shown outside boxes, block IDs and thread IDs shown inside the left and right boxes, respectively.

safely load three 32-bit pixels into shared memory while maintaining an SM occupancy of 50%. Fig. 6 demonstrates an example pixel access pattern. Once all pixel data has been loaded to shared memory, barrier synchronization is used to ensure all other threads within a block have fetched their data. Finally, each thread performs edge and silhouette detection on the pixels that it fetched, accessing neighbouring pixels from shared memory as needed, and stores the end results in shared memory for the subsequent weight calculation stage.

The goal of weight calculation is to compare the temporary frame, which resides in global memory, to each particle image in the particle grid, which is partitioned across shared memory, and generate a weight for each particle with pixel-level accuracy. A divide-and-conquer approach is used to first generate a weight for each pixel in the particle grid, then sum the pixel weights to generate a weight for each thread block, and finally sum the block weights of all blocks that tile a given particle image to generate a weight for that particle (Fig. 7).

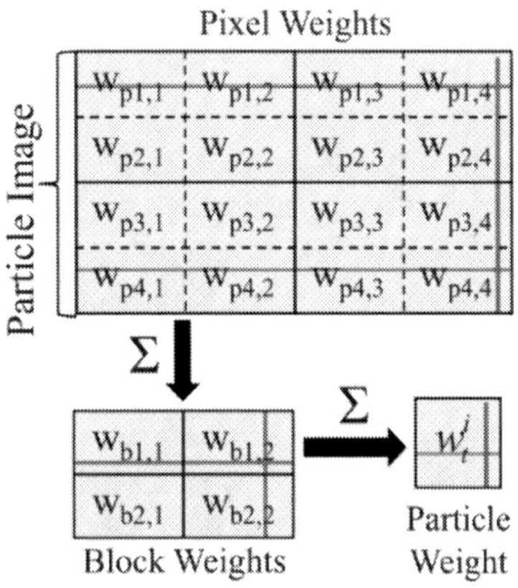


**Figure 7.** Particle Weight calculation

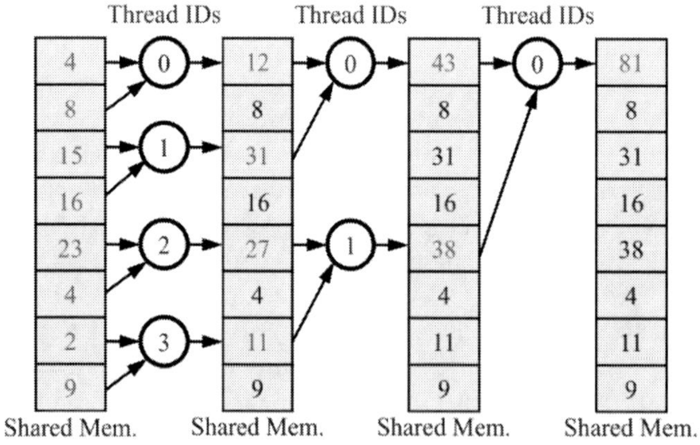


**Figure 8.** Weight summation parallel reduction algorithm

Since shared memory does not persist across kernel launches, weight calculation is done by the same kernel that performs edge and silhouette detection on the particle grid to circumvent unnecessary global memory accesses. Each thread loads three registers with the frame pixels corresponding to the previously fetched particle grid pixels and pixel weights $w_{pixel}$ are calculated according to

$$w_{pixel} = \epsilon_{max} - |p_{frame} - p_{particle}| \tag{2}$$

where $p_{frame}$ is the floating point value of a frame pixel, $p_{particle}$ is a particle pixel, and $\epsilon_{max}$ is the maximum possible error (i.e., residual) the pixels can achieve. Pixel weights are written back to shared memory.

The most computationally complex aspect of weight calculation is the summation of pixel weights. This step is performed by a parallel reduction algorithm similar to [23]. Each row of a thread block reduces its corresponding shared memory row to a single value. This done in five iterations, the first three of which are partially shown in Fig. 8. In the first iteration, 16 threads of a row each sum two adjacent shared memory cells and store the results in shared memory. The next iteration uses eight threads to sum the results of the previous iteration, and so on. The algorithm is applied to three shared memory rows by each row of the thread block, yielding a single column of 24 weights for the block, without any bank conflicts. A single thread then serially adds the values in this column to produce a single weight for the entire block and stores it in global memory.

Finally, a new kernel is launched (serving as a global synchronization barrier) to sum the weights of all blocks that tile each particle and writing the resulting particle weights to page-locked host memory. A major drawback of GPU computing is the significant latency associated with data transfers between the GPU and CPU. The described approach circumvents excessive data transfers by performing model rendering, projection, feature extraction, and particle image evaluation entirely on the GPU, requiring only one floating point value per particle to be transfered across the PCI-E bus.

## 4. Results and Discussion

The system was tested on an Intel Core2 Duo E8400 3.0GHZ PC with 4 GB DDR3 RAM, running Windows 7, CUDA 3.1, and DirectX 9 and equipped with an Asus ENGTX295 NVIDIA GeForce GTX295 graphics card (2 GPUs, each with 240 streaming processors and 896MB GDDR3 memory). A rigid (6 DOF) 3-D model of a hand comprised of 389 polygons was created using Blender [24] and used as a test object. An Allied Vision Technology GC660C 119FPS 659x493 GigE CCD camera was used to record a five second (600 frame) video sequence at a resolution of 320x240, during which the hand moves extremely quickly and through all degrees of freedom, with several occurrences of simultaneous translation and rotation. The hand in the sequence was tracked at 5 possible particle image resolutions with $N$ ranging from 50 to 4050, and a weighted average of 25 particles was used for state estimation. A demonstration of tracking is shown in Fig. 9 and accuracy and frame rate results are shown in Fig. 10. Because ground truth is not defined for this video sequence, accuracy is measured by comparing the silhouettes of the tracking target and the 3-D model configured in the estimated pose; specifically, accuracy is defined as the ratio of correctly classified pixels (tracking target vs background pixels) to the total number of pixels in the bounding box that contains both silhouettes. Note that some resolutions were not possible at certain values of $N$ due to limitations of Direct3D.

As expected, the frame rate of the system decreases as particle image resolution or the number of particles increase. Conversely, accuracy increases with an increase in either of these parameters; however, diminishing returns are realized once a resolution of 96x72 and a particle count of approximately 1000 are reached. These settings can be viewed as the ideal operating parameters for tracking, simultaneously delivering both speed and accuracy (31 fps with 95.2% accuracy). For applications that demand a higher frame rate but less accuracy, reducing the

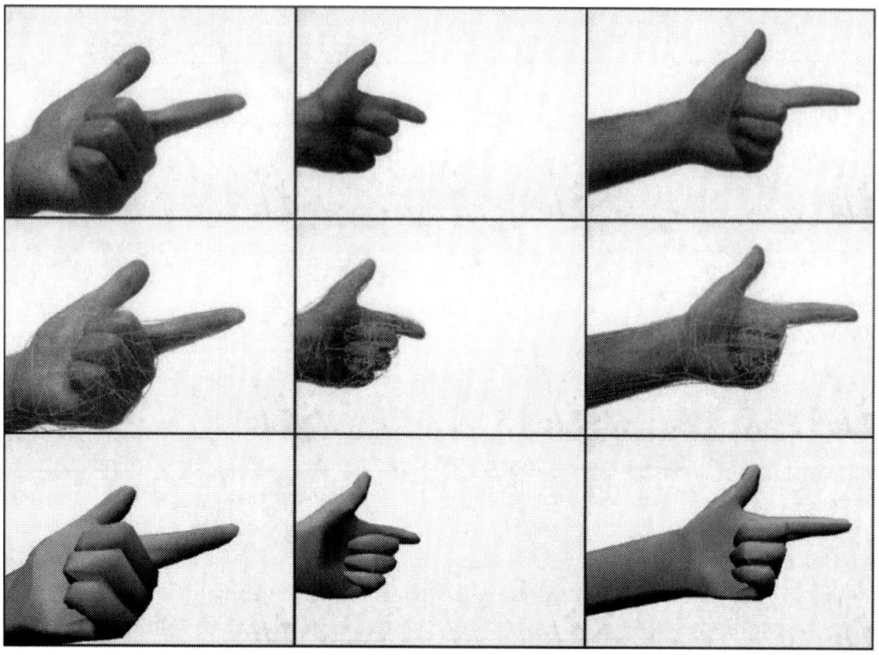

**Figure 9.** Tracking demonstration for three example frames: from top to bottom, the three rows show the original video frame, the frame with a wire-frame overlay, and a rendering of the 3-D model in the estimated pose of the tracking target, respectively.

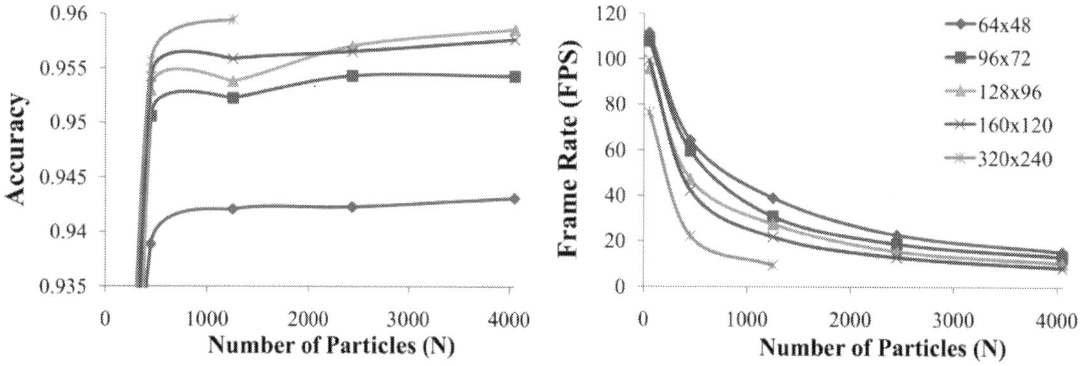


**Figure 10.** Frame rate and accuracy results at 5 particle image resolutions and the number of particles varying between 50 and 4050.

particle count to 500 nearly doubles the framerate while decreasing accuracy by just 0.2%. A qualitative analysis of the system demonstrates that tracking is both accurate and robust, with a nearly one-to-one mapping between the pose of the tracking target and the estimate. Rotation proves more difficult to estimate than translation as there are often multiple solutions when dealing with monocular video. Additionally, abrupt changes in direction at high velocities can cause the tracking target to momentarily be lost; however, recovery normally takes less than 5 frames, as the system dynamics model used by the particle filter compensates. Additional views would certainly increase accuracy, but at a significant increase in computational complexity.

The results of GPU acceleration are also of interest. Table 2 compares the amount of time taken for each stage of the tracking system in both GPU and CPU implementations for the ideal operating parameters described above. Timing results were captured by GPU timers, which are accurate within half a microsecond. The feature detection and weight calculation stage is the target of GPU-acceleration and demonstrated a speedup of almost 18 times over an equivalent CPU implementation. It is important to note that both versions use the GPU for model rendering as it would be illogical to attempt to use the CPU for this task in any scenario. This explains why the CPU implementation slightly outperformed the GPU in this stage, as the GPU resources are dedicated exclusively to model rendering.

**Table 2.** Tracking time per frame (1250 particles, 96x72 resolution, Sobel edge detection)

| Stage | GPU Time (ms) | CPU Time (ms) | Speedup |
|---|---|---|---|
| Render Particle Images | 11.90 | 9.67 | 0.81x |
| Acquire Frame | 5.68 | 2.50 | 0.44x |
| Map Resources | 2.65 | 3.64 | 1.37x |
| **Edge Detection & Weight Calc.** | **8.78** | **157.23** | **17.90x** |
| Particle Filter | 1.38 | 1.04 | 0.75x |
| Other | 1.93 | 1.84 | 0.95x |
| Total | 32.34 | 175.92 | 5.44x |

The speedup is proportional to both the particle image resolution and number of particles and ranges between 2 and 20 times depending on these parameters. One major limitation of the GPU-acceleration lies in Direct3D, which is not able to independently control multiple GPUs. As a result, while the memory spaces of both GPUs are utilized to increase the maximum size of the particle grid, only one GPU is used to render particle images and execute CUDA kernels. Additionally, some aspects of feature extraction and weight calculation are not ideal candidates for a GPU implementation; for example, padding the edges of particle images for convolution with the Sobel kernels requires significant conditional logical that leads to warp divergence and the parallel reduction of all pixel weights to a single value is inherently difficult to parallelize. Nonetheless, during each second of tracking, the GPU generates, processes, and reduces nearly 1GB of pixel data to less than 40KB of particle weight data. This demonstrates the GPU's utility as a powerful, massively parallel processor that enables previously-unobtainable, real-time performance of the 3-D model-based approach to tracking.

## 5. Conclusion

This work has demonstrated that robust and accurate real-time 3-D model-based tracking is feasible using modern consumer-grade GPUs. Despite some limitations in maximally utilizing multiple GPUs and the challenges associated with parallelizing some inherently serial aspects of the particle weighting algorithm, speedups up to 20x over an equivalent CPU implementation were realized. Future work includes adapting the system to support articulated object tracking (e.g., a hand with a flexible finger and thumb for virtual 'gripping') and exploring more specific applications, such as augmented reality. With the recent release of NVIDIA's Fermi architecture [25], designed primarily with the GPU-computing community in mind, the utility of GPU-acceleration in high-performance computing tasks, such as model-based tracking and countless other applications, will certainly continue to grow.

## Acknowledgments

The authors would like to thank M Kinsner and T Kirubarajan for their guidance, as well as NSERC Canada and McMaster University for funding this research.

## References

[1] Lepetit V and Fua P 2005 *Foundations and Trends in Computer Graphics and Vision* pp 1–89

[2] Stenger B, Thayananthan A, Torr P H S and Cipolla R 2006 *IEEE Trans. Pattern Analysis and Machine Intelligence* **28** 1372–1384

[3] NVIDIA Corporation 2010 *CUDA Programming Guide* Santa Clara, CA 3rd ed

[4] Sherrod A 2006 *Ultimate Game Programming With DirectX* (Boston, MA: Charles River Media)

[5] Arulampalam M S, Maskell S, Gordon N and Clapp T

[6] Tan G, Guo Z, Chen M and Meng D 2009 *ICS '09: Proceedings of the 23rd international conference on Supercomputing* pp 380–389

[7] Pock T, Unger M, Cremers D and Bischof H 2008 *Computer Vision and Pattern Recognition Workshops, 2008. CVPRW '08. IEEE Computer Society Conference on* pp 1 –8

[8] Halfhill T R 2008 *Microprocessor Report*

[9] Kirk D B and mei W Hwu W 2010 *Programming Massively Parallel Processors: A Hands-on Approach* (Burlington, MA: Morgan Kaufmann Publishers)

[10] Breitenstein M, Kuettel D, Weise T, Van Gool L and Pfister H 2008 *CVPR'08: IEEE Conference on Computer Vision and Pattern Recognition* pp 1–8

[11] Lehment N H, Arsic D, Kaiser M and Rigoll G 2010 *Computer Vision and Pattern Recognition Workshops (CVPRW), 2010 IEEE Computer Society Conference on* pp 87 –92

[12] Lenz C, Panin G and Knoll A 2008 *VMV'08: International Workshop on Vision, Modeling and Visualization*

[13] Lozano O and Otsuka K 2008 *ICASSP '08: IEEE International Conference on Acoustics, Speech and Signal Processing* pp 713 – 716

[14] Liu K Y, Tang L, Li S Q, Wang L and Liu W 2009 *ICME'09: Proceedings of the 2009 IEEE international conference on Multimedia and Expo* pp 1813–1816

[15] Cabido R, Concha D, Pantrigo J J and Montemayor A S 2009 *ISPAN '09: Proceedings of the 2009 10th International Symposium on Pervasive Systems, Algorithms, and Networks* (Washington, DC, USA: IEEE Computer Society) pp 757–762

[16] Pezzementi Z, Voros S and Hager G D 2009 *ICRA'09: Proceedings of the 2009 IEEE international conference on Robotics and Automation* (Piscataway, NJ, USA: IEEE Press) pp 1225–1232 ISBN 978-1-4244-2788-8

[17] Kanter D 2008 NVIDIA's GT200: Inside a parallel processor URL `http://www.realworldtech.com/page.cfm?ArticleID=RWT090808195242`

[18] NVIDIA Corporation 2010 *CUDA Best Practices Guide* Santa Clara, CA 3rd ed

[19] Microsoft Corporation 2003 *Microsoft DirectX 9.0c SDK Documentation.*

[20] Luna F 2003 *Introduction to 3D Game Programming with DirectX 9.0* (Plano, TX: Wordware Publishing)

[21] Isard M and Blake A 1998 *Inter. Journal of Comp. Vis.* **29** 5–28

[22] NVIDIA Corporation 2008 *CUDA GPU Occupancy Calculator* Santa Clara, CA

[23] Harris M 2007 Optimizing parallel reduction in cuda Tech. rep. NVIDIA Corporation

[24] Roosendaal T and Selleri S 2009 *The Official Blender 2.3 Guide* (Austria: Springer-Verlag/Wien)

[25] NVIDIA Corporation 2009 NVIDIAs Next Generation CUDA Compute Architecture: Fermi Tech. rep. NVIDIA Corporation

# A modular CUDA-based framework for scale-space feature detection in video streams

**M. Kinsner**[1], **D. Capson**[1] **and A. Spence**[2]

[1] Deparment of Electrical and Computer Engineering, McMaster University, Hamilton, Ontario, Canada

[2] Deparment of Mechanical Engineering, McMaster University, Hamilton, Ontario, Canada

E-mail: `kinsnemh@mcmaster.ca`, `capson@mcmaster.ca`, `adspence@mcmaster.ca`

**Abstract.** Multi-scale image processing techniques enable extraction of features where the size of a feature is either unknown or changing, but the requirement to process image data at multiple scale levels imposes a substantial computational load. This paper describes the architecture and emerging results from the implementation of a GPGPU-accelerated scale-space feature detection framework for video processing. A discrete scale-space representation is generated for image frames within a video stream, and multi-scale feature detection metrics are applied to detect ridges and Gaussian blobs at video frame rates. A modular structure is adopted, in which common feature extraction tasks such as non-maximum suppression and local extrema search may be reused across a variety of feature detectors. Extraction of ridge and blob features is achieved at faster than 15 frames per second on video sequences from a machine vision system, utilizing an NVIDIA GTX 480 graphics card. By design, the framework is easily extended to additional feature classes through the inclusion of feature metrics to be applied to the scale-space representation, and using common post-processing modules to reduce the required CPU workload. The framework is scalable across multiple and more capable GPUs, and enables previously intractable image processing at video frame rates using commodity computational hardware.

## 1. Introduction

In specific computer and machine vision applications, the size or scale of features to be detected is either unknown in advance or changes significantly during imaging. These changes can be caused by a varying distance between camera and target, or by blurring effects within poorly focused or motion blurred images. Multi-scale image processing techniques enable detection of these features, but the requirement to process data at many scale levels within a single image imposes a substantial computational load. In applications where image processing at video frame rates is required, such as in some machine vision metrology systems, processing using commodity hardware requires computational acceleration. Recent advances in General Purpose Graphics Processing Units (GPGPUs) have enabled new classes of multi-scale algorithms for video frame rate processing within these applications.

This paper describes a modular framework for GPGPU-accelerated multi-scale image processing, with specific application to scale-space feature detection. Recent results are presented from implementation of the framework on an NVIDIA CUDA-enabled GPU, with detectors implemented for two classes of scale-space features.

The remainder of this paper is organized as follows: Section 2 provides a brief introduction to scale-space techniques and the associated detection of Gaussian blob and ridge features; Section 3 describes the modular framework for scale-space feature detection on a GPU; Section 4 describes the GPGPU implementation; Section 5 presents results, and the paper concludes in Section 6.

## 2. Scale-space Image Processing

### *2.1. Generation of the scale-space*

Scale-space theory provides a framework suitable for feature detection where little is known about the scale (size) of relevant features, and where the scale varies considerably within and between images in a data set. A parameter $t$ is introduced, known as the scale parameter, and is used as a filter variance parameter to progressively filter the original image at successive variance levels [1]. The kernel used to perform the filtering operation must meet specific requirements to generate a linear scale-space representation, and in the continuous case, the unique kernel with these properties is the Gaussian [2]. A basic scale-space is defined on an image $f$ as

$$L(x, y; t) = g(x, y; t) * f \tag{1}$$

where $g$ represents a two-dimensional Gaussian kernel with variance $t$. When discretized in practical systems, the Gaussian kernel does not necessarily produce the correct scale-space properties, so alternative discrete kernels have been proposed [3, 4], the later of which is used for this work. More specifically, a separable convolution kernel with smoothing parameter step size of $\Delta t = \frac{1}{3}$, is [4]:

$$k = [\frac{1}{6} \quad \frac{2}{3} \quad \frac{1}{6}] \tag{2}$$

Multiple discrete scale levels are chosen and the image is filtered at these chosen variances, collectively forming a scale-space representation of the image. Figure 1 depicts an original image frame, and progressive filtering at two variance levels. As the filter variance increases, finer features in the image are filtered away, providing a progressively smoothed and simplified representation.

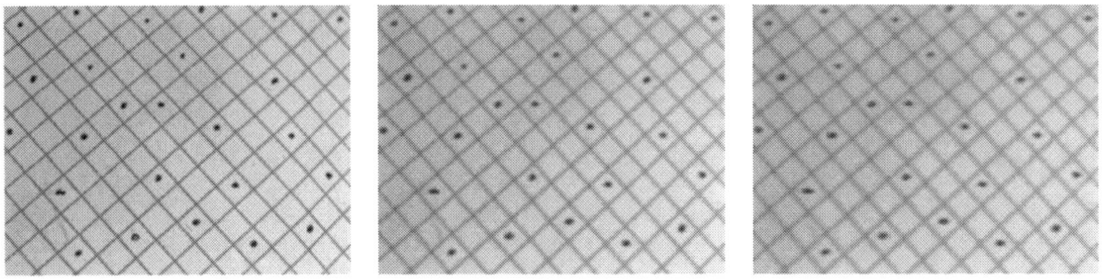

(a) Original image (b) Filtered with $t = 39$ pixels$^2$ (c) Filtered with $t = 66$ pixels$^2$

**Figure 1.** Progressive filtering of a sample image

### *2.2. Feature detection*

Once a scale-space representation of the original image has been computed, feature metrics and associated post-processing are applied. The details of processing depend upon the characteristics of the feature class, with two such classes described following. Selection and demonstration of blob and ridge feature detectors is motivated by a machine vision application where blob-like fiducials must be tracked across frames, and grid lines (ridges) must be extracted in the presence of substantial depth-of-field blur.

*2.2.1. Blob Detection* Blob features can be described through many metrics, but one measure useful for scale-space extraction is a scale-normalized Laplacian (scale-normalized discrete second derivative approximation) operator [5, 6]. The operator is applied to the scale-space image representation at all scale levels, and a spatially local maximum search is then performed (also across scale levels). A global sorting operation can then locate the strongest metric maxima, where the strongest responses are taken to be the best defined blobs. This method of blob detection allows features of varying sizes and spatial extents to be detected, intrinsically adjusting for the effects of changing camera range-to-target and blur.

*2.2.2. Ridge Extraction* Ridge extraction is more complex than blob detection in the sense that ridges are not restricted to local regions of an image, and must be extracted as continuous structures. The metrics defining a ridge are typically more arithmetically intense than primitive feature detectors (such as blob detection), and require the use of multiple metrics from which zero iso-surfaces are interpolated and intersected to define ridge segments [7]. A series of ridge extraction metrics have been proposed in the literature, one of which has been implemented in this work [7]. A primary metric is designed to detect ridge characteristics in the image, but in many cases the same physical ridge is detected at multiple scale levels in the image representation. A second metric is then applied to the scale-space that defines the optimal scale at which to detect the feature, typically based on some quality such as resolution or confidence in spatial localization. The two metrics form zero iso-surfaces, which are intersected to form the ridges. Ridge segments are interpolated within a voxel bounded by neighbouring pixels at neighbouring filter variance levels, and although continuous across multiple voxels, the ridge segments are computed independently of one another. A linking stage is then required to follow connected segments of a ridge, establishing the length and properties, and producing a connected output.

### *2.3. Other Features*
Many other feature metrics have been proposed in the literature including edge, junction, and corner detectors. Common post-processing tasks and re-use of the scale-space representation lead to the conclusion that a modular architecture is suitable, wherein detectors for additional feature classes can be added with minimal effort.

## 3. Modular Architecture
In the context of image feature detection, scale-space techniques involve three primary phases: (a) generation of the scale-space image representation; (b) application of feature detection metrics; and (c) post-processing to extract features from the multi-scale metric data. Generation of the scale-space representation is common across various feature detectors, and as such can be viewed as the initial stage in a feature detection framework.

Detection of features involves application of a feature-specific metric (multiple metrics in some cases), followed by post-processing to reduce the data into a compact form usable by output applications with minimum additional computation. Post-processing tasks including non-maximum suppression, local extrema search, and value-based sorting are common to multiple classes of detectors, and are therefore amenable to reuse in a modular architecture. Figure 2 shows such an architecture, where all feature detectors reuse the output from a common scale-space generation module, and blob and junction detectors use common maxima search and sorting modules. The structure of the metric computation phase is common across all features, with the only difference being the metric equation implemented.

Passing of intermediate data between processing modules can be accomplished in two ways. First, intermediate data output by one module can be stored back to global memory, allowing a following module to explicitly load into shared memory the data that it requires. This method

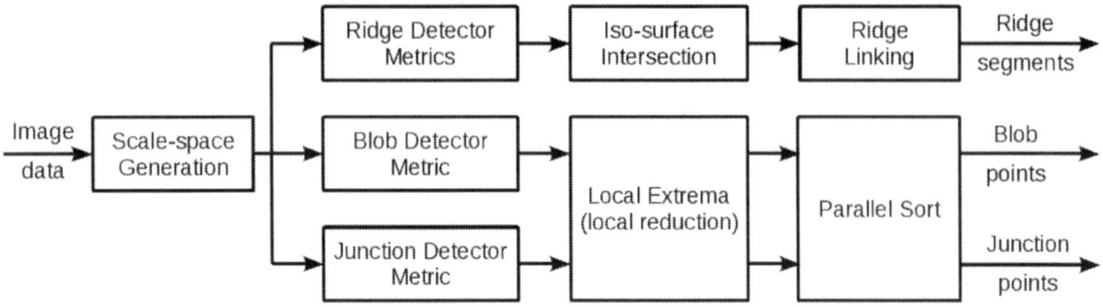


**Figure 2.** Modular feature detectors

causes the intermediate data to be buffered in the slower global memory, but does not impose any constraints on the data block size or structure to be used by subsequent modules.

A second method of intermediate data passing is to incorporate the modules together into a single GPU kernel, but separating the modules as code libraries. The isolated code modules are therefore combined at the compile phase, at the cost that the developer must pay more attention to the local data storage structures. This technique is less flexible, but can save data store/load overhead when a subsequent module requires an identical or similar structure of data in the multiprocessor's shared memory.

## 4. GPGPU Implementation

### *4.1. GPGPU Overview*

Recent advances in GPGPU technology by major manufacturers have extended the conventional GPU graphics pipeline into a more general purpose computing platform, highly efficient for many arithmetically intense applications. Algorithms must be parallelized for efficient GPU implementation, and further must be structured to execute within the computing resource constraints of a group of Single Instruction Multiple Data (SIMD)-like cores. Arithmetically intense algorithms typically experience the best acceleration on GPGPUs, especially where groups of threads exhibit non-divergent behaviour in the SIMD processing stream.

This work focusses on one NVIDIA GPGPU product line, utilizing the programming interface known as the Compute Unified Device Architecture (CUDA). A commodity GTX 480 graphics card is chosen as a realistic GPU, currently available in standard computing systems used for machine vision metrology.

### *4.2. Scale-space Algorithms on GPGPU*

Scale-space algorithms exhibit properties making them suitable for GPGPU implementation, specifically spatial locality, arithmetic intensity, and operations that can employ parallel reduction techniques. Spatial locality of data exists during the scale-space generation and metric computation phases of feature detection, allowing independent portions of the data set to be assigned to independent GPU multiprocessors. A lack of required communication between the independent data provides for efficient scaling properties across larger or additional GPUs.

The scale-space generation and metric computation tasks exhibit arithmetic intensity, with many arithmetic operations taking place per data load operation from global memory. This characteristic allows memory fetches to be hidden behind computation operations, and leads to an implementation that is not tightly memory bound.

Post-processing tasks such as maxima search and data summary operations can be structured to conform to parallel programming patterns, including parallel reduction and parallel sorting.

### *4.3. Scale-space Generation*

The kernel defined by (2) is used to generate the scale-space image representation, but it produces a filter variance step of $\Delta t = \frac{1}{3}$. Discrete scale levels must be chosen when generating the scale-space, and for this work, eight scale levels are defined at $t = 1, 4, 7, 10, 13, 16, 19, 22$ where $t$ is the filter variance parameter in pixels$^2$. Repeated application of (2) is required in each image dimension to achieve the necessary filter variance, so to achieve a flexible implementation, an intermediate "unit step kernel" $k_{unit}$ is generated which provides an incremental filter operation by a variance of $\Delta t = 1$.

$$k_{unit} = [\frac{1}{216} \quad \frac{3}{54} \quad \frac{51}{216} \quad \frac{11}{27} \quad \frac{51}{216} \quad \frac{3}{54} \quad \frac{1}{216}] \tag{3}$$

The "unit step kernel" $k_{unit}$ is then repeatedly applied to the image data to achieve the required scale-space filter variances, with each set of repeated applications called a "step kernel" in Figure 3. The convolutions are separable, so separate operations are performed to filter the image data in the horizontal and vertical image axes.

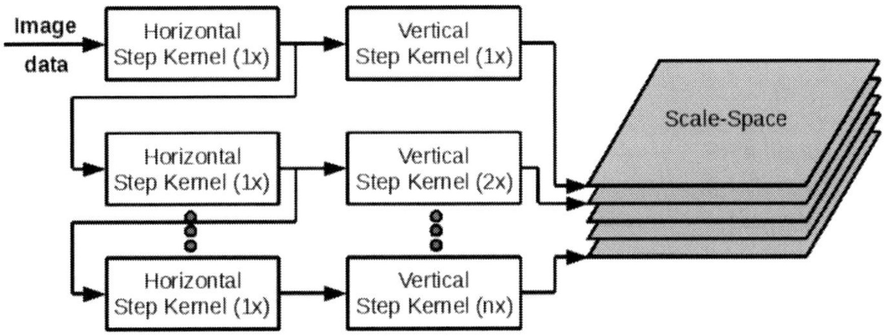


**Figure 3.** Scale-space generating convolution work flow

Output from the convolution operations forms the image scale-space representation, which is subsequently used to generate the various feature detection metrics.

### *4.4. Computation of Feature Metrics*

Specific metrics are computed for each type of feature to be extracted, as described in Section 2. The metric equations typically contain many spatial derivative approximations in the scale-space [7], and these are computed using central difference approximations as the core. To maintain uniformity across different orders of differentiation in a given image dimension, higher order derivative kernels are formed through repeated convolution of a first order central difference kernel. Using this approach, the spatial extent of data required to form a derivative approximation expands with the order of differentiation, so overlapping border data is loaded into the GPU multiprocessors to avoid fetching bordering data from global memory in a non-coalesced fashion.

Metric values are computed for each pixel of the image, at each scale level in the scale-space. Output from the metric computation calculations is therefore of the same dimension as the scale-space representation (for each metric computed), and this data forms the raw input to the feature extraction operations.

### *4.5. Blob Detection*

The work flow for blob feature detection is presented in Figure 4. After application of the blob detection metric to the scale-space representation, an extrema search is performed within a patch of data that is contained in a GPU multiprocessor. This search is structured as a parallel reduction operation using coalesced memory accesses for efficiency, and produces the maximum blob response within a given scale-space image patch. These local maxima can then be searched by the CPU to find a few maximum blob responses, or sorted on the GPU such that the CPU only needs to access the initial items in a sorted list to locate the strongest blob responses. When performed on the GPU, a parallel radix sort [8] is used to order the local maxima by response strength.

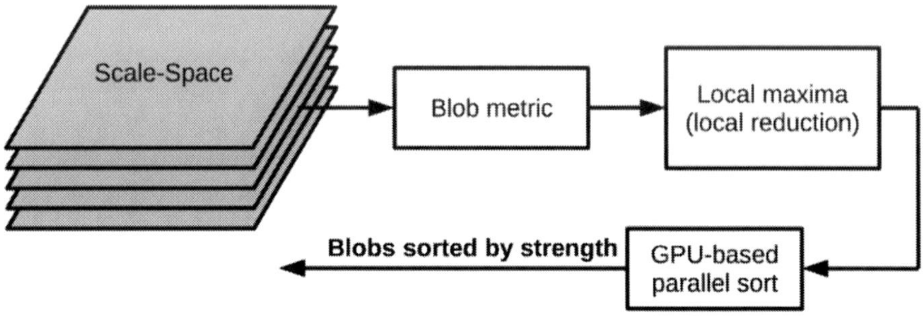


**Figure 4.** Blob detection work flow

*4.6. Ridge Detection*

Scale-space ridge detection on an NVIDIA GTX 280 has been described in a previous publication [9], so details of the ridge detection process such as iso-surface intersection and ridge summary data are not repeated here. The overall ridge detection process is shown in Figure 5. Arithmetically intense metrics are computed upon the scale-space image representation, and interpolated zero iso-surfaces are intersected to form ridge segments. The segments are then linked and examined within local blocks of the scale-space, and where valid portions of a ridge are found, summary data is produced that can later be used by the CPU to rapidly locate and analyse long ridges. The iso-surface intersection and summary data operations are independently performed on subsets of the scale-space data that is local to a specific GPU multiprocessor. These operations are also common to other scale-space algorithms such as edge detectors (given alternative edge detection metrics), so like blob detection post-processing, these modules can be reused and directly applied to alternative feature detectors.

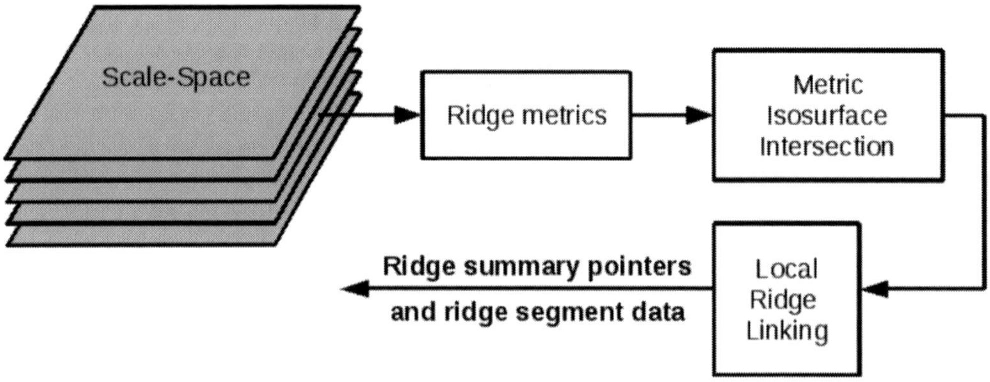


**Figure 5.** Ridge detection work flow

## 5. Results

Sample output from the GPGPU-accelerated scale-space feature detectors is presented in Figures 6 and 7. Timing results are presented in Table 1 from implementation of the blob and ridge detectors on the most recent generation NVIDIA GTX 480 graphics card, running CUDA 3.0. The image resolution was 1024 by 768 pixels, and eight discrete scale-space scale levels were used, with scale variances of $t = 1, 4, 7, 10, 13, 16, 19, 22$. The total GPU processing time, reported in the final row of Table 1, includes the operations between and the overhead of kernel calls. This aggregate processing rate is better than 15 frames per second, which is the required frame rate for the machine vision camera used in testing.

Table 2 presents resource utilization for the various GPU kernels, and represents a significant increase in resource availability from previous generations of NVIDIA hardware. Registers are the floating point memory locations local in scope to a single thread, while shared memory is shared between groups of threads executing within a single multiprocessor (single thread block). Occupancy is the ratio of active threads to the maximum number of threads supported by a multiprocessor, as constrained by the hardware resources available and the individual utilization of each thread. For further description of the GPU architecture, the reader is referred to [10, 11]. Previous hardware generations provided fewer hardware resources, especially in terms of register availability in the presence of many parallel threads. Increased hardware resources simplify implementation of arithmetically intense algorithms, reducing the need to manually decompose operations for control of register usage.

Memory copies between the host system and GPU mass memory are not a bottleneck to performance. Data transfer times are variable, but small compared with the aggregate GPU processing time per frame. Current hardware further provides the capability to asynchronously transfer data between host and device, allowing data transfer operations to occur in parallel with kernel execution.

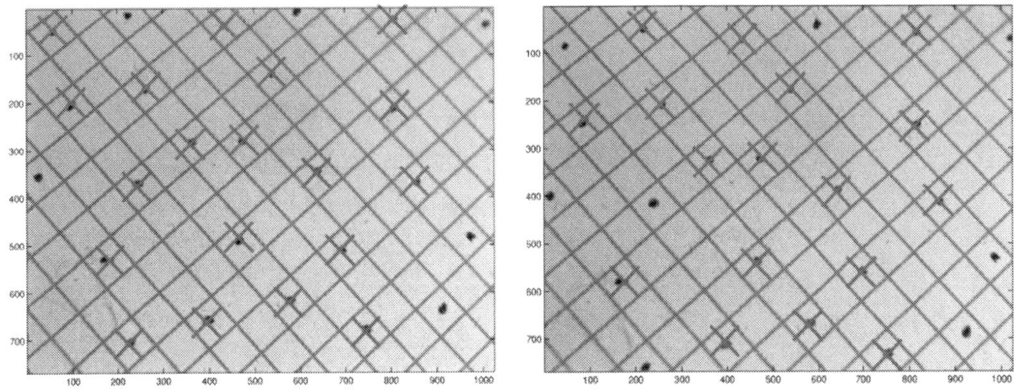


(a) Sample frame 1 (b) Sample frame 2

**Figure 6.** Sample images with strongest scale-space blob metric response locations shown by red markers

## 6. Conclusions

This paper has described a framework for GPGPU-accelerated scale-space feature detection intended to operate at video frame rates. The architecture is modular, allowing additional feature detectors to re-use both the image scale-space representation, and also common post-processing and analysis modules. By design the framework can be easily extended to new feature classes, and is scalable across multiple and more capable GPUs. Extraction of ridge

Journal of Physics: Conference Series **256** (2010) 012005 doi:10.1088/1742-6596/256/1/012005

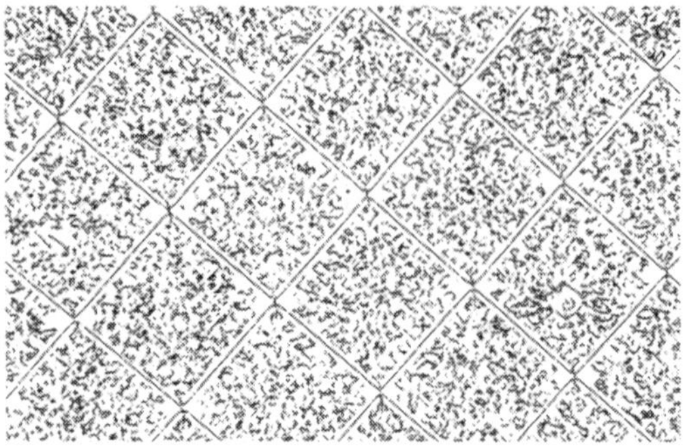

(a) Ridge segments - straight grid lines and unfiltered noise between the lines

(b) Ridge segments - height representing the scale-space scale at which a segment was detected

(c) Ridge segments - straight grid lines and unfiltered noise between the lines

(d) Ridge segments - height representing the scale-space scale at which a segment was detected

**Figure 7.** Scale-space ridge detector output line segments from a single sample image, with unfiltered noise detected between the straight grid lines

**Table 1.** GPU Timing Results

| GPU Kernel | GPU Time/Frame [ms] |
|---|---|
| Horizontal Convolution | 0.6 |
| Vertical Convolution | 3.75 |
| Ridge Metric Computation | 15.56 |
| Ridge Iso-surface Intersection | 21.58 |
| Ridge Linking | 1.5 |
| Blob Metric and Local Maximum | 1.3 |
| Blob Strength Radix Sort | 3.6 |
| Total processing time (with overhead) | 48.76 |

and blob features is achieved at faster than 15 frames per second on video sequences from a real machine vision system, utilizing an NVIDIA GTX 480 graphics card for acceleration. This processing speed enables a set of algorithms previously intractable at video frame rates, providing simultaneous extraction of multiple feature classes from a non-trivial scale-space representation. The proposed framework can further be used to implement other multi-scale feature detection algorithms, and arithmetically intense image processing techniques in general.

**Table 2.** GPU Resource Utilization

| GPU Kernel | Registers | Shared Memory [bytes] | Occupancy [%] |
|---|---|---|---|
| Horizontal Convolution | 7 | 4224 | 100 |
| Vertical Convolution | 17 | 3208 | 100 |
| Ridge Metric Computation | 54 | 4928 | 33 |
| Ridge Iso-surface Intersection | 25 | 8128 | 58 |
| Ridge Linking | 15 | 2240 | 88 |
| Blob Metric and Local Maximum | 10 | 6976 | 100 |

**Acknowledgments**
Funding from SHARCNET and NSERC is gratefully acknowledged.

**References**

[1] Witkin A P 1983 *Proceedings of the 8th International Joint Conference on Artificial Interlligence* pp 1019–1022
[2] Babaud J, Witkin A P, Baudin M and Duda R O 1986 *IEEE Trans. Pattern Anal. Mach. Intell.* **8** 26–33 ISSN 0162-8828
[3] Lindeberg T 1993 *Journal of Mathematical Imaging and Vision* **3** 349–376
[4] Lim J Y and Stiehl H 2003 *Scale Space Methods in Computer Vision* (Springer Berlin / Heidelberg) pp 132–147
[5] Lindeberg T 1998 *International Journal of Computer Vision* **30** 79–116
[6] Lindeberg T 1993 *International Journal of Computer Vision* **11** 283–318
[7] Lindeberg T 1996 *Proceedings of Computer Vision and Pattern Recognition* pp 465–470
[8] Satish N, Harris M and Garland M 2009 *Parallel and Distributed Processing Symposium, International* **0** 1–10
[9] Kinsner M, Spence A and Capson D 2010 *Journal of Computer Aided Design and Applications* **7**
[10] NVIDIA 2010 *NVIDIA CUDA Programming Guide* URL `http://developer.nvidia.com/object/cuda.html`
[11] NVIDIA 2010 *Tuning CUDA Applications for Fermi* URL `http://developer.nvidia.com/object/cuda.html`

# Mapping the MPM maximum flow algorithm on GPUs

**Steven Solomon and Parimala Thulasiraman**

Department of Computer Science, University of Manitoba, Winnipeg, Manitoba, Canada

E-mail: umsolom9@cs.umanitoba.ca, thulasir@cs.umanitoba.ca

**Abstract.** The GPU offers a high degree of parallelism and computational power that developers can exploit for general purpose parallel applications. As a result, a significant level of interest has been directed towards GPUs in recent years. Regular applications, however, have traditionally been the focus of work on the GPU. Only very recently has there been a growing number of works exploring the potential of irregular applications on the GPU. We present a work that investigates the feasibility of Malhotra, Pramodh Kumar and Maheshwari's "MPM" maximum flow algorithm on the GPU that achieves an average speedup of 8 when compared to a sequential CPU implementation.

## 1. Introduction

Recently, Graphics Processing Units (GPUs) have become powerful co-processors for general purpose parallel applications. As a result of the GPU's high degree of parallelism and computational power, significant interest has been directed toward the platform for parallel application development. A vast portion of the literature, however, focuses on applications for solving regular problems. Regular problems exhibit structured and predictable data access patterns and typically allow a developer to exploit data-parallelism in an intuitive manner. As a result, regular applications generally map well to the GPU architecture.

Irregular problems, however, have not been as extensively studied on the GPU. Irregular problems use pointer-based data structures such as graphs or trees. Maximum flow computations or unstructured grid problems are two such examples of irregular problems. Due to the composition of the data, these problems exhibit unstructured and unpredictable data access patterns. As a result, irregular problems are significantly more difficult to implement efficiently in a parallel manner. Ruetsch and Micikevicius [1] detail an example of the performance degrading effects of inefficient memory usage on the GPU. Efficiently mapping an irregular problem to the GPU becomes difficult as a result of the unpredictable data access patterns.

We investigate an irregular, graph-based algorithm that solves the maximum flow problem. In general, graph theory applications make use of the graph data structure. A graph is composed of a set of vertices that are connected by edges. These edges can be visualized as a road connecting one intersection (vertex) to another. An edge is said to be *directed* if it allows movement in only one direction, or *undirected* if this restriction does not exist. A *path* between two vertices $a$ and $b$ traverses some arbitrary number of edges (starting from $a$) until $b$ is reached. A *shortest* path between two vertices $a$ and $b$ is a path of minimum length from $a$ and $b$ (that is, a path containing the minimum number of edge-traversals required to move from $a$ to $b$).

A flow network expands the concept of the graph by introducing a flow capacity for each edge that describes how much flow an edge can support. In network theory, the flow can be visualized as data traversing some link between nodes (vertices) in a network graph.

For our implementation, we focus on the use of general, or random, graphs. A general graph emphasizes the irregular nature of graph data as vertices are connected by edges to other vertices in unstructured, *random* patterns. This approach differs from the use of more structured grid graphs by removing the possibility of optimizations which take into account the structured nature (as was shown by Hussein et al. [2]).

The maximum flow problem has applications in a variety of areas. For example, Hussein et al. [2] use a maximum flow algorithm to perform image processing (in the form of image segmentation). The maximum flow problem can also be applied to transportation problems (which further have applications in network theory), such as the transshipment problem [3].

In this paper, we consider the mapping and implementation of Malhotra, Pramodh Kumar and Maheshwari's [4] maximum flow algorithm (MPM algorithm) on NVIDIA GPUs using CUDA. In the literature, Goldberg and Tarjan's [5] push-relabel maximum flow algorithm has been the target for parallelization on different computer architectures and this is no different for the GPU. Section 3 will discuss some of the current implementations in the literature.

To the best of our knowledge, there exists no current work on parallelizing the MPM algorithm on the GPU. We believe that though the push-relabel algorithm is the best-known sequential maximum flow algorithm, it is not the best suited algorithm for the GPU. The synchronous nature of the MPM algorithm allows the problem to be mapped efficiently on the GPU for both structured and general graphs.

The following Section provides a brief introduction into the GPU architecture and CUDA programming model. Section 3 discusses the current related work for parallel irregular algorithms as well as maximum flow algorithms. A description of the maximum flow problem as well as the MPM algorithm is provided in Section 4, and Section 5 details our methodology used to map the MPM algorithm on the GPU. Section 6 provides some performance results and lists some hypotheses to explain sources of potential performance degradation in the algorithm. Finally, Section 7 provides our conclusions and briefly discusses some potential future work.

## 2. CUDA Programming Model

The Compute Unified Device Architecture, or CUDA, provides developers with a framework for general purpose GPU development. This section will briefly cover the exposed GPU architecture as well as the general threading model provided by CUDA. NVIDIA's CUDA Programming Guide [6] provides further details on any information presented in this section. As we are working with a GPU from the GT200 series, all information included in this section pertains only to that architecture.

### *2.1. GPU Architecture*

The GPU is composed of two major components: (i) The GPU core, and, (ii) The off-chip memory. A series of Streaming Multiprocessors (SMs) compose the GPU core. Each of these SMs contains eight Scalar Processors (SPs). In our work, we use a GTX 260 GPU with 27 SMs and, thus, a total of 216 SPs. Each of the SMs executes following the Single Instruction Multiple Threads (SIMT) model. A modification of the Single Instruction Multiple Data (SIMD) model, SIMT focuses on threads executing identical instructions on different pieces of data.

For data storage, the GPU provides various memories. The largest, and slowest, memory available is the aforementioned off-chip memory, or "global memory". This memory is read/write accessible to all threads executing across all of the SMs in the GPU. Unfortunately, accessing global memory results in a steep access latency penalty. NVIDIA [6] measures the approximate latency as 400 to 600 cycles for each access. Further, no automatic caching of data read from global memory takes place. As a result, this latency hit occurs with each access.

At the SM level, we consider the "shared" memory. Shared memory is as fast as accessing a register [6] if no bank conflicts exist[1]. Clearly, the speed of shared memory results in it being a tremendous asset for a GPU application. Using shared memory to cache data accessed in global memory can result in significant performance improvements for a GPU algorithm.

### *2.2. Threading Model*

Threads in CUDA are organized into a series of hierarchical groups. At the lowest level is the thread warp. A thread warp is composed of 32 threads executing on the same SM. Each thread in a warp must be executing the exact same instruction as every other thread in the same warp. Threads whose execution paths have diverged are marked as inactive during instructions that do not pertain to their path. This is, in effect, the SIMT model as discussed in Section 2.1. As a result, the performance impact of branch-heavy code should become apparent. The worst case, for example, involves each thread in a warp taking a unique execution path through a block of code. In this example, the warp's threads will end up executing the block's instructions in serial, rather than in parallel.

One level higher, all warps executing on the same SM are organized into a thread block. Each thread within a thread block is provided an ID unique only among other threads in the same thread block. Further, each thread block itself is provided an ID that uniquely identifies it from all other thread blocks. At the highest level exists the thread grid. The thread grid is composed of all thread blocks participating in the execution of the kernel.

Shared memory is exclusively allocated to each thread block executing within an SM. For example, if there are four thread blocks scheduled on a single SM, each thread block can only use 1/4 of the total available shared memory for itself[2]. This allocated shared memory is read/write accessible only to threads within the same thread blocks. Other thread blocks cannot access shared memory data allocated by another thread block within the same SM.

## 3. Related Work

There exists a number of sequential algorithms that solve the maximum flow problem. Ford and Fulkerson [7] described one of the first maximum flow algorithms that use the concept of a flow augmenting path. This algorithm was later improved on by Edmonds and Karp [8]. Later, Dinic [9] improved on these concepts further and introduced the idea of a layered network.

Malhotra, Pramodh Kumar and Maheshwari's [4] "MPM algorithm" provides an efficient, parallelizable way of finding maximum flows within a layered network. There are several phases to this algorithm which will be explained in Section 4. We consider the MPM algorithm as the algorithm is very iterative and synchronous — concepts that we believe allow it to map well to the GPU architecture.

Finally, Goldberg and Tarjan [10, 11] moved away from the layered network concept and proposed a "push-relabel" algorithm that solves the maximum flow problem. Unlike the previous algorithms mentioned above, the push-relabel algorithm is very asynchronous in nature. As a result, we believe it may not be the best-suited maximum flow algorithm for the GPU.

A large amount of research in high performance computing has targeted the GPU in recent years. Most of the literature, however, has focused on accelerating very regular algorithms on the GPU. Typically, these regular algorithms display tremendous speedup on the GPU when compared to sequential CPU code as well as alternative parallel architectures. For example,

[1] Shared memory is split in to 16 32-bit wide banks, multiple requests for data from the same bank arriving at the same time are serialized.

[2] All thread blocks allocated for a task are automatically initialized to use the same resources as every other thread block. Thus, each thread block must use an equal portion of shared memory.

Fujimoto [12] provides an implementation of fast, efficient dense matrix-vector multiplication on the GPU. Fujimoto's results show a speedup of approximately 32 on the GPU compared to a sequential CPU implementation.

Unfortunately, the current breadth of literature for irregular algorithms on the GPU remains limited when compared to regular algorithms. The literature does, however, contain some existing work on maximum flow algorithms for the GPU. The authors of these papers focus on implementations of Goldberg and Tarjan's [5] push-relabel algorithm, as it is the maximum flow algorithm with the best-known sequential time complexity.

Hussein et al. [2] first showed a GPU implementation of Goldberg and Tarjan's [11] push-relabel algorithm. The authors focused on grid graphs, however, which reduces the potential of their algorithm. As a result, Hussein et al. had the ability to develop optimizations that specifically target the structured nature of grid graphs. These optimizations included the lockstep breadth-first traversal.

Following the work of Hussein et al. [2], Vineet and Narayanan [13] detailed their own implementation of the push-relabel algorithm on the GPU. The authors, too, looked only at grid graphs for their implementation. One optimization they investigated involved a method to skip inactive vertices from being processed in a GPU kernel. Vineet and Narayanan also investigated how packing data together impacted the performance of their GPU algorithm. Surprisingly, they showed that their data compaction schemes generally resulted in worse performance than simply leaving the data as-is.

Harish et al. [14] provided a more recent example of a push-relabel GPU algorithm. Unlike Hussein et al. [2] and Vineet and Narayanan [13], Harish et al. [14] considered random graphs. The authors implemented a scan operation that aided in the identification of only those vertices that will be active in the next iteration. Their algorithm then only assigns threads to cover those vertices that are marked as active. Harish et al. further investigate the performance difference between high-degree and low-degree graphs. They concluded that high-degree graphs typically show improved performance over a sequential implementation.

Much of the motivation for our work with the MPM algorithm stems from Kulkarni et al.'s [15] Galois Project. The authors' designed the Galois Project to look at automatic parallelization for irregular algorithms. Kulkarni et al. [16] show that a mapping towards iterative, synchronous algorithms is ideal for parallel irregular algorithms. The MPM algorithm itself is a very iterative, very synchronous algorithm. Hence, this provides an opportunity to observe how an algorithm with these properties maps and performs on the GPU architecture.

## 4. A Maximum Flow Algorithm for the GPU

The maximum flow problem begins with a graph, $G$, containing a set of vertices, $V$, connected by directional edges, $E$, with no self-loops. Each edge, $e$, maintains two values:

(i) $c(e)$ – The capacity for flow. That is, how many units of flow this edge can support.

(ii) $f(e)$ – The current amount of flow through this edge.

When $f(e) = c(e)$ the edge $e$ is said to be saturated, as no further flow can be sent across this edge. Clearly, the amount of flow reaching a vertex is bound by the amount of flow that can enter, or leave that same vertex across the incoming or outgoing edges. Finally, we define the residual capacity of an edge to be $\tilde{c}(e) = c(e) - f(e)$, which describes how much remaining flow edge $e$ can still support.

There exists two uniquely identified vertices in $G$. One vertex, labelled as the *source*, contains only outgoing edges. As the name implies, it is the source of flow in the graph. Another vertex in the graph, the *sink*, contains only incoming edges. In the maximum flow problem, we want to find the maximum amount of flow that can be pushed from the source to the sink. A correct solution results in an outflow from the source that is greater than or at least equal to any other possible solution. Further, a correct solution must satisfy the following rule: the inflow at the sink must be equal to the outflow from the source. That is, every unit of flow leaving the source must, eventually, reach the sink by flowing across a series of valid edge choices.

The MPM algorithm accepts a graph $G = (V, E, s, t)$, where $V$ is a set of vertices, $E$ is a set of weighted edges (the weight represent $c(e)$ for each edge), $s$ is the source vertex and $t$ is the sink vertex.

A layered network structure acts as the base concept behind the MPM algorithm. The layered network, $G_L$, is constructed from $G$ in a manner similar to a breadth-first traversal of a graph. Starting at $s$, we traverse all edges where $\tilde{c}(e) > 0$ (that is, edges that can still support at least one unit of flow) and assign the adjacent vertices a distance label of 1. We then repeat the process for all of the vertices with a distance label of 1, marking the next set of adjacent vertices with a distance label of 2. This process continues until $t$ has been assigned a distance label. Once assigned a distance label, the label is not changed, even if the same vertex is "hit" again during a later edge traversal. Vertices and edges that have been traversed during this breadth-first traversal of $G$ are added to $G_L$.

Once $t$ has been assigned a distance label, any vertices with no outgoing edges must be pruned from $G_L$ (excluding $t$ itself). By pruning a vertex we remove the vertex itself from $G_L$ as well as any edges connecting to this vertex. This pruning effect may then propagate to other vertices as a result of the removal of other edges/vertices. This pruning step ensures that the layered network contains a series of shortest paths: any edge chosen at any layer in $G_L$ is guaranteed to form one step in a shortest path between $s$ and $t$.

Following the pruning step, we must then choose the vertex with the minimum potential, $v_m$, in $G_L$. The potential of a vertex is defined as the minimum of the in or out-potential, where the in-potential is the sum of all $\tilde{c}(e)$ values for each edge moving into this vertex, and the out-potential is the sum of all $\tilde{c}(e)$ values for each edge moving out of this vertex. The potential of $v_m$ is then pushed as flow to $t$ and pulled as flow from $s$. What this step involves is simply pushing this flow across arbitrary forward edges from $v_m$ to the next layer, and across arbitrary backward edges from $v_m$ to the previous layer in $G_L$. This process continues until the flow on either side of $v_m$ has reach $s$ and $t$.

Once flow has reached $s$ and $t$, the algorithm prunes $G_L$ once again. Clearly, at least one vertex, $v_m$, must be removed as flow equal to the potential of $v_m$ was sent through $G_L$ (and, thus, all of the outgoing and/or incoming edges will be saturated). Following the pruning, the algorithm searches for the next $v_m$ and continues sending flow throughout $G_L$.

The process of pushing/pulling flow in $G_L$ terminates when a path from $s$ to $t$ can no longer be found in $G_L$. When this event occurs, the $f(e)$ values for edges in $G$ are updated with the new, updated values from $G_L$. A new layered network is constructed and the algorithm restarts using the updated data for $G$. The algorithm itself terminates when a layered network can no longer be constructed that contains both $s$ and $t$. At this point, the algorithm returns, and a maximum flow possible for $G$ is reported.

### *4.1. Example*

We provide a brief example of a full iteration of the MPM algorithm. Figure 1 provides each step of the algorithm. (a) shows the original graph, $G$. The capacity of each edge is listed above each edge (on the left-hand side for edge $(4, t)$), and the current flow being sent through an edge is listed below each edge. (b) shows the construction of the layered network, $G_L$. In order to start pushing and pulling flow, as shown in (c), the vertex of minimum potential in $G_L$ is first determined. Clearly, vertex 2 has a minimum potential of 1. Hence, 1 unit of flow must be pushed from and pulled to vertex 2. Following this pushing and pulling of flow, edge $(2, t)$ becomes saturated. As a result, it is pruned from $G_L$. Now, vertex 2 no longer has any outgoing edges and, thus, its potential is 0. It, too, is removed from $G_L$ along with edge $(s, 2)$.

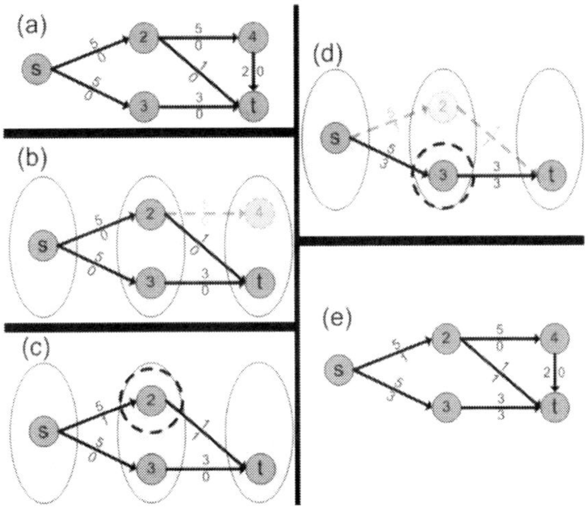


**Figure 1.** An iteration of the MPM algorithm.

In (d), the next vertex with minimum potential is chosen. The vertex chosen is vertex 3, with a potential of 3 (Note that vertex $t$ was also a valid choice in this scenario, as it, too, has a potential of 3). 3 units of flow are pushed and pulled from vertex 3 to the sink and source. Following this operation, edge $(3, t)$ becomes saturated. Similar to the example in the previous paragraph, this edge is removed and the pruning effect cascades to vertex 3 and edge $(s, 3)$. At this point in time, a path no longer exists between $s$ and $t$ in $G_L$. Thus, this iteration is complete and the amount of flow moved along each edge in $G_L$ is merged with the current state of $G$. The algorithm then "restarts", using the current state of $G$ as the input data.

## 5. Solution Methodology

We implement the MPM algorithm on an NVIDIA GTX 260 graphics card using CUDA. As a result of the massive parallelism possible on the GPU, we assign a single vertex in the graph to each thread in the GPU. This allows us to exploit very fine-grained parallelism but still allows our implementation to work on very large graph sizes (in terms of vertex count, $|V|$). To mimic the synchronicity of the original sequential algorithm, we decompose the algorithm into a series of sub-algorithms or phases. These phases are as follows:

(i) layered network construction (a modified breadth-first traversal of $G$),
(ii) pruning of the layered network,
(iii) searching for the vertex with the minimum potential, and,
(iv) pushing and pulling of flow.

Our implementation divides each phase into a separate GPU kernel. This separation of phases into kernels allows for explicit synchronization between all threads at each step in the algorithm[3].

We focus on the mapping and implementation of the layered network construction phase, an extension to our previous work in [17] where we investigate and analyze the performance of a breadth-first search GPU algorithm. The layered network construction phase for the MPM algorithm is, essentially, a modified breadth-first search. The breadth-first search must be modified, as the layered network construction phase must take into account the $\tilde{c}(e)$ value of an edge while performing the traversal. This requires a greater degree of computations and memory accesses in order to accomplish.

[3] CUDA does not allow for synchronization between all threads during the execution of a kernel, only threads within a thread block.

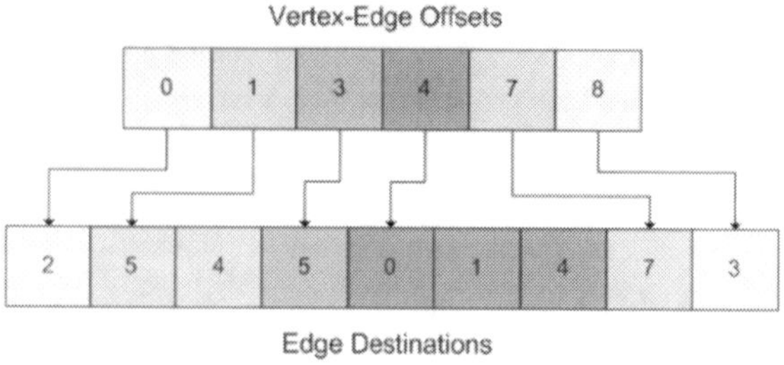


**Figure 2.** Adjacency-list representation of a graph using two arrays.

Rather than create a layered network, $G_L$, which would incur an added degree of performance degradation due to memory accesses, we use the GPU copy of $G$ to manage all aspects of the MPM algorithm. We maintain lightweight state arrays that represent the active vertices and edges in $G_L$ and allow the algorithm to work on the local copy of $G$ directly. The layered network construction phase marks each edge as active in the edge state array during the traversal of the graph. Each traversed vertex in the vertex state array is marked with the distance label of that vertex. This method has the added benefit of not requiring an update to $G$ after each pushing/pulling of flow phase, as $G$'s data is being modified directly.

Furthermore, each edge, rather than storing both the $c(e)$ and $f(e)$ values, simply maintains a record of the $\tilde{c}(e)$ value. This serves to reduce both the memory accesses and computations required when confirming an edge is still useful (as both $c(e)$ and $f(e)$ would be required otherwise). As we still need the $f(e)$ values for the final solution (in order to determine the total amount of flow across each edge), we transfer the $\tilde{c}(e)$ values back to the CPU when the algorithm has completed. The CPU maintains an original copy of $G$ with the original $c(e)$ values. From this, we can compute the final $f(e)$ values for each edge.

In order to improve the overall memory performance of our algorithm, we use a dual-array adjacency list to store $G$. Figure 2 shows the two arrays used. Harish et al. [18, 14] describe this technique as being memory efficient on the GPU. We load vertex-edge offset *start* values into the shared memory assigned for each thread block. Thread $i$ (covering some vertex $j$) in each thread block then reads the *end* offset value (that is, the *start* offset value for the next vertex, $j+1$) from the shared memory data that is written by thread $i+1$. The total number of redundant global memory reads required to read in the start and end offsets for each vertex's edge sets is reduced with this method.

As a brief example, consider the third element in the "vertex-edge offset" array in Figure 2. The value of 3 tells us that the edges for the third vertex begin at element 3 (assuming a zero-indexed array) in the edge destination array. By looking at the next element in the vertex-edge offset array, which is equal to 4, we know that elements up to (but not including) 4 in the edge destination array also define edges for vertex 3. Hence, vertex 4 has a single edge: $e = (3, 5)$.

Algorithm 1 shows the CPU loop used to drive the MPM algorithm. Each of the phases of the MPM algorithm run on the GPU, with the CPU managing the kernel invocations and termination condition checking. As a result, we require only $4-8$ bytes of data transferred from the GPU to the CPU during each iteration of a phase in order to check termination conditions. Initially, the graph is constructed by the CPU, and then transferred to the GPU. Large-scale transfers back to the CPU only take place when the MPM algorithm has terminated.

The end result is a very synchronous, very iterative algorithm. Synchronization exists between kernel invocations (essentially iterations and phases), and each phase as well as the algorithm itself is entirely iterative in nature.

**Algorithm 1** MPM Algorithm Framework

**Require:** $G = (V, E, s, t)$
  **while** A path exists from $s$ to $t$ in $G$ **do**
    $G_L \leftarrow G$ partitioned into a layered network
    **while** A path exists from $s$ to $t$ in $G_L$ **do**
      Find vertex $v$ with min. vertex potential
      Push/pull flow from $v$ to $s$, $t$
      Prune $G_L$
    **end while**
  **end while**

## 6. Results

All of the results provided in this Section were gathered from the parallel GPU algorithm running on a GTX 260, and a sequential CPU algorithm running on an Intel Core 2 Duo at 3.0Ghz. The sequential code was written to be optimized for sequential execution but was not optimized specifically for the CPU architecture.

We provide Figure 3 as an example of the execution time for one phase of the algorithm — the construction of the layered network. We test the execution time on random graphs with an average vertex degree of 5. The measurement includes only the time required to construct a layered network on the GPU. It is clear that the GPU outperforms the CPU by a significant margin for larger graphs. The GPU algorithm displays a speedup of approximately 10 for the largest graph size tested.

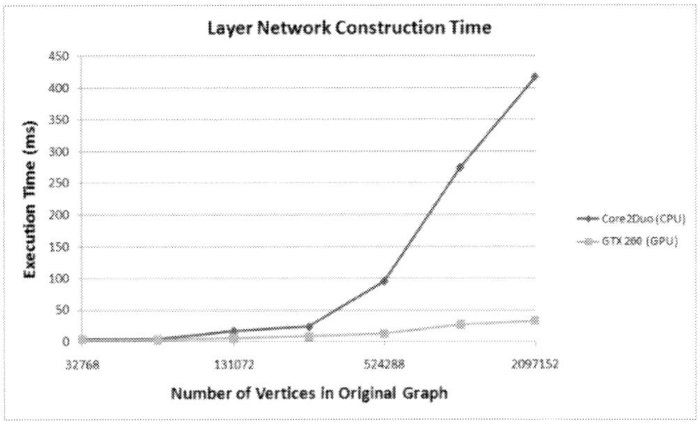


**Figure 3.** Layered Network Construction Execution Times

We further test the MPM algorithm as a whole on graphs with a vertex degree of 10. The results, provided in Figure 4, show that the GPU implementation provides a greater degree of performance than the sequential CPU implementation. The speedup for the GPU MPM algorithm as a whole is, on average, 8, when compared to the sequential CPU version.

An immediate issue that is likely reducing performance is the nature of the memory accesses. Phases (1), (2) and (4) (as listed in Section 5) require a significant level of unpredictable memory accesses during the algorithm in order to update important data related to the state of each vertex and edge. As described by Ruetsch and Micikevicius [1], the performance of a GPU algorithm can be significantly impacted by the performance of memory accesses. As a result, the high amount of unpredictable memory accesses will likely be the root cause of potential performance degradation in the algorithm. Unfortunately, due to the irregularity of these phases, the act of improving memory performance becomes tremendously difficult if not impossible.

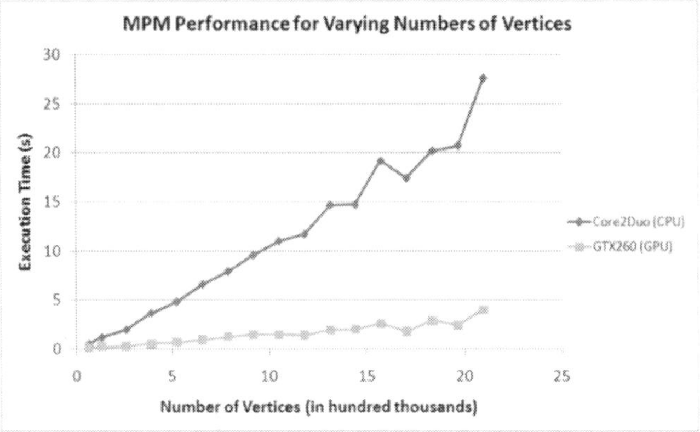


**Figure 4.** MPM Algorithm Execution Times

Another issue is the load balancing between thread warps. Currently, all threads covering all vertices are launched with each kernel invocation, regardless of whether or not every thread will be active in the coming iteration. This results in a load imbalance amongst threads in a thread warp. Perhaps not all of the vertices covered by threads in a warp will be active in the next iteration, and, due to this, less parallelism is being exploited. Consider an example where only 20 out of 32 threads in a warp have active vertices. In this scenario, only 20 out of the 32 threads will be executing. 12 threads will be idling, wasting resources that could be put to use doing *useful* work. Of course, the act of preparing and maintaining a list of vertices that will be active in the next iteration may result in an equal level of performance degradation in and of itself. As discussed in the next Section, we leave experimentation of this idea as future work.

## 7. Conclusions and Future Work

We have presented the details of a GPU implementation of the MPM algorithm. Our results show that the GPU algorithm can achieve, on average, approximately 8 times speedup over a sequential CPU implementation. Furthermore, one phase of the algorithm, the construction of a layered network, shows a speedup of 10 when compared to a sequential CPU implementation. Despite the architectural inefficiencies and difficulties associated with parallel irregular algorithms on the GPU, the GPU MPM algorithm can still achieve greater performance than a CPU algorithm.

We believe there still exists some future work to be performed on our implementation in order to improve the performance further. One such modification that may result in improved performance is the reduction in threads that are not performing useful work. This will serve to reduce branch divergence as well as ensure that each thread within each SM is performing useful work, rather than sitting idle. The experimentation of this hypothesis is left as future work.

## References

[1] Ruetsch G and Micikevicius P 2009 Optimizing matrix transpose in CUDA Published in the CUDA SDK 2.3 package
[2] Hussein M, Varshney A and Davis L 2007 On implementing graph cuts on CUDA *First Workshop on General Purpose Processing on Graphics Processing Units* (Boston, MA, USA)
[3] Fleischer L 1998 Faster algorithms for the quickest transshipment problem with zero transit times *Proceedings of the ninth annual ACM-SIAM symposium on Discrete algorithms (SODA 98)* (San Francisco, California, United States: Society for Industrial and Applied Mathematics) pp 147–156
[4] Malhotra V M, Kumar M P and Maheshwari S N 1978 *Information Processing Letters* **7** 277–278
[5] Goldberg A V and Tarjan R E 1988 *Journal of the ACM* **35** 921–940
[6] NVIDIA 2010 *Cuda 3.0 Programming Guide* NVIDIA

[7] Ford L and Fulkerson D 1956 *Canadian Journal of Mathematics* **8** 399–404
[8] Edmonds J and Karp R 1972 *Journal of the Association for Computing Machinery* **19** 248–264
[9] Dinic E A 1980 *Soviet Math. Dokl.* **11** 1277–1280
[10] Goldberg A 1987 *Efficient Graph Algorithms for Sequential and Parallel Computers* Ph.D. thesis Laboratory of Computer Science, MIT
[11] Goldberg A and Tarjan R 1988 *Journal of the Association for Computing Machinery* **35** 921–940
[12] Fujimoto N 2008 Faster matrix-vector multiplication on GeForce 8800GTX *IEEE International Symposium on Parallel and Distributed Processing (IPDPS 2008)* (Miami, Florida, USA: IEEE) pp 1–8
[13] Vineet V and Narayanan P J 2008 CUDA cuts: Fast graph cuts on the GPU *IEEE Computer Society Conference on Computer Vision and Pattern Recognition Workshops* (Anchorage, AK: IEEE) pp 1–8
[14] Harish P, Vineet V and Narayanan P 2009 Large graph algorithms for massively multithreaded architectures Technical Report IIIT/TR/2009/74 International Institute of Information Technology Hyderabad Hyderabad, India
[15] Kulkarni M, Pingali K, Walter B, Ramanarayanan G, Bala K and Chew L P 2007 Optimistic parallelism requires abstractions *Proceedings of Programming Language Design and Implementation 2007* (San Diego, CA, USA: ACM) pp 211–222
[16] Kulkarni M, Burtscher M, Inkulu R, Pingali K and Cascaval C 2009 How much parallelism is there in irregular applications? *Proceedings of the 14th ACM SIGPLAN Symposium on Principles and Practice of Parallel Programming* (Raleigh, NC, USA: ACM) pp 3–14
[17] Solomon S and Thulasiraman P 2010 Performance study of mapping irregular computations on GPUs *The 11th IEEE International Workshop on Parallel and Distributed Scientific and Engineering Computing (PDSEC-10)* (Atlanta, USA: IEEE)
[18] Harish P and Narayanan P 2007 Accelerating large graph algorithms on the GPU using CUDA *14th International Conference on High Performance Computing (HiPC 2007)* (Goa, India: Springer-Verlag) pp 197–208

Journal of Physics: Conference Series **256** (2010) 012007 doi:10.1088/1742-6596/256/1/012007

# Accelerated Synchrotron X-ray Diffraction Data Analysis on a Heterogeneous High Performance Computing System

**J. Qin and M. A. Bauer**

Computer Science Department, The University of Western Ontario, London, ON. N6A 5B7 Canada

Email: qin.jinhui@gmail.com , bauer@uwo.ca

**Abstract**. The analysis of synchrotron X-ray Diffraction (XRD) data has been used by scientists and engineers to understand and predict properties of materials. However, the large volume of XRD image data and the intensive computations involved in the data analysis makes it hard for researchers to quickly reach any conclusions about the images from an experiment when using conventional XRD data analysis software. Synchrotron time is valuable and delays in XRD data analysis can impact decisions about subsequent experiments or about materials that they are investigating. In order to improve the data analysis performance, ideally to achieve near real time data analysis during an XRD experiment, we designed and implemented software for accelerated XRD data analysis. The software has been developed for a heterogeneous high performance computing (HPC) system, comprised of IBM PowerXCell 8i processors and Intel quad-core Xeon processors. This paper describes the software and reports on the improved performance. The results indicate that it is possible for XRD data to be analyzed at the rate it is being produced.

## 1. Overview of XRD data analysis

In order to understand and predict properties of materials, detailed information about the structure of the material at the atomic-scale level is needed. Synchrotrons can be used to generate X-rays in order to probe materials at the atomic level. One approach is to use X-ray diffraction (XRD) to do this. The Laue X-ray diffraction (XRD) technique has been used to determine the atomic-scale structures of materials based on the Bragg's Law [1]. When a material that consists of a periodic atomic structure, such as a crystal, is irradiated with X-rays, it produces a diffraction pattern showing numerous sharp spots, called Bragg diffraction peaks. By measuring and analyzing the positions and intensities of these peaks, one can determine the arrangement of atoms in the crystalline material. Not only can Laue diffraction provide geometrical information about the crystalline structure, it can also be extended to measure the intensity of X-ray diffraction to determine the strain tensor in sample materials [2]. The strain tensor information will then help scientists/engineers to determine possible flaws in materials.

During an XRD experiment, a material sample is scanned by X-rays at a number of points over an area of the sample. A CCD detector captures the diffractions and produces a 2D image for each scan point (see Figure 1 for an example of an XRD image). Image sizes are dependent on the CCD detector. For an area XRD scan, a typical step size for moving the scan point by point is a couple of microns. The data generated in XRD experiments can consist of a large number of images. A number of existing software packages for XRD data analysis have been developed to analyze these images. These include the 3D X-ray Micro-diffraction Analysis Software Package [3] at the Advanced Photon Source (APS), which was developed by scientists at the Oak Ridge National Laboratory (ORNL), and the X-ray Micro-diffraction Analysis Software (XMAS) [4] at the Advanced Light Source (ALS) in Lawrence Berkeley National Laboratory (LBNL). The common feature of these two packages is that they both are Windows-based software, implemented in the Interactive Data Language (IDL) [5]. Both can process a large amount of XRD data sequentially.

Figure 1 An example of an XRD image

Analysis of XRD data analysis typically consists of applying three computational methods on each of the XRD images sequentially. The three procedures are *peak searching*, *indexing* and *strain analysis*. The *peak searching* process attempts to extract useful information about intensity points (peaks) from an image to be used as input for the next two procedures. The *indexing* procedure takes the output from the *peak searching* procedure and generates the structure information about the sample material, e.g. the orientation of a crystalline lattice plane from which a diffraction spot (i.e. a peak that appeared in the XRD image) is generated. Based on the indexing results and peak information, the strain analysis procedure then produces strain tensor information of the sample. Based on the indexing results and strain tensor information, an orientation map and a strain map can be generated for the entire scanned area from which all XRD data were collected. Those maps become the most interesting part for scientists and engineers.

The main purpose of peak searching procedure is to identify the *locations and intensities* of those peaks captured on an XRD image. The accuracy of peak search results become critical to the quality of indexing and strain analysis. However, images generated from X-ray diffraction are often noisy, and peaks shown within an image are often "blobs" of different sizes and shapes. A number of factors, often referred to as the Lorentz factors [6], affect the quality of images from X-ray diffraction, including scattering, angular-velocity factor, polarization, etc. These make the peak searching procedure complicated.

In order to produce more precise peak information for indexing and strain analysis, the peak searching procedure can be split into three subtasks: a threshold finding process, which is used to remove background noise; a "blob" searching process, which searches for all the blobs with intensities above the threshold; and a blob fitting process, which applies some 2D curve fitting functions on each blob area to determine a reasonable location and intensity of a peak.

Basically, the source data from an XRD experiment is a sequence of digital image files. The XRD data analysis software actually deals with a large amount of data, typically, digital images in the hundreds or even thousands. Existing analysis software processes these images individually and sequentially. This is a very time consuming process, normally requiring days to complete the processing of an entire set of data from one XRD experiment. As noted, synchrotrons are typically used to generate X-rays to probe the structure of materials. Synchrotron time is valuable and it is often difficult to get a scheduled beam time. Data analysis using existing software means that researchers are typically doing the analysis after their time on the synchrotron. More timely analysis could help researchers make decisions on subsequent experiments and gain significant insight into the materials that they are studying. Performance improvement in the analysis software could make the analysis more useful.

In this paper, we report on the development of parallel XRD data analysis software which has been developed on a heterogeneous HPC system comprised of Intel and IBM Cell processors. The software developed and the system it runs on makes it possible for XRD data to be processed in "near-real time", that is, nearly as fast as it is being produced. A description of the target HPC system is provided in Section 2. The development approach is presented in Section 3, including the challenges and an overview of the design and implementation of the software. Performance evaluations are reported in Section 4, and Section 5 provides some conclusions and future work. This work was done as part of the Science Studio project [7][1].

## 2. Target HPC system

Our target system, called *prickly*, is one of the clusters in SHARCNET [9]. *Prickly* is located at the University of Western Ontario (UWO). It is a *heterogeneous* HPC system consisting of one head node for hosting user logins and a chassis with 12 Linux cluster blades providing total 160 computing cores. Among the 12 blades, four blades are Intel blades and the other eight are IBM QS22 Cell blades. On each of the Intel blades, there are *two* quad-core Xeon E5420 processors running at 2.5GHz with 8GB of memory. Each of the IBM Cell blades contains *two* PowerXCell 8i processors, so called Cell processors, running at 3.2GHz with 16GB of memory. Gigbit Ethernet provides the inter-connection between any two blades on *prickly*, including Cell blades and Intel blades.

Unlike traditional multi-core processors which are homogenous, such as those on Intel blades, the Cell processor itself has *heterogonous* multi-cores [10]. It employs two types of cores optimized for different kind of tasks. Each Cell processor has nine cores, i.e. *one* PowerPC Processor Element (PPE) and *eight* Synergistic Processor Elements (SPEs). The PPE is just a traditional 64-bit Power processor with 512K cache and acts as a large-scale processor core to run the operating system and performs control-intensive tasks. In contrast, the SPEs are much simpler, but devote more resources to perform computationally intensive tasks. Since each Cell blade has two Cell processors, in total, there are *sixteen* SPEs on each Cell blade. The sixteen SPEs are independent, 128-bit vector processors, which can execute Single Instruction Multiple Data (SIMD) instructions. Each SPE has its own local storage (256KB) for instructions and data. The local storage can be considered as the cache for each SPE, except that users' programs need to decide and manage what should be taken from main memory and stored locally during the execution. The SPE access to the memory is achieved through its Direct Memory Access Controller (DMAC). The DMAC can work concurrently with SPE executions, which hides the latency caused by memory accesses. However, user programs need to decide how to overlap

[1] The Science Studio project has been funded by CANARIE [8].

the data movement between the main memory and a local store in order to make the computations done on an SPE as efficient as possible.

The Element Interconnect Bus (EIB) provides four 128-bit data transmission channels for the inter-communication among PPE, SPEs, main memory and I/O. The EIB can support up to 307GB/s bandwidth between any two bus units. Therefore, with EIB, each SPE can not only work alone, but also be chained together to perform data processing with an intensive workload, such as stream processing.

The PPE is intended to run the operating system and coordinate computations. A key performance advantage comes from Cell's sixteen de-coupled SPE SIMD engines with dedicated resources for computations, including large register files and DMA channels. By incorporating advanced multiprocessing technologies, the Cell processor is especially suitable for high performance workloads. While the Cell's special architecture offers many advantages for high performance computations, the architecture also makes programming on Cell a non-trivial task.

## 3. Application development approach

We had two main objectives in porting the XRD data analysis application onto the target HPC system, i.e. *prickly*. First, we wanted to create an implementation of the three major analysis procedures where the processing tasks were pipelined in order to accelerate the processing of an XRD image. Second, we also wanted an implementation so that multiple XRD images could also be processed in parallel.

### 3.1. Challenges

There were two major challenges involved in this porting process. First, the original software was written in IDL [5]. IDL is a proprietary scripting language provided with extensive subroutine libraries for processing/visualizing images in the fields of physics, astronomy and medical imaging. Our target system ran Linux and could only compile and execute programs in C/C++/Fortran. The software needed to be rewritten in C in order to be able to adapt it for execution on Cell processors. The source code of the XRD analysis software was originally developed at ORNL and made use of many of the library functions within the IDL package. With neither the source code of the library functions nor detailed computation algorithms for XRD data analysis procedures, rewriting the software in C became a challenging task.

Another challenge was programming on Cell itself. Because of the Cell's heterogeneous architecture, the PPE and SPE execute different instruction sets. Programs written for the PPE and SPEs need to be compiled by different compilers. The PPE manages SPE processes as POSIX threads and so threaded programming skill is required. As each SPE has its own local store for holding instructions and input/output data, the data needs to be moved back and forth between the local store and the main memory with explicit DMA commands. Moreover, in order to hide the memory access latency in order to make efficient programs, programmers must handle the local store management, such as data locality, overlapping data transformation and inter-unit communication within SPE's computations. Because of the limited space (256K) for a local store on an SPE, only tasks that can fit within that local store are good candidates to be considered for a SPE, otherwise, advanced overlay management needs to be considered [10].

### 3.2. Redesign of the software

In order to attain an accelerated XRD data analysis on the target HPC system, the software needed to be redesigned to be parallel and take an advantage of the Cell's multi-core processing. Considering the challenges of programming on Cell, it was more efficient to first identify the performance "bottleneck" of the entire process and then target an implementation around that "critical" part. As a result, instead of simply porting all three procedures onto Cell, we focused on the critical part first.

As introduced in Section 1, there are three major procedures involved in XRD data analysis. Our measurements indicated that more than 80% of the processing time was spent in the peak searching procedure; therefore, it was initially targeted as the "critical" computation to be considered for porting to the Cell.

The peak searching procedure involves finding a threshold, blob searching, and curve fittings on each of the blobs. Among all three subtasks in peak searching, curve fitting is the most intensive one. During curve fitting on each blob, it applies *two* 1-D fittings (i.e., one for X direction and one for Y direction) and *one* 2-D fitting for a box area around each blob. Figure 2 illustrates blobs identified in an XRD image with a certain intensity threshold. Each fitting process actually entails solving a multi-variable, non-linear least square minimization problem. It involves iterations to update the state of corresponding variables continually until a certain criteria are met. Specifically, the 1-D fitting involves solving four variables; these results become the initial states for the 2-D fitting. In turn, the 2-D fitting involves solving for six variables. The existing software carries out the curve fittings sequentially for each of the blobs in an image, which is very time consuming and becomes the bottleneck of the entire XRD data analysis.

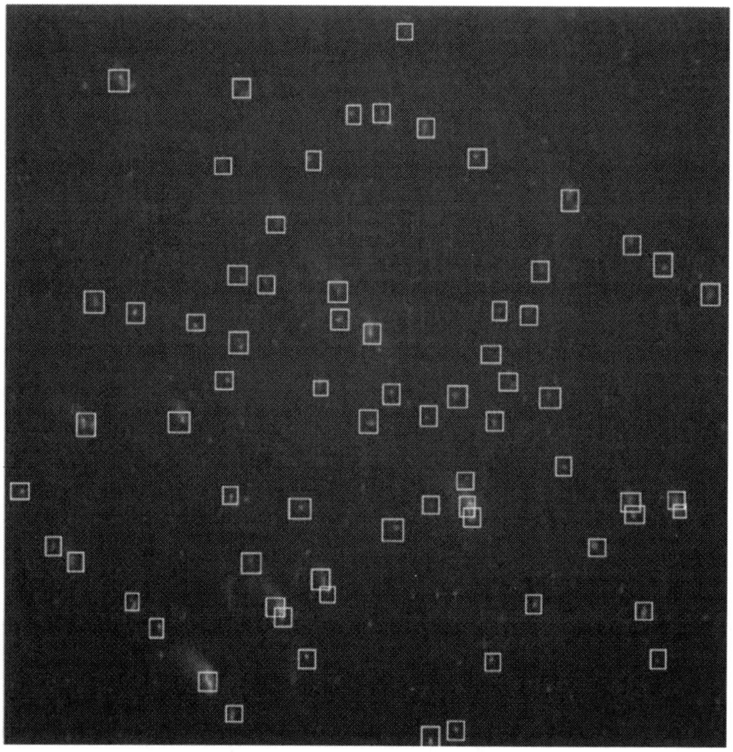

Figure 2. An XRD image with identified blobs that need curve fittings

Considering the computational power of a Cell's SPE, with a limited local store, it works well for a process with relatively small size but needs to run many times. Fortunately, the curve fitting is applied to each blob, which is in a relatively smaller area than the entire image area. The computation of the fitting process is also relatively intense and needs to be applied to every blob in an image. Therefore, the curve fitting process was selected as the processing task for the Cell's SPEs. Then, once a collection of blobs has been identified, the fitting process can be done on the Cell's multiple SPEs in parallel, i.e. multiple blobs can be fitted by multiple SPEs simultaneously.

Even though the size of the data accessed during each curve fitting process is relatively small, it still took some effort to implement such a task on Cell's SPE because of its limited local store. As has been mentioned, the curve fitting process involves actually solving a non-linear, multi-directional least square fitting problem. The GNU Scientific Library (GSL) [11] provides a minimization package in C for solving such problems. However, the size of GSL, or even only the size of the minimization package, is too large to fit into a SPE's local store. In order to fit the desired code on the SPE, we manually extracted each of functions needed for the minimization computation and then included them as part of the source code to be compiled on Cell's SPEs. Combined with the two other sub-tasks i.e., threshold discovery and blob identification, on Cell's PPE, after the curve fitting for blobs had been successfully ported onto the Cell's SPEs, the peak searching procedure could then be carried out entirely on a Cell processor.

To analyze a large set of XRD images, and in order to achieve a better throughput, there are two key challenges that must be overcome. First, we need to find out what needs to be run only once and which result can be shared by all the other images from the same set of experiments, i.e., to identify those computations that do not need to be applied to every image. For example, the background information of images from a same set of experiments can be very similar because they all have the same experimental settings. Therefore, the threshold finding process only needs to be applied to one of the images, and the threshold can be passed to and shared by all the other images.

Second, for a parallel application on such a heterogeneous HPC system, it is a multi-threaded and multi-process program. Therefore, we need to consider the overhead in generating such two kinds of parallel computational elements. For example, in order to have $n$ images processed in parallel requires $n$ parallel processes. In processing each of the images, doing curve fitting to $m$ blobs within each image in parallel requires $m$ parallel threads. For a given set of computation resources, the execution time for processing a large set of images varies when the configurations are different, i.e. the settings of $m$ and $n$. Efficient processing depends on the computational workload and the difference in overhead between multi-thread and multi-process. We explore this difference in settings in the experiments presented in Section 4.

### 3.3. Parallel XRD data analysis

The three major procedures involved in XRD data analysis have been designed to execute on different types of nodes on *prickly*, a blade center with two different types of blades. The peak searching procedure has been ported to Cell nodes, and the indexing and strain analysis procedures have been ported to the Intel Xeon nodes.

In order to achieve high performance in processing a large number of XRD images, we consider a parallel computational approach involving processing multiple images in parallel and fitting multiple blobs in parallel for each image. Since curve fitting is the most intensive task it has been assigned to the Cell's SPEs and threshold finding and blob searching are kept as tasks for Cell's PPEs. It is assumed that the intensity levels or background information are similar among images collected from the same XRD experiment and therefore the threshold finding process only needed to be performed once on one of the images, e.g. normally the first image. The rest of the images can then share this same threshold during the rest of the analysis. The design of such a parallel XRD data analysis program on our target system is illustrated in Figure 3.

Using Figure 3 as a guide, the processing proceeds as follows. Initially, $n$ images are loaded to be processed in parallel. Each of these images is initially processed on one of the PPEs in order to identify blobs; a list of blobs is produced. Curve fitting is then done on each list of blobs on an SPE. The processing of the blobs from an image results in a list of possible peaks (PeakList in the Figure 3). The list of peaks is then passed to the indexing computation which results in a grain orientation map and the index data. The index data is then used in the strain computation which produces a strain map. Because of relatively less computational workload when compared with the peak searching procedure, the indexing procedure and strain analysis procedure are currently kept as sequential processes for each image and done on Intel blades. The size of each of these processes is too large to easily fit into the limited local storage on Cell's SPE as they stand. Moreover, the Cell's PPE is structured specifically for handling control intensive tasks instead of computational tasks [10], therefore, we decide to allocate the process of indexing and strain analysis to the Intel blades instead of the Cell blades on *prickly*.

The communication and data transformation between any two nodes on *prickly* is achieved through MPI [12]. MPI is a language independent protocol. It is commonly used in computer clusters and supercomputers to support point-to-point and collective communications. For the computational elements allocated to the Cells' SPEs, management is achieved through Linux POSIX Threads (Pthread). All the nodes share the same file system.

In summary, the XRD data analysis developed is a mixed multi-threaded and multi-process application which can achieve an accelerated processing performance as we shall describe in the next section.

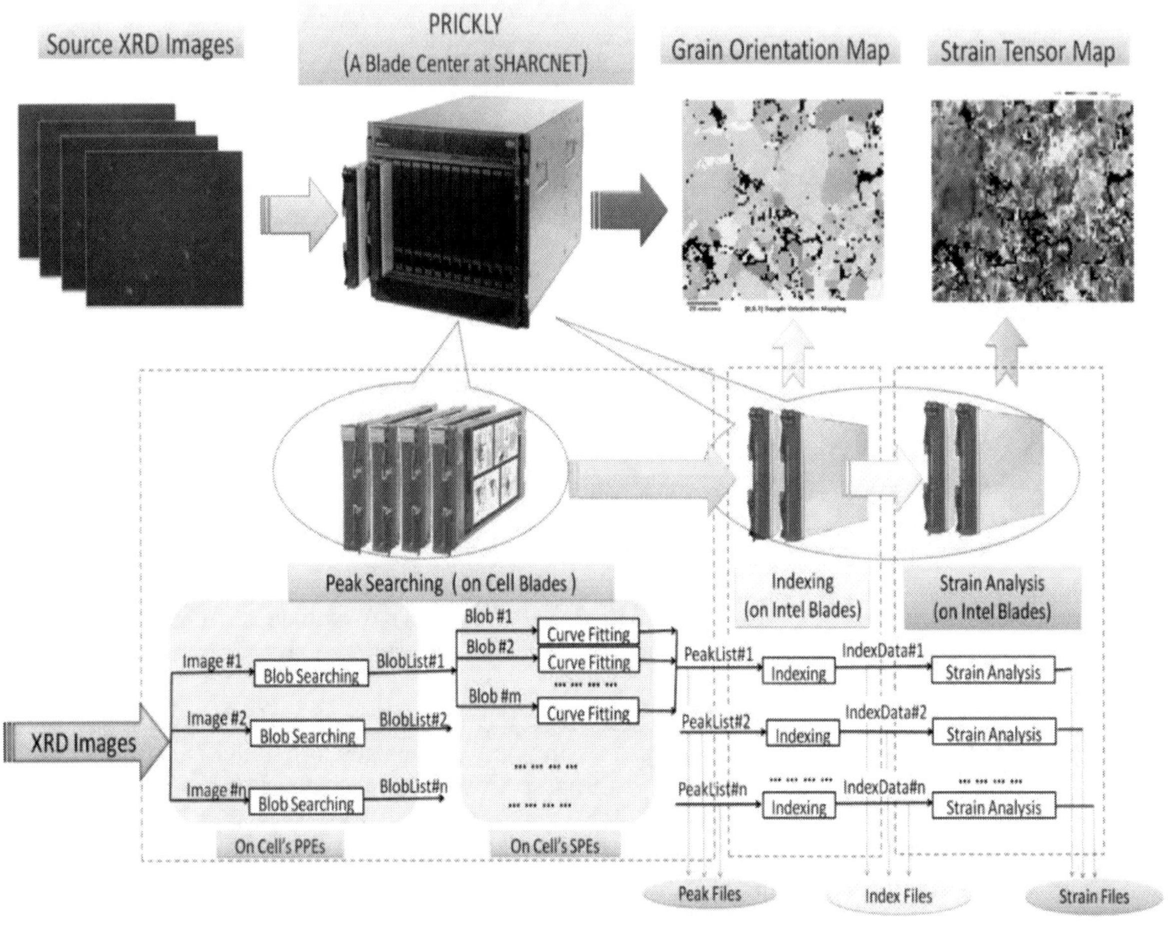


Figure 3. The configuration of high performance XRD data analysis on Prickly

## 4. Performance evaluation

The design in Figure 3 has been implemented and deployed on *prickly*. In order to evaluate the processing performance and study the impact of various parallel settings on the processing, we have carried out a number of experiments considering images of different sizes and considering different numbers of processing elements on *prickly*. Since XRD images could have different sizes depending on the CCD detector that had been used for collecting the images, we considered two different sizes of XRD images. The XRD images were collected from experiments from the Advanced Photon Source (APS) [13] and others were from experiments done at the Canadian Light Source (CLS) [14]. An XRD image from APS has 1042X1042 pixels and is about 2MB/image; while an image from CLS has 2084X2084 pixels and is about 8MB/image. That means the workload for processing these two types of images is different.

### 4.1. XRD analysis on a desktop machine

We measured the time needed to process a set of images sequentially with the original IDL software on a desktop machine, i.e., each processing step was done sequentially and one image after the other. We compared the performance of using the original IDL software and using the non-parallelized C version of the software on the same desktop machine. The desktop machine used was an Intel dual core 2.4GHz processor with 3GB memory. The average processing speed was measured in seconds/image and is presented in Table 1. As expected, the software in C is faster than the original IDL version. The speedup is relatively modest for both the APS images and the CLS images, respectively, 1.47 and 1.25.

Table 1. Desktop performance of XRD analysis on sequential images

| Image size (pixels/image) | Total number of images | Total processing time (sec.) | | Average speed (sec./image) | | Speedup (C vs. IDL) |
|---|---|---|---|---|---|---|
| | | IDL | C | IDL | C | |
| 1042X1042 (APS) | 64 | 276.0 | 188.34 | 4.31 | 2.94 | 1.47 |
| 2084X2084 (CLS) | 64 | 919.0 | 735.26 | 14.36 | 11.49 | 1.25 |

### 4.2. XRD analysis on *prickly*

The C version of the software was then ported to *prickly*, as illustrated in Figure 3, where the three procedures have been chained together to work in a pipeline. Multiple images can then be processed in parallel simultaneously, which means that multiple pipelines can be created for XRD image analysis. Since the system has total of 4 Intel blades and 8 Cell blades, we examined the performance of XRD data analysis on *one* pair of Cell-Intel blades and on *multiple* pairs of Cell-Intel blades.

In using one of the Cell-Intel nodes on *prickly*, i.e. 1 Cell blade and 1 Intel blade, as described in Figure 3, peak searching is done on the Cell blade; while indexing and strain analysis are done on the Intel blade. Since each Cell blade has total 16 SPEs, if $n$ images are processed in parallel, i.e. $n$ pipelines, and $m$ SPEs are allocated for each image or each pipeline, $m$ and $n$ are constrained to be values such that $m \times n=16$. Different settings for $m$ and $n$ were tested. For the workload on the Intel blade, if $n$ pipelines are initiated for processing $n$ images in parallel, $n$ processes are created on the Intel blade, and each of these $n$ processes works on the indexing and strain analysis on one of $n$ images. The operating system takes care of workload distribution among the *eight* cores on the Intel blade. We measured the average processing speed of various settings. The speedup of each test case is compared to the desktop speed using IDL and C software. Tables 2 and 3 present the measured results on one pair of computation nodes on *prickly*. Table 2 reports the results for the APS images and Table 3 reports the results for the CLS images.

Table 2. Results of processing APS images on *one* pair of Cell-Intel nodes on *prickly*

| Total number of images | Number of images in parallel (number of pipelines) | Number of SPEs for each image | Total processing time (seconds) | Average speed (sec./image) | Speedup vs. IDL (speed 4.31 sec./image) | Speedup vs. C (speed 2.94 sec./image) |
|---|---|---|---|---|---|---|
| 64 | 1 | 16 | 40.54 | 0.63 | 6.84 | 4.67 |
| 64 | 2 | 8 | 27.42 | 0.43 | 10.02 | 6.84 |
| 64 | 4 | 4 | 22.64 | 0.35 | 12.31 | 8.4 |
| 64 | 8 | 2 | 16.41 | 0.26 | 16.58 | 11.31 |
| 64 | 16 | 1 | 14.05 | 0.22 | 19.59 | 13.36 |

The results presented in Table 2 illustrate that for processing images of the size of the APS images, the more images that are processed in parallel, the better throughput, i.e. processing 16 images resulted in speedups of 19.59 compared to IDL on the desktop and 13.36 compared to C on the desktop.

In processing the larger size images, i.e. those from CLS, the results of Table 3 suggest that processing 4 images in parallel and with 4 SPEs allocated for each image can produce the best throughput. The processing of multiple images in parallel was achieved through multi-process programming, while the parallel blob fitting on the Cell was achieved through multi-threaded programming. In general, process creation cost is much larger than thread creation cost. The overall performance gain of a parallel application is dependent on balancing the computational workload and the trade-off in setting up the parallel processing elements. The result of reducing the number of SPEs allocated for each image, i.e., to 2 or 1 in Table 3, in lieu of having more parallel pipelines to process more images in parallel is not sufficient to overcome the blob processing done on each of the larger images within the SPEs. Consequently, using 4 pipelines and 4 SPEs results in the best performance for images of this size.

Table 3. Results of processing CLS images on *one* pair of Cell-Intel nodes on *prickly*

| Total number of images | Number of images in parallel (number of pipelines) | Number of SPEs for each image | Total processing time (seconds) | Average speed (sec./image) | Speedup vs. IDL (speed 14.36 sec./image) | Speedup vs. C (speed 11.49 sec./image) |
|---|---|---|---|---|---|---|
| 64 | 1 | 16 | 182.07 | 2.84 | 5.06 | 4.05 |
| 64 | 2 | 8 | 115.95 | 1.81 | 7.93 | 6.35 |
| 64 | 4 | 4 | 98.27 | 1.55 | 9.26 | 7.41 |
| 64 | 8 | 2 | 107.12 | 1.67 | 8.60 | 6.88 |
| 64 | 16 | 1 | 107.50 | 1.68 | 8.55 | 6.84 |

It is also possible to use multiple Intel and Cell blades to process a sequence of images. To examine the performance implications of this, we also ran a number of experiments of XRD data analysis using *multiple* pairs of Cell-Intel nodes. Based on the results presented in Tables 2 and 3, *1* SPE per APS image and *4* SPEs per CLS image were used. Tables 4 and 5 present the results.

Table 4. Results of processing APS images on *multiple* pairs of Cell-Intel nodes on *prickly*

| Total number of images | Pairs of Cell-Intel node(s) | Number of images in parallel (number of pipelines) | Total processing time (sec.) | Average speed (sec./image) | Speedup vs. IDL (speed 4.31sec./image) | Speedup vs. C (speed 2.94 sec./image) |
|---|---|---|---|---|---|---|
| 64 | 1 | 16 | 14.05 | 0.22 | 19.59 | 13.36 |
| 64 | 2 | 32 | 9.20 | 0.14 | 30.78 | 21.00 |
| 128 | 3 | 48 | 9.58 | 0.07 | 61.57 | 42.00 |
| 128 | 4 | 64 | 9.28 | 0.07 | 61.57 | 42.00 |

The results presented in Tables 4 and 5 illustrate that the performance of XRD data analysis has been boosted significantly when more computational resources are used. For the smaller sized XRD images collected at APS, as presented in Table 4, when 48 processing pipelines were setup on *three* pairs of Cell-Intel blades, the average processing speed can reach as high as 0.07 sec./image, which is 61.57 times speedup compared with 4.31sec./image of using existing IDL software on a desktop machine. For larger size XRD images collected at CLS, when 16 processing pipelines were setup on *four* pairs of Cell-Intel blades, the average processing speed (see Table 5) can reach as high as 0.59 sec./image, which is 24.34 times speedup compared with 14.36 seconds/image when using the IDL software on a desktop machine.

Notably, the measured *speedup* from both sets of experiments presented in Tables 4 and 5 do not result in a linear improvement as more nodes are added. One of the factors affecting the speedup is the increased overhead of MPI in setting up more pipelines exchanging information across the two types of nodes on *prickly*. Another factor that affects achieving a linear speedup is the data transfer and communication cost of the Gigabit Ethernet; as more processes are added there is an increase in communication. A more thorough investigation of exactly which factors play the most prominent role in limiting the speedup when processors are added could be one of the research tasks in future.

Table 5. Results of processing CLS images on *multiple* pairs of Cell-Intel nodes on *prickly*

| Total number of images | Pairs of Cell-Intel node(s) | Number of images in parallel (number of pipelines) | Total processing time (sec.) | Average speed (sec./image) | Speedup vs. IDL (speed 14.36 sec./image) | Speedup vs. C (speed 11.49 sec./image) |
|---|---|---|---|---|---|---|
| 64 | 1 | 4 | 98.27 | 1.55 | 9.26 | 7.41 |
| 64 | 2 | 8 | 61.45 | 0.96 | 14.96 | 11.97 |
| 64 | 3 | 12 | 44.71 | 0.70 | 20.51 | 16.41 |
| 64 | 4 | 16 | 37.80 | 0.59 | 24.34 | 19.47 |

## 5. Conclusion and future work

In this paper we reported on the development of accelerated XRD data analysis software running on a heterogeneous HPC system, i.e. the cluster *prickly* on SHARCNET. Using the computational power of *prickly*, especially the Cell processors, the accelerated XRD data analysis can achieve a speedup of up to 60 times faster than a desktop performance of the original IDL software package, depending on the size of images and the number of computation nodes used on *prickly*. The design and implementation has not yet been optimized to fully use the computing resources provided by *prickly* and so there is likely still room to further improve the computational performance if there is such a need in the future. Other designs using the computing resources of *prickly* could also be examined as alternative schemes for computation. For example, it would be interesting to compare and examine the performance of a parallel implementation without using any of the Cell blades and only using the quad-core Xeon processors on the Intel blades.

The current implementation has achieved the functionality for a high performance data analysis of XRD data. Our next step is to implement the process of data transmission followed by our analysis in near real time, that is, as it is collected during a synchrotron experiment. This would result in a near real time data collection and analysis process. This is currently underway using the VESPERS beamline [15] at the Canadian Light Source (CLS), combined with CANARIE [8]'s dedicated lightpath for data transmission from CLS to UWO. The expected performance will allow researchers to examine initial results from beamline experiments, make decisions regarding the experiment (e.g. stop it and reposition the beam if the scan seems unproductive), and to make decisions regarding subsequent experiments. Such a service model could also be adapted to other synchrotron and HPC facilities.

## References

[1]. M. M. Woolfson, *An Introduction to X-Ray crystallography*, 2nd edition, Cambridge University Press, 1997.

[2]. J. Chuang and G. Ice, "Automated indexing for texture and strain measurement with broad-band pass x-ray micro-beams", *Journal of Applied Physics*, No. 9, Vol. 86, 1999
[3]. XRD data analysis at APS : http://www.aps.anl.gov/Sectors/33_34/microdiff/downloads/
[4]. XRD data analysis at ALS: http://xraysweb.lbl.gov/microdif/user_resources.htm
[5]. K. P. Bowman, *An Introduciton to Programming with IDL: Interactive Data Language*, Elsevier Inc. 2006
[6]. A.J.C. Wilson, *Elements of X-Ray Crystallography*, Addison-Wesley Pub. Co. ,1970
[7]. Science Studio website: http://sciencestudioproject.com/
[8]. CANARIE website: http://www.canarie.ca/
[9]. SHARCNET website: https://www.sharcnet.ca/
[10]. M. Scarpino, *Programming the Cell Processor: For Games, Graphics, and Computation*, Printice Hall, 2008
[11]. M. Galassi, J. Davies, J. Theiler, B. Gough, G. Jungman, P. Alken, M. Booth and F. Rossi, *GNU Scientific Library Reference Manual - Third Edition (v1.12),* Network Theory Ltd. 2009
[12]. P. Pacheco, *Parallel Programming with MPI,*Morgan Kaufmann Publishers, Inc. 1997
[13]. APS website: http://www.aps.anl.gov/About/ALD/
[14]. CLS website: http://www.lightsource.ca/index.php
[15]. VESPERS website: http://vespersbeamline.com/index.php

# Polymer Dynamic Field Theory on Graphics Processing Units

**I. Wright**
Perimeter Institute, 31 Caroline Street North, Waterloo, Ontario, Canada N2L 2Y5
E-mail: iwright@perimeterinstitute.ca

**R. A. Wickham**
Department of Physics, University of Guelph, Guelph, Ontario, Canada N1G 2W1
E-mail: rwickham@uoguelph.ca

**Abstract.**
This paper explores the application of graphics processing units (GPUs) to a study of the dynamics of diblock copolymer (DBCP) melts. DBCP melts exhibit ordered mesophases with potential applications in nanotechnology, but are often found out of equilibrium. The length scales involved have previously rendered the numerical modelling of these materials intractable in certain regimes. We adapt a code originally written in parallel using the Message Passing Interface to GPUs using the NVIDIA® CUDA™ architecture. We gain a factor of 30 performance improvement over the original code at large system size. We then use this performance improvement to study DBCP melts in two computationally time-intensive regimes: droplet nucleation close to phase coexistence, and dynamics under confinement in a small cylindrical nanopore.

## 1. Diblock Copolymers

A diblock copolymer consists of two distinct bonded polymeric chains, in a configuration such as

`A-A-A-A-B-B-B-B.`

Polymer chains have more coiled than stretched microstates, so entropic forces favour a coiled chain configuration, resulting in a disordered phase in a diblock copolymer melt at high temperature. An effective repulsion between monomers `A` and `B` counteracts this, preferring a stretched configuration with maximum separation between unlike monomers. These two effects compete, leading to the formation of various phases of ordered microdomains of `A` and `B` through a process known as self-assembly (see Figure 1). Observed structures of these microdomains include lamellar domains, hexagonally-packed cylindrical domains and body-centred cubic lattices of spherical domains. Self-assembly makes these materials attractive for applications where the creation and control of structures on nanometer scales is desirable – for example, as lithographic templates for nanowires and photonic crystals.

While perfectly periodic equilibrium phases are well-described by theory, the processing of DBCP melts requires an understanding of the dynamics of the formation of these mesophases.

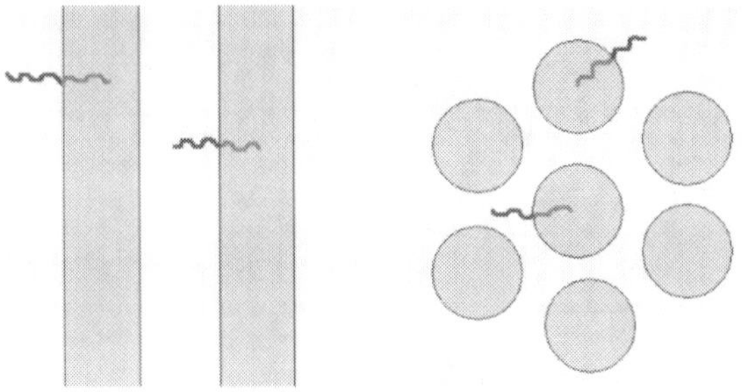

**Figure 1.** An illustration of the location of an individual polymer within a melt in the lamellar and cylindrical mesophases, respectively.

For example, if one crosses a phase boundary while processing a block copolymer, a non-uniform phase can result due to the nucleation of a droplet of the stable structure out of the original (metastable) structure.

In addition, the equilibrium behaviour of DBCP melts and their dynamics are altered in the presence of a surface field, creating novel microstructures in confined spaces, including concentric lamellae, helical and gyroid structures. The dynamics of the formation of these microstructures is largely unexplored to date.

## 2. The Brazovskii Model for Diblock Copolymers

The behaviour of DBCP melts is found to be dependent on only two parameters [1]. The first, $f$, is the quotient $\frac{N_A}{N}$, where $N_A$ is the number of A monomers in a polymer chain and $N$ is the total number of monomers in the chain (the polymerization index). The second parameter is the product $\chi N$, where $\chi$ is the Flory-Huggins parameter, a temperature dependent parameter characterizing the monomer-monomer interaction strength.

We use the Landau-Brazovskii model [2] to describe DBCP melts in the weak segregation limit. We define an order parameter $\phi(\mathbf{r})$ to describe the thermal average of the deviation of the melt from homogeneity,

$$\phi(\mathbf{r}) = \langle(1-f)\rho_A(\mathbf{r}) - f\rho_B(\mathbf{r})\rangle, \tag{2.1}$$

where $\rho_A(\mathbf{r})$ and $\rho_B(\mathbf{r})$ are the densities of monomers A and B respectively. In a high temperature, disordered phase, $\phi(\mathbf{r})$ is zero everywhere, because the ratio of A to B monomers at any point in the melt is dependent only on $f$. As the melt is cooled and ordered structure forms, $\phi(\mathbf{r})$ deviates from zero and becomes periodic.

We encode the competing actions of the entropic force and the monomer-monomer interaction in the many-chain Edwards Hamiltonian [3]. Following the method of Leibler [1] and Ohta and Kawasaki [4], we use the tools of statistical mechanics to expand the free energy of the melt in powers of the order parameter. We use the de Gennes random phase approximation [5] to write the expansion for interacting monomers in terms of the Green's functions derived for non-interacting monomers moving in a mean field describing interactions.

We truncate this free energy expansion at fourth order and take the weak segregation limit by expanding around a single dominant scattering wavevector $q_0$ in reciprocal space, corresponding to the periodicity of the melt. This approximation is valid in a regime close to the mean-field critical point. The resulting expression for the free energy in real space can be written in the form used by Wickham, Shi and Wang [6]:

$$F = \frac{1}{V}\int d\mathbf{r}\left\{\frac{\xi^2}{2}[(\nabla^2+1)\phi]^2 + \frac{\tau}{2}\phi^2 - \frac{\gamma}{3!}\phi^3 + \frac{1}{4!}\phi^4\right\}. \tag{2.2}$$

We have absorbed the dependence on the wavevector $q_0$ through a change of variables. Here $V$ is the volume of the melt, while $\xi$, $\tau$ and $\gamma$ are dependent on the parameters $f$ and $\chi N$. In fact, only $\tau$ is dependent on $\chi N$, and is thus the only variable dependent on the temperature of the melt. Also, $\gamma$ depends on the asymmetry of the copolymer composition, and is zero when $f = 0.5$.

## 3. Simulating the Dynamics of DBCP Melts

The code we have adapted to run on graphics processing units (GPUs) was originally developed by Robert Wickham's group at the University of Guelph. To model the kinetics of a DBCP melt, we evolve the system in time governed by a diffusion equation,

$$\begin{aligned}\frac{\partial\phi}{\partial t} &= -\nabla^2\frac{\delta F}{\delta\phi} \\ &= -\nabla^2\left\{\xi^2[(\nabla^4+2\nabla^2+1)\phi] + \tau\phi - \frac{\gamma}{2}\phi^2 + \frac{1}{6}\phi^3\right\}. \end{aligned}\tag{3.1}$$

The field thus evolves towards spatial configurations that minimize the free energy of the system.

To implement the kinetics numerically, equation (3.1) is integrated at every discretized timestep, using a real space finite difference approach. We initialize a field $\phi(x, y, z)$ in a mesh in three dimensions. We then either impose periodic boundary conditions to simulate an infinite melt in the study of nucleation, or impose a cylindrical mask on the melt in the $x$ and $y$ directions to model confinement.

We implement a second-order accurate Laplacian operator, calculated by averaging the gradient between each point on the mesh and its nearest neighbours,

$$\nabla^2 O \simeq \frac{1}{6}(\Delta O_x + \Delta O_y + \Delta O_z), \tag{3.2}$$

where, for example

$$\Delta O_x = O(x-1, y, z) + O(x+1, y, z) - 2O(x, y, z) \tag{3.3}$$

and $O$ is any operand. This function is applied to $\phi$ to calculate $\nabla^2\phi$, then again to $\nabla^2\phi$ to calculate $\nabla^4\phi$. Finally, the operator is applied to $\frac{\delta F}{\delta\phi}$ to calculate the right hand side of equation (3.1). Because each application of the Laplacian operator is dependent on nearest neighbours, the final right hand side of equation (3.1) is equivalent to the application of a third-to-nearest neighbour mask, in accordance with the prescriptions of a finite difference approach.

To model confinement we impose Neumann boundary conditions at the confining surface, setting those gradients that cross the boundary to zero. We then use a fourth order Runge-Kutta integration technique [7] to integrate the right-hand side of equation (3.1).

Because the discretized Laplacian of the field at a point in the mesh is dependent only on the values of its nearest neighbours and all other operations performed on each point in the mesh are independent of neighbours, it is natural to turn to parallelization to evolve the field. The original code of the Wickham group used the Message Passing Interface (MPI) to achieve parallelization by dividing the melt into slices along the z-axis, and computing each slice on a different processor, with MPI calls passing boundary values between processors.

While this is an improvement over serially iterating over every value of the field, the fact that the evolution of each mesh point depends only on its nearest neighbours makes much broader parallelization possible. We turn to graphics processing units (GPUs) and the NVIDIA CUDA architecture to achieve this parallelization.

## 4. Implementation on Graphics Processing Units

### *4.1. The CUDA Architecture*

Unlike traditional Central Processing Units (CPUs), Graphics Processing Units (GPUs) devote most of their architecture to computation, with massive numbers of arithmetic logic units (ALUs) and less space devoted to caching and flow control operations [8]. These many ALUs allow many units of computation to be executed in parallel. The NVIDIA® CUDA™ architecture defines an instruction set that takes advantage of this parallelization, by allowing the programmer to replace serial loop structures with *kernels*, which represent an operation that can be performed in parallel over an entire data set. These kernels are executed as *threads*, arranged into *warps* run as a single computational entity on multiple ALUs.

To take advantage of this parallelization, we identified those sections of the original code where an operation was performed for each element in the data set, characterised by a 3-dimensional `for`-loop. We then wrote kernels to replace those sections of the code, operating on an initial field configuration transferred from traditional memory to GPU device memory at the beginning of the computation.

We left the MPI structure of the code intact, creating a hybrid CUDA-MPI code. Some latency is introduced in transferring data from the GPU to be sent to neighbouring processors in MPI codes, but the ability to parallelize both within GPUs and across many cores outweighs this deficiency.

### *4.2. Implementation of the Laplacian*

Threads are arranged into *blocks*. Each block has an area of shared memory associated with it, which can access global device memory in parallel in a process known as *coalesced memory access*. In cases where the same data is accessed by many threads, shared memory can provide impressive performance improvements. We explored this approach in tandem with a global memory approach for the implementation of the Laplacian.

Following work done by Leist, Playne and Hawick [9], who investigated a similar system, we implemented two kernels for the Laplacian operator. The first accessed data for nearest neighbours from global memory, while the second took advantage of shared memory and coalesced memory access to call nearest neighbour information without making multiple calls to global device memory. A third kernel, using texture memory, was discarded in the early stages of development as too complex to implement practically in concurrence with the original MPI structure of the code. Leist, Playne and Hawick [9] implement a texture memory kernel in two dimensions and present results indicating that investigation into texture memory may produce further performance improvements, but we leave this for future work.

One subtlety involved with coalesced memory access is the treatment of border threads. A block can hold a maximum of 512 threads [8], indexed in any number of dimensions. In order to take advantage of CUDA's parallelization, kernels must be free of *branches*, where threads in a warp take different execution paths such as through an `if`-statement. Thus, border threads cannot be treated differently from internal threads. As such, we follow Leist, Playne and Hawick in implementing non-updating border threads, which hold values to allow their nearest neighbours inside the block to update without a kernel branch.

In both cases, we used the maximum block size of 512 in order to exploit GPU parallelization to its fullest extent. When reading from global memory, where the position of adjacent threads is less important, we mapped to a $16 \times 32$ block of threads. When reading from shared memory, we initialized in three dimensions to a $16 \times 8 \times 4$ block size, including the non-updating border threads. The latter specification was implemented to ensure that one dimension of the block was of a size sufficient to take advantage of all 16 consecutive threads that may be read simultaneously.

The authors of Ref. [9] consider a two-dimensional system ringed by a one-dimensional square of border threads. We consider a three-dimensional system and a two-dimensional surface of border threads, resulting in further latency caused by the extra threads introduced. This problem could be alleviated through further work, following methods documented in Ref. [10][1], where a two-dimensional tile is used as the basic block.

## 5. Performance Results

We benchmarked the performance of the old MPI code and compared it with results from the GPU code. The performance improvements are shown in Figure 5. We can see from Figure 5 that at mesh dimension $N = 384$, where $N = \sqrt[3]{L_x L_y L_z}$ and $L_k$ is the number of elements in the mesh in the $k$-dimension, the performance improvement is a factor of 30, a result that facilitates much of the other work presented in this paper.

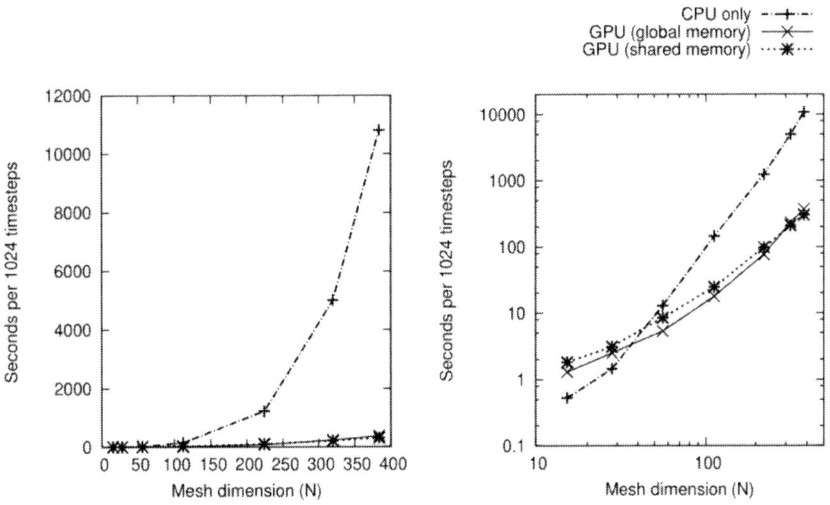


**Figure 2.** Linear and logarithmic plots of the performance gains of the various GPU kernels, compared against the CPU implementation. Both implementations use MPI parallelization on 8 cores. Mesh dimension $N = \sqrt[3]{L_x L_y L_z}$.

[1] We acknowledge the referees of this paper for their help in directing us to this reference.

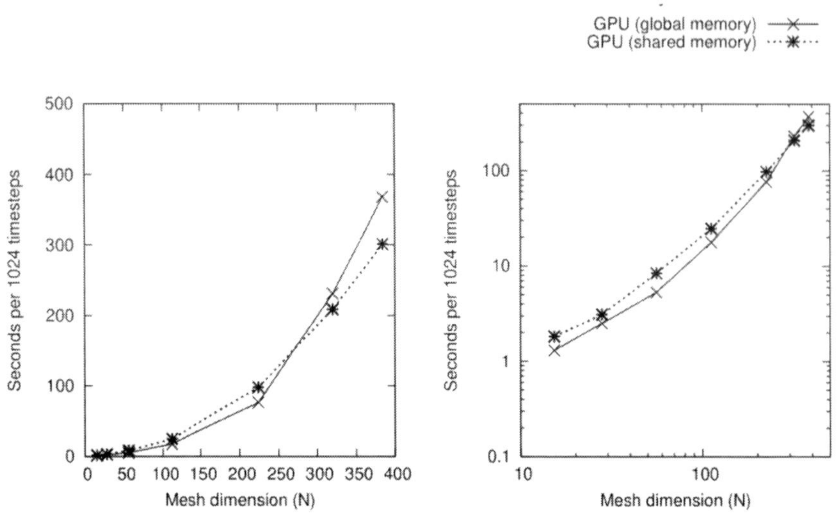


**Figure 3.** A comparison of the performance of the kernels using global and shared memory. Conditions are the same as in Figure 5

Leist, Playne and Hawick's studies in two dimensions found that a kernel using shared memory was faster than one using global memory for any value of $N^3$ greater than approximately $2 \times 10^4$ [9]. However, our global memory kernel was more efficient than the shared memory kernel for any $N^3$ less than $2 \times 10^7$ (see Figure 3). We suspect that this discrepancy is due to the three-dimensional nature of our problem requiring many more non-updating border threads and thus creating more latency in the code.

## 6. Droplet Nucleation

The performance improvements discussed in the previous section are important when a large system size is required. A situation where this is the case is nucleation. In a first-order phase transition such as the lamellar-cylinder phase transition in DBCP melts, there is a region between the phase boundary and the spinodal line where the phase is metastable (see Figure 4). If fluctuations in the field nucleate a sufficiently large droplet of the stable phase, characterised by critical size, the droplet will overcome the energy barrier preventing the phase transition and grow, resulting in the entire system transitioning to the stable phase. Both disorder-order transitions [11, 12] and order-order transitions [13, 6] have been studied theoretically. Our research focuses on order-order transitions.

We manually nucleate a droplet of the stable cylindrical phase in a lamellar melt initialized at a temperature where it is metastable (see Figure 5). Very close to the coexistence boundary and away from the spinodal regime, the size of the critical droplet, which will grow and effect the phase transition, is much larger than the characteristic length associated with the critical wavevector $\frac{2\pi}{q_0}$. These two differing length scales result in the need for a large system size to model nucleation accurately.

Figure 6 shows both results obtained using MPI [14] and new results obtained using the MPI-GPU hybrid code. The hybrid code demonstrably tests nucleation theory in regions much closer to coexistence than previously investigated. The size of the error bars on the new data can be shrunk by further work with the code, as they represent the region in which it may not be clear after $10^6$ timesteps whether the droplet has grown or shrunk.

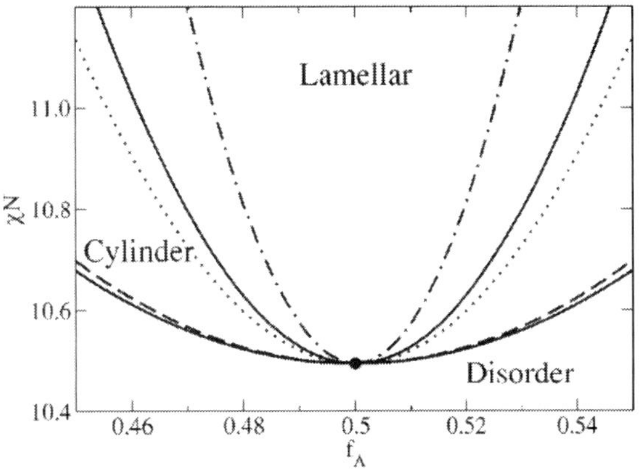


**Figure 4.** The phase diagram for the lamellar-cylinder transition as derived by Wickham, Shi and Wang [6]. Solid lines - phase boundaries; dashed line - disorder-order spinodal; dotted line: lamellar-to-cylinder spinodal; dashed-dotted line: cylinder-to-lamellar spinodal. Other phase boundaries, such as for the bcc and gyroid phases, are not shown. The central point is the mean-field critical point, where a symmetric DBCP will undergo a second-order phase transition.

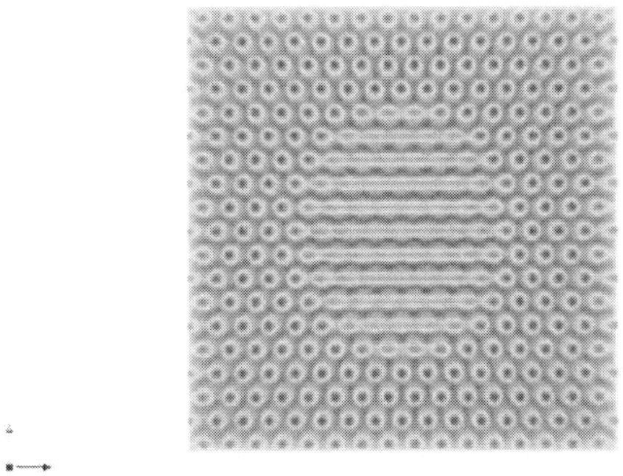

**Figure 5.** A cross-section of a cylindrical bulk where a lamellar droplet has just been nucleated in our simulation. In this case $R \approx 6L_0$.

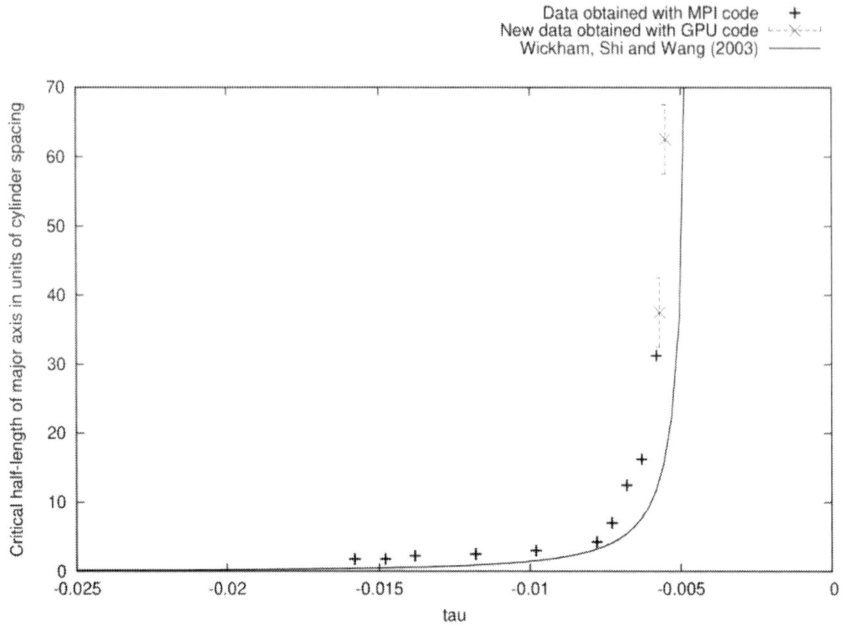


**Figure 6.** Half-length of major axis of critical droplet, measured in cylinder spacings, as a function of $\tau$.

## 7. Dynamics in a Cylindrical Nanopore

If we place a diblock copolymer melt into contact with a surface, it is possible that the substance of which the surface is composed may interact unequally with monomers A and B. This would favour configurations of the system where one species of monomer is closer to the surface. The combination of confinement and the presence or absence of a surface field produces novel structures, such as concentric lamellar structures, helices and double helices, particularly at small pore radius.

The equilibrium behaviour of these morphologies has been derived in detail [15, 16, 17, 18, 19, 20], and is dependent on block asymmetry (the value of $f$), surface field strength and the incommensurability of the diameter of the confined space with the natural periodicity of structures formed in the bulk. Pore diameters $D_p = (n + 0.5)L_0$, where $n$ is an integer and $L_0$ is the bulk structure period, are expected to deviate most starkly from bulk structures. In addition, narrow pores with high curvature are expected to create large deviations from bulk behaviour.

The dynamics of the formation of these structures has been largely unexplored. Many of the most interesting novel structures, such as helical structures, appear at very small pore radius. A coarse mesh at this small system size introduces unwanted effects, so a very fine mesh is required. To maintain the stability of the code we thus used a very short timestep, which would have made simulation intractable without the performance improvements of the GPU code. We use the new code to study two features of interest - the growth of structure from a disordered phase, and the formation of chiral structures such as helical structures from an achiral phase.

We use the same form of the surface field used in Ref. [21]. If $R_p$ is the radius of the pore, $\lambda$ is the decay length of the interaction and $\sigma$ is the interaction cutoff length, then the interaction strength $H(\mathbf{r})$ is defined to be

$$H(\mathbf{r}) = \begin{cases} H_S \frac{\exp\{(\sigma+|\mathbf{r}|-R_p)/\lambda\}-1}{\exp\{\sigma/\lambda\}-1} & R_p - \sigma \leq |\mathbf{r}| \leq R_p \\ 0 & \text{otherwise.} \end{cases} \tag{7.1}$$

We follow Li and Wickham in setting $\sigma = 0.5R_g$ and $\lambda = 0.25R_g$, where $R_g$ is the radius of gyration of the copolymer. This prescription sets the strength of the monomer-pore interaction to be of roughly the same order as the monomer-monomer interaction strength.

*7.1. The Growth of Structure*

We first set the surface field strength $H_s = 0.05$, and the pore diameter to be larger than the characteristic period of lamellae in the bulk $L_0 = \frac{2\pi}{q_0}$. We observe that the structure of concentric lamellae grows from the pore wall inwards, as shown in Figure 7, with outer rings showing much stronger segregation than inner rings. This is important in nanotechnology applications where many concentric rings are desired, as each ring appears to take longer to form than the previous one. After $10^5$ timesteps, the largest pore radius we simulated, at $\frac{R_p}{L_0} = 5.1$ ($R_p$ the pore radius), is still not completely formed (See Figure 7c).

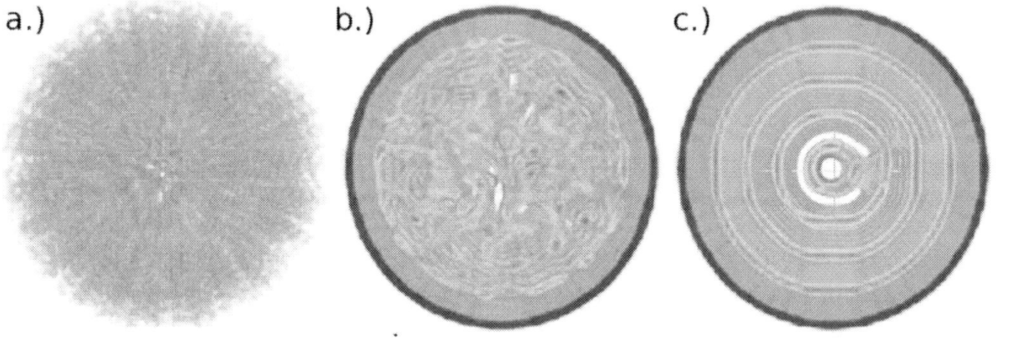


**Figure 7.** The growth of concentric lamellar structure for $H_s = 0.05$, $f = 0.45$, $\chi N = 15$, $\frac{R_p}{L_0} = 5.1$. Stronger segregation is indicated towards the red end of the spectrum, while weaker segregated areas are shown in blue. Timesteps are t=0; t=1000; t=50 000

*7.2. Chirality*

We next chose a finer mesh and examined structures found when $1 < \frac{D_p}{L_0} < 3$ ($D_p$ the pore diameter). Many of these structures are chiral, and we were particularly interested in the transition from a uniform, disordered phase to a chiral phase. The route to chiral structure is in this case through a gyroid-like phase, which then transforms asymmetrically into linked disc structures, which in turn accrete into the helical phases observed in the final states: a double helix in the case of $\frac{D_p}{L_0} = 2.4$; a single helix for $\frac{D_p}{L_0} = 1.6$ (see Figure 8); and interestingly more slowly forming disc-like structures at other radii, which we expect at long times to form one of the equilibrium structures predicted by Li and Wickham [19].

These results are particularly interesting because the theoretical bulk prediction at this temperature is of lamellar structure, while helical structures are generally thought to appear through the distortion of a cylindrical phase by the presence of the nanopore [19, 21]. This may be explained by a stress placed on the system by the imposition of periodic boundary conditions along the z-axis, which may stabilize the helical phase.

To explore chiral structure in more depth, we increased the asymmetry of the polymers in the melt by decreasing $f$, and initialized in an achiral bcc sphere phase, evolving in the presence of a nanopore at a temperature where a cylindrical phase is predicted in the bulk. The results of this simulation are shown in Figure 9. It can be seen that the bcc nearest neighbours mark

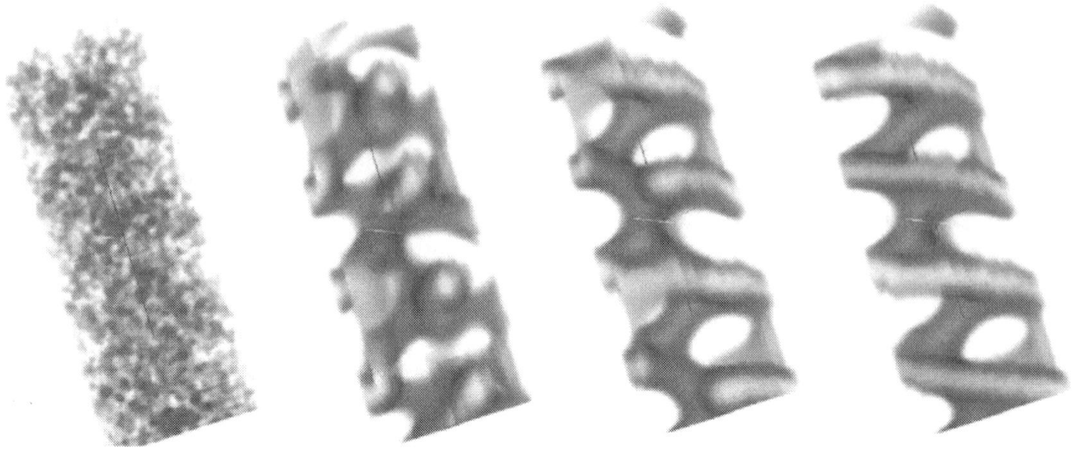

**Figure 8.** The growth of chiral structure for $H_s = 0.05$, $f = 0.45$, $\chi N = 15$, $\frac{D_p}{L_0} = 1.6$, t = 0; $1 \times 10^5$; $2 \times 10^5$; $1 \times 10^6$

out the final helical structure. Chirality appears to be chosen at random, with each handedness equally likely to be found over multiple runs.

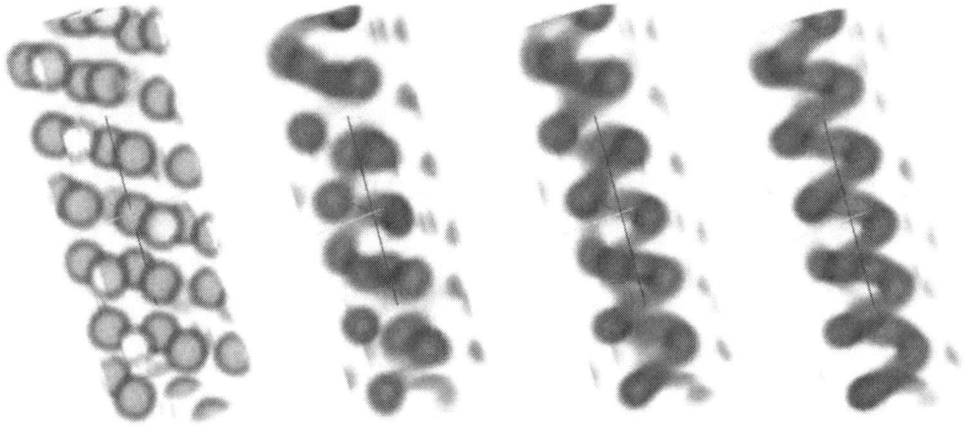

**Figure 9.** The growth of helical structure from an initial bcc sphere phase. $H_s = 0.05, f = 0.4, \chi N = 11.8$ $\frac{D_p}{L_0} = 1.6$, t = 0; $5 \times 10^4$; $1 \times 10^5$; $2 \times 10^5$ .

## 8. Conclusion

In summary, we have demonstrated the applicability of graphics processing units (GPUs) to polymer dynamic field theory, finding a factor of 30 performance improvement over an existing MPI code at large system size. We have used this performance improvement to study two aspects of the dynamics of diblock copolymer melts in regimes that were previously computationally intractable: droplet nucleation close to phase coexistence and the dynamics of structure growth and chirality in the presence of a cylindrical nanopore.

Future work on the simulation code itself should focus on reducing latency due to the passing of boundary information from one GPU to another, which is currently a hurdle in the performance of the code when run on a large number of GPUs. This would be achieved by implementing non-updating border threads on each GPU core.

In addition, the performance of the code would allow for a full study of the dynamics of DBCP melts under nanopore confinement across regimes of temperature, asymmetry and pore radius.

This work was supported by the Perimeter Scholars International Program, and by NSERC.

## References

[1] Leibler L 1980 *Macromolecules* **13** 1602–1617
[2] Brazovskii S 1975 *Sov. Phys. JETP* **41** 6725–6760
[3] Edwards S 1966 *Proc. Phys. Soc. (London)* **88** 265
[4] Ohta T and Kawasaki K 1986 *Macromolecules* **19** 2621–2632
[5] de Gennes P 1970 *J. Phys. (Paris)* **31** 235
[6] Wickham R, Shi A and Wang Z 2003 *J. Chem. Phys.* **118** 10293–10305
[7] Butcher J 2003 *Numerical methods for ordinary differential equations* (John Wiley and Sons)
[8] NVIDIA® Corporation CUDA™ 2.0 Programming Guide Available at: http://www.nvidia.com/ [last accessed April 2010]
[9] Leist A, Playne D and Hawick K 2009 *Concurrency Computat.: Pract. Exper.* **21** 2400–2437
[10] Micikevicius P 3d finite difference computation on gpus using cuda Available at: http://developer.download.nvidia.com/CUDA/CUDA_Zone/papers/gpu_3dfd_rev.pdf [last accessed August 2010]
[11] Fredrickson G and Binder K 1989 *J. Chem. Phys* **91** 7265–7275
[12] Hohenberg P and Swift J 1995 *Phys. Rev. E* **52** 1828–1845
[13] Nonomura M and Ohta T 2001 *J. Phys.: Condens. Matter* **13** 9089–9112
[14] Spencer R and Wickham R 2010 Unpublished
[15] He Z, Song M, liang H and Pan C 2001 *J. Chem. Phys* **114** 10510–10513
[16] Sevink G, Zvelindovsky A, Fraaije J and Huinink H 2001 *J. Chem. Phys.* **115** 8226–8230
[17] Y Wu *et al* 2004 *Nature Materials* **3** 816–822
[18] Feng J, Liu H and Hu Y 2006 *Macromol. Theory Simul.* **15** 675–685
[19] Li W and Wickham R 2006 *Macromolecules* **39** 8492–8498
[20] Ma M, Thomas E, Rutledge G, Yu B, Li B, Jin Q, Ding D and Shi A C 2010 *Macromolecules* **43** 3061–3071
[21] Li W and Wickham R 2009 *Macromolecules* **42** 7530–7536

Journal of Physics: Conference Series **256** (2010) 012009 doi:10.1088/1742-6596/256/1/012009

# Fast polynomial multiplication on a GPU

**Marc Moreno Maza and Wei Pan**
University of Western Ontario, London ON, Canada
E-mail: `moreno@csd.uwo.ca; wpan9@csd.uwo.ca`

**Abstract.** We present CUDA implementations of Fast Fourier Transforms over finite fields. This allows us to develop GPU support for dense univariate polynomial multiplication leading to speedup factors in the range $21 - 37$ with respect to the best serial C-code available to us, for our largest input data sets. Since dense univariate polynomial multiplication is a core routine in symbolic computation, this is promising result for the integration of GPU support into computer algebra systems.

## 1. Introduction

Polynomials and matrices are the fundamental objects on which most computer algebra algorithms operate. In the past 15 years, significant efforts have been deployed by different groups of researchers for delivering highly efficient software packages for computing symbolically with polynomials and matrices, like LINBOX, MAGMA, and NTL [16, 1, 25]. However, most of these works are dedicated to serial implementation, in particular in the case of polynomials. Only a few studies [18, 21, 22] report on parallel implementation (targeting multicores) of polynomial arithmetic. None of the computer algebra software packages available today takes advantage of graphics processing units (GPUs) in support of libraries for polynomial arithmetic. The work reported in this paper aims at filling this gap.

This contrasts sharply with the state of affairs in numerical linear algebra and in digital signal processing. For instance, the commercialized software system MATLAB with its Parallel Computing Toolbox [17] and GPU Toolbox [13] provides programming support for different parallelism paradigms (data-parallelism, MPI, multithreading) and parallel architectures (GPUs, multicores, clusters) together with many library functions taking advantage of this support. In digital signal processing, in particular for the computation of Fast Fourier Transforms (FFTs), the use of hardware acceleration technologies, notably GPUs, has been investigated in several works [10, 20, 12, 27].

In this paper, we present a GPU implementation of fast polynomial multiplication. We focus on dense univariate polynomials over prime fields for the following reasons. First, many algorithms in symbolic computation tend to densify intermediate data, even if the input and output are sparse. Second, multivariate polynomial multiplication can be reduced to univariate multiplication through the so-called Kronecker's substitution. Third, computation with polynomials over non-prime fields can be reduced to the prime field case by means of modular techniques. We refer to the landmark book *Modern Computer Algebra* [11] for an extensive presentation of these ideas.

This reduction to dense univariate polynomial over coefficient fields $\mathbb{Z}/p\mathbb{Z}$, where $p$ is a prime, allows us to rely on FFT techniques, which is the basis of fast polynomial arithmetic [4]. However, as detailed in Section 2.3, FFT computations over finite fields present specific challenges. For this reason, techniques for FFTs with floating point number coefficients are not sufficient for supporting polynomial multiplication over finite fields. This motivates the work reported in this paper.

Most serial implementations of FFT over finite fields, see [8] and the references therein, rely on the radix-2 Cooley-Tukey Formula [6]. On multicores, the row-column FFT algorithm is used successfully, see [21, 22]. In the case of GPUs, to which this paper is devoted, it is natural to revisit the popular FFT formulas of Cooley-Tukey and Stockham [26] in the context of finite fields. We review these formulas in Section 2.2. As in [23, 5] we take advantage of the formalism of tensorial calculus to generate code and identify our GPU kernels. The Cooley-Tukey and Stockham formulas differ only in the way that intermediate computations are stored. We concentrated our efforts on these two formulas, despite of the existence of other formulas for computing FFTs, for the following reasons. First, the radix-2 Cooley-Tukey formula is well understood in the context of finite fields. Second, in numerical computing, the Stockham formula seems to be well-suited for GPU implementation [10].

In this work, we present our detailed implementations of the Cooley-Tukey and Stockham FFT formulas, aiming at utilizing the horsepower of Graphics Processing Units (GPUs). The organization of the paper is as follows. In Section 2, we first formalize FFTs in terms of Kronecker (tensor) product, then we discuss an efficiency-critical operation in a finite field, namely modular multiplication. Sections 3 and Section 4 focus on our CUDA [2] implementation of the Cooley-Tukey and Stockham FFTs. We present experimental results in Section 5 and draw conclusions in the end.

## 2. The Kronecker product and FFT over finite fields

This section reviews the Fast Fourier Transform (FFT) in the language of tensorial calculus, see [28] for an extensive presentation. This formalism facilitates code generation as explained in [5, 9], and in particular it helps identifying GPU kernel specifications. We also highlight the specific features of FFTs over finite fields and refer to [22] for details. Throughout this paper, we denote by $\mathbb{K}$ a field. In practice, this field is often a prime field $\mathbb{Z}/p\mathbb{Z}$ where $p$ is a prime number greater than 2.

### *2.1. Basic operations on matrices*

Let $n, m, q, s$ be positive integers and et $A, B$ be two matrices over $\mathbb{K}$ with respective formats $m \times n$ and $q \times s$. The tensor (or Kronecker) product of $A$ by $B$ is an $mq \times ns$ matrix over $\mathbb{K}$ denoted by $A \otimes B$ and defined by

$$A \otimes B = [a_{k\ell}B]_{k,\ell} \quad \text{with} \quad A = [a_{k\ell}]_{k,\ell} \tag{1}$$

The direct sum of $A$ and $B$ is an $(m+q) \times (n+s)$ matrix over $\mathbb{K}$ denoted by $A \oplus B$ and defined by

$$A \oplus B = \begin{bmatrix} A & 0 \\ 0 & B \end{bmatrix}. \tag{2}$$

More generally, for $n$ matrices $A_0, \ldots, A_{n-1}$ over $\mathbb{K}$, the direct sum of $A_0, \ldots, A_{n-1}$ is defined as $\oplus_{i=0}^{n-1} A_i = A_0 \oplus (A_1 \oplus (\cdots \oplus A_{n-1}) \cdots)$. The stride permutation matrix $L_m^{mn}$ permutes an input vector $\mathbf{x}$ of length $mn$ as follows

$$\mathbf{x}[im + j] \mapsto \mathbf{x}[jn + i], \tag{3}$$

for all $0 \leq j < m$, $0 \leq i < n$. If $\mathbf{x}$ is viewed as an $n \times m$ matrix, then $L_m^{mn}$ performs a transposition of this matrix.

### *2.2. Discrete Fourier Transform*

We fix an integer $n \geq 2$ and an $n$-th primitive root of unity $\omega \in \mathbb{K}$. The $n$-point Discrete Fourier Transform (DFT) at $\omega$ is a linear map from the $\mathbb{K}$-vector space $\mathbb{K}^n$ to itself, defined by $\mathbf{x} \longmapsto \mathrm{DFT}_n\,\mathbf{x}$ with the n-th DFT matrix

$$\mathrm{DFT}_n = [\omega^{k\ell}]_{0 \leq k,\, \ell < n}. \tag{4}$$

In particular, the DFT of size 2 corresponds to the butterfly matrix

$$\mathrm{DFT}_2 = \begin{bmatrix} 1 & 1 \\ 1 & -1 \end{bmatrix}. \tag{5}$$

The well-known Cooley-Tukey Fast Fourier Transform (FFT) [6] in its recursive form is a procedure for computing $\mathrm{DFT}_n\,\mathbf{x}$ based on the following factorization of the matrix $\mathrm{DFT}_n$, for any integers $q, s$ such that $n = qs$ holds:

$$\mathrm{DFT}_{qs} = (\mathrm{DFT}_q \otimes I_s) D_{q,s} (I_q \otimes \mathrm{DFT}_s) L_q^{qs}, \tag{6}$$

where $D_{q,s}$ is the diagonal twiddle matrix defined as

$$D_{q,s} = \bigoplus_{j=0}^{q-1} \mathrm{diag}(1, \omega^j, \ldots, \omega^{j(s-1)}), \tag{7}$$

Formula (8) illustrates Formula (6) with $\mathrm{DFT}_4$:

$$\begin{aligned}
\mathrm{DFT}_4 &= (\mathrm{DFT}_2 \otimes I_2) D_{2,2} (I_2 \otimes \mathrm{DFT}_2) L_2^2 \\
&= \begin{bmatrix} 1 & 0 & 1 & 0 \\ 0 & 1 & 0 & 1 \\ 1 & 0 & -1 & 0 \\ 0 & 1 & 0 & -1 \end{bmatrix} \begin{bmatrix} 1 & 0 & 0 & 0 \\ 0 & 1 & 0 & 0 \\ 0 & 0 & 1 & 0 \\ 0 & 0 & 0 & \omega \end{bmatrix} \begin{bmatrix} 1 & 1 & 0 & 0 \\ 1 & -1 & 0 & 0 \\ 0 & 0 & 1 & 1 \\ 0 & 0 & 1 & -1 \end{bmatrix} \begin{bmatrix} 1 & 0 & 0 & 0 \\ 0 & 0 & 1 & 0 \\ 0 & 1 & 0 & 0 \\ 0 & 0 & 0 & 1 \end{bmatrix} \\
&= \begin{bmatrix} 1 & 1 & 1 & 1 \\ 1 & \omega & -1 & -\omega \\ 1 & -1 & 1 & -1 \\ 1 & -\omega & -1 & \omega \end{bmatrix} = \begin{bmatrix} 1 & 1 & 1 & 1 \\ 1 & \omega^1 & \omega^2 & \omega^3 \\ 1 & \omega^2 & \omega^4 & \omega^6 \\ 1 & \omega^3 & \omega^6 & \omega^9 \end{bmatrix}.
\end{aligned} \tag{8}$$

Assume that $n$ is a power of 2, say $n = 2^k$. Formula (6) can be unrolled so as to reduce $\mathrm{DFT}_n$ to $\mathrm{DFT}_2$ (or a base case $\mathrm{DFT}_m$, where $m$ divides $n$) together with the appropriate diagonal twiddle matrices and stride permutation matrices. This unrolling can be done in various ways. Before presenting one of them, we introduce a notation. For integers $i, j, h \geq 1$, we define

$$\Delta(i,j,h) = (I_i \otimes \mathrm{DFT}_j \otimes I_h) \tag{9}$$

which is a square matrix of size $ijh$. For $m = 2^\ell$ with $1 \leq \ell < k$, the following formula holds:

$$\mathrm{DFT}_{2^k} = \left( \prod_{i=1}^{k-\ell} \Delta\left(2^{i-1}, 2, 2^{k-i}\right) \left(I_{2^{i-1}} \otimes D_{2,2^{k-i}}\right) \right) \Delta\left(2^{k-\ell}, m, 1\right) \left( \prod_{i=k-\ell}^{1} \left(I_{2^{i-1}} \otimes L_2^{2^{k-i+1}}\right) \right). \tag{10}$$

Therefore, Formula (10) reduces the computation of $\mathrm{DFT}_{2^k}$ to composing $\mathrm{DFT}_2$, $\mathrm{DFT}_{2^\ell}$, diagonal twiddle endomorphisms and stride permutations. This is the basis of the implementation presented in Section 3. Another recursive factorization of the matrix $\mathrm{DFT}_{2^k}$ is

$$\mathrm{DFT}_{2^k} = (\mathrm{DFT}_2 \otimes I_{2^{k-1}}) D_{2,2^{k-1}} L_2^{2^k} (\mathrm{DFT}_{2^{k-1}} \otimes I_2), \tag{11}$$

from which one can derive the Stockham FFT [26] as follows

$$\mathrm{DFT}_{2^k} = \prod_{i=0}^{k-1} (\mathrm{DFT}_2 \otimes I_{2^{k-1}})(D_{2,2^{k-i-1}} \otimes I_{2^i})(L_2^{2^{k-i}} \otimes I_{2^i}). \tag{12}$$

This is the basis of the implementation presented in Section 4.

### 2.3. FFTs over finite fields

As mentioned in the introduction, for FFT computations, the case where $\mathbb{K}$ is a finite field offers specific challenges in comparison to the case where $\mathbb{K}$ is the field $\mathbb{C}$ of complex numbers. This explains why FFT techniques used for numerical computation do not adapt straightforwardly to symbolic computation.

On the algebraic side, the most obvious difference is that the prime field $\mathbb{Z}/p\mathbb{Z}$ admits an $n$-th primitive root of unity if and only if $n$ divides $p-1$. In addition, radix 2 FFTs are often preferred to others in symbolic computation, since many algorithms, such as those for polynomial factorization, require the prime $p$ to be small, say $p = 3, 5, 7, \ldots$. Since the radix must be invertible in $\mathbb{Z}/p\mathbb{Z}$, this essentially imposes the restriction to radix 2 FFTs. See [11] for details on these algebraic considerations.

On the implementation side, multiplying two elements $a, b$ of $\mathbb{Z}/p\mathbb{Z}$ is obviously a key routine. Unlike the case of single and double precision floating point arithmetic, the operation $(a, b, p) \longmapsto (ab) \mod p$, for $a, b, p \in \mathbb{Z}$, is not provided directly by hardware. This operation is thus an efficiency-critical low-level software routine that the programmer has to take care of. When $p$ is a machine word size prime, which is the assumption in this paper, two techniques are popular in the symbolic computation community.

The first one takes advantage of hardware floating point arithmetic, see [7]. We call `double_mul_mod` our implementation of this technique, for which our CUDA code is shown below. The fourth argument `pinv` is the inverse of `p` which is precomputed in floating point.

```
__device__ int double_mul_mod(int a, int b, int p, double pinv) {
   int q  = (int) ((((double) a) * ((double) b)) * pinv);
   int res = a * b - q * p;
   return (res < 0) ? (-res) : res;
}
```

In our implementation, double precision floating point numbers are encoded on 64 bits and make this technique work correctly for primes $p$ up to 30 bits.

The second technique, called the Montgomery reduction [19], relies only on hardware integer arithmetic. We summarize this elegant trick. Consider a positive integer $R \geq p$ such that $\gcd(R, p) = 1$. Hence there exists integers $R^{-1}, p'$ such that we have:

$$RR^{-1} - pp' = 1 \text{ and } 0 < p' < R. \tag{13}$$

Consider an integer $x$, satisfying $0 \leq x < p^2$, and for which we want to compute $x/R \mod p$. Let $c$ and $d$ (resp. $e$ and $f$) be the quotient and remainder in the Euclidean division of $x$ by $R$ (resp. $dp'$ and $R$). Then, it is easy to prove that there exists an integer $q$ such that $x + fp = qR$ holds, that is, satisfying $q \equiv x/R \mod p$. If $p > 2$ then $R$ can be chosen to be a power of 2. Therefore, with this choice, computing $x/R \mod p$ amounts to 2 multiplications, 2 additions and 3 shifts. Now, in order to compute products in $\mathbb{Z}/p\mathbb{Z}$, one "represents" any residue class $a \mod p$ by $aR \mod p$. Then, applying the previous trick to $x = (aR)(bR) \mod p$ one obtains efficiently $(ab)/R \mod p$, that is, the representative of $(ab) \mod p$. An improved version of this trick was proposed in [15].

## 3. Implementation of the Cooley-Tukey FFT

We stick to the notations and hypotheses introduced in Section 2.2. Our purpose is to describe our CUDA implementation of Formula (10). The idea behind this formula is that the base case $\mathrm{DFT}_m$ can be implemented efficiently, for $m$ small enough, typically $m = 16$. This formula can be interpreted as the composition of three computational steps:

**S$_1$:** $\mathbf{x} \longmapsto \prod_{i=k-\ell}^{1} (I_{2^{i-1}} \otimes L_2^{2^{k-i+1}})\mathbf{x}$,

**S$_2$:** $\mathbf{x} \longmapsto \Delta\left(2^{k-\ell}, m, 1\right)\mathbf{x}$,

**S$_3$:** $\mathbf{x} \longmapsto \prod_{i=1}^{k-\ell} \Delta\left(2^{i-1}, 2, 2^{k-i}\right)\left(I_{2^{i-1}} \otimes D_{2,2^{k-i}}\right)\mathbf{x}$.

According to the definition $\Delta(2^{k-\ell}, m, 1) = I_{2^{k-\ell}} \otimes \mathrm{DFT}_m$, the step $\mathbf{S}_2$ essentially reduces to execute a sequence of base $\mathrm{DFT}_m$, each of which operates on a subarray of $\mathbf{x}$, independently. Therefore, we focus hereafter on $\mathbf{S}_1$ and $\mathbf{S}_3$. Note that in steps $\mathbf{S}_1$ and $\mathbf{S}_3$ we need to double-buffer the array to avoid synchronizations among different CUDA thread blocks, see [3] for details. That is, at the same time, we have two arrays $X$ and $Y$ of length $n$, one of which is the input and the other is the output, and they switch their role after a kernel call on the input.

### 3.1. *Implementation of step* $\mathbf{S}_1$

Step $\mathbf{S}_1$ consists of a sequence of calls to the following GPU kernel, with $s$ ranging from 1 to $\frac{n}{2m}$. Its specification is

```
/**
 * Compute Y = (I_s x L_2^{n/s})X
 *
 * @X, input array of length n
 * @Y, output array of length n
 */
void list_transpose_kernel(int *Y, int *X, int n, int s);
```

Performing the product of $I_s \otimes L_2^{n/s}$ by a vector $\mathbf{x}$ of length $n$ is equivalent to

(i) dividing $x$ evenly into $s$ subarrays,

(ii) regarding each subarray as a $\frac{n}{2s} \times 2$ matrix and transposing it.

Therefore, step $\mathbf{S}_1$ essentially consists of $s$ matrix transpositions of size $\frac{n}{2s} \times 2$. Following the spirit of [24] for matrix transposition, we realized an efficient subroutine to transpose a list of matrices. Note that we could not directly adapt their code since each matrix has only two columns. Without padding the input data with zeros, our implementation is still able to utilize the shared memory of CUDA devices effectively. For simplicity, we present our implementation with the following example.

**Example 3.1** Let $M$ be a $16 \times 2$ matrix. We set the thread block size to $16 \times 2$ with indices $(i, j)$ for $0 \leq i < 16$ and $j = 0, 1$. Then we first read $M$ into an array $M_s$ of size 32 residing in the shared memory space as follows

```
int i = threadIdx.y * 16 + threadIdx.x;
M_s[i] = M[i];
```

That is, the above segment of code transforms $M$ into the shared array $M_s$ via two coalesced reads, without changing the data layout. Still, we look at the shared array $M_s$ as a $16 \times 2$ matrix, then we achieve the transposition by writing the data back to the global memory column-wise as follows

```
int i = threadIdx.y * 16 + threadIdx.x;
M[i] = M_s[threadIdx.x * 2 + threadIdx.y];
```

The first 16 threads (a half warp) $\{(i, 0) \mid 0 \leq i < 16\}$ read in $M_s[0], M_s[2], \ldots, M_s[30]$, and write to $M[0], M[1], \ldots, M[15]$. On the other hand, the second half warp of threads $\{(i, 1) \mid 0 \leq i < 16\}$ read in $M_s[1], M_s[3], \ldots, M_s[31]$ and write to $M[16], M[17], \ldots, M[31]$. Again all the writes to global memory are coalesced.

The above example can be generalized to transpose a list of $m \times 2$ matrices with only coalesced reads and writes for any $m \geq 16$, which satisfies the specification of the kernel *list_transpose_kernel*.

### *3.2. Implementation of step* $\mathbf{S}_3$

We are going to map the formula

$$(I_{2^{i-1}} \otimes \mathrm{DFT}_2 \otimes I_{2^{k-i}})(I_{2^{i-1}} \otimes D_{2,2^{k-i}}), \quad 1 \le i \le k - \ell \tag{14}$$

to a GPU kernel function. The following relation is easy to prove:

$$(I_{2^{i-1}} \otimes \mathrm{DFT}_2 \otimes I_{2^{k-i}})(I_{2^{i-1}} \otimes D_{2,2^{k-i}}) = I_{2^{i-1}} \otimes \big((\mathrm{DFT}_2 \otimes I_{2^{k-i}})D_{2,2^{k-i}}\big) \tag{15}$$

Hence step $\mathbf{S}_3$ consists of a sequence of calls to the following GPU kernel, with $q$ ranging from $\frac{n}{2}$ to $m$. It specification is

```
/**
 * Compute Y = (I_{n/2q} x (DFT_2 x I_q) D_{2, q}) X
 *
 * @X, input array of length n
 * @Y, output array of length n
 */
void list_butterfly_kernel(int *Y, int *X, int n, int q);
```

We notice that $(\mathrm{DFT}_2 \otimes I_q)D_{2,q}$ is, in fact, the *classical butterfly* operation, which can be realized as,

```
for (i = 0; i < q; ++i) {
    Y[i]   = X[i] + X[q+i] * W[i];
    Y[q+i] = X[i] - X[q+i] * W[i];
}
```

with $W[i] = \theta^i$ and $\theta$ is a $(2q)$-th primitive root of unity. The formula $(\mathrm{DFT}_2 \otimes I_q)D_{2,q}$ will be applied to a segment of data of length $2q$. Hence, with $n/2$ threads, one can realize *list_butterfly_kernel* which implements the formula $I_{2^{i-1}} \otimes \big((\mathrm{DFT}_2 \otimes I_{2^{k-i}})D_{2,2^{k-i}}\big)$ for each $i$.

## 4. Implementation of the Stockham FFT

Recall the Stockham FFT formula (12):

$$\mathrm{DFT}_{2^k} = \prod_{i=0}^{k-1} (\mathrm{DFT}_2 \otimes I_{2^{k-1}})(D_{2,2^{k-i-1}} \otimes I_{2^i})(L_2^{2^{k-i}} \otimes I_{2^i}).$$

For each fixed $0 \le i < k$, it consists of three computational steps:

$\mathbf{A}_1$: $\mathbf{x} \longmapsto (L_2^{2^{k-i}} \otimes I_{2^i})\mathbf{x}$,
$\mathbf{A}_2$: $\mathbf{x} \longmapsto (D_{2,2^{k-i-1}} \otimes I_{2^i})\mathbf{x}$,
$\mathbf{A}_3$: $\mathbf{x} \longmapsto (\mathrm{DFT}_2 \otimes I_{2^{k-1}})\mathbf{x}$.

Similar to the implementation of the Cooley-Tukey FFT, we double-buffer these steps.

### *4.1. Implementation of step* $\mathbf{A}_1$

We describe how to map the formula $L_2^{n/s} \otimes I_s$ to a GPU kernel, where $n$ is the FFT size and $s$ is the stride size. Let $M$ be an $(n/s - 1) \times 2s$ matrix stored in the row-major layout. The effect of this stride permutation on $M$ is to perform the following reordering:

$$M = \begin{bmatrix} \mathbf{S}_0 & \mathbf{S}_1 \\ \mathbf{S}_2 & \mathbf{S}_3 \\ \mathbf{S}_4 & \mathbf{S}_5 \\ \vdots & \vdots \\ \mathbf{S}_{(n/s-2)} & \mathbf{S}_{(n/s-1)} \end{bmatrix} \Longrightarrow T = \begin{bmatrix} \mathbf{S}_0 & \mathbf{S}_2 & \mathbf{S}_4 & \cdots & \mathbf{S}_{(n/s-2)} \\ \mathbf{S}_1 & \mathbf{S}_3 & \mathbf{S}_5 & \cdots & \mathbf{S}_{(n/s-1)} \end{bmatrix}$$

where $\mathbf{S}_i$ denotes the elements of $M$ with indices $is \cdots (is+s-1)$. When we regard $M$ as an $(n/s-1)\times 2$ matrix (each stride $\mathbf{S}_i$ is a single element of $M$), the output matrix $T$, of size $2\times(n/s-1)$, is the matrix transposition of $M$.

We describe how to realize this stride permutation in a CUDA kernel. Let $\tau$ be the number of threads in a thread block, typically $\tau = 128$. To use the shared memory space efficiently, $\tau$ should be a multiple of 16. Under the above setting, the number of threads blocks required is given by $\lambda = \frac{n}{\tau}$. We need to distinguish the following two cases

(1) $s \geq \tau$, that is, $\delta = \frac{s}{\tau}$ blocks are needed to move a stride of length $s$,

(2) $s < \tau$, that is, a thread block moves $\delta = \frac{\tau}{s}$ strides of data.

The reason to have such a case discussion is that the relation between $\tau$ and $s$ determines the behavior of each thread block, specified in detail as follows.

Case $s \geq \tau$: Given a thread block index $0 \leq i < \lambda$, we define

$$i_q = \mathrm{quo}(i, \delta) \quad \text{and} \quad i_r = \mathrm{rem}(i, \delta).$$

Here $i_q$ determines the stride index which thread block $i$ is working on and $i_r$ determines the offset inside this stride. If $i_q$ is a multiple of 2 then $\mathbf{S}_{i_q}$ appears in the first row of the output matrix, otherwise it appears in the second row. Hence the offset for the thread block $i$ is given by the following formula:

$$\mathrm{rem}(i_q, 2) * \frac{n}{2} + \mathrm{quo}(i_q, 2) * s + i_r * \tau. \tag{16}$$

In this case, each thread block does a direct copy, that is, no data shuffle is needed.

Case $s < \tau$: Given a thread block index $0 \leq i < \lambda$, there are two offsets:

$$i * \mathrm{quo}(\tau, 2) \quad \text{and} \quad i * \mathrm{quo}(\tau, 2) + \frac{n}{2}. \tag{17}$$

In this case, the behavior inside each thread block is not just a direct copy. Since there are $\delta$ strides inside each thread block, those strides with even index use the first offset to move data, while the other strides use the second offset to move data (this may be viewed as an in-block data shuffle).

### 4.2. *Implementation of steps* $\mathbf{A}_2$ *and* $\mathbf{A}_3$

According to its definition, $D_{2,2^{k-i-1}}$ is a diagonal matrix of size $2^{k-i}$ and thus $D_{2,2^{k-i-1}} \otimes I_{2^i}$ is again a diagonal matrix of size $n$, with each diagonal element repeated $2^i$ times. Hence step $\mathbf{A}_2$ simply scales $\mathbf{x}$ with powers of the primitive root of unity $\omega$. On the other hand, step $\mathbf{A}_3$ is a list of basic butterflies with stride size $n/2$. This step is accesses data in a very uniform manner. In the following section, we discuss the performance implications of steps $\mathbf{A}_2$ and $\mathbf{A}_3$.

## 5. Experimentation

We have realized in CUDA 2.2 both Cooley-Tukey FFT and Stockham FFT, and conducted a series of benchmarks using a Geforce GTX 285 graphics card on a desktop with the processor Intel Core 2 Quad CPU Q9400 @ 2.66GHz and 6 GB main memory. This graphics card has the compute capability 1.3, consists of 30 multiprocessors, each of which has 8 cores for integer and single-precision floating-point arithmetic operations. However, each multiprocessor has only 1 double-precision floating-point unit. This is an important characteristic for our `double_mul_mod` routine implementing the map $(a, b, p) \mapsto (ab) \mod p$, as described Section 2.3. We note that for the most recent NVIDIA graphics cards having the compute capability 2.x, the ability to perform double-precision floating-point operations has been greatly enhanced.

The experimentation is described in the following three subsections. Section 5.1 is dedicated to modular multiplication, and more precisely, to a comparative implementation of the map $(a, b, p) \mapsto (ab) \bmod p$ on both CPU and GPU. Section 5.2 presents the results for our GPU implementation of the FFT formulas of Cooley-Tukey and Stockham, as described in Sections 3 and 4. Section 5.3 compares the performance of FFT-based univariate polynomial multiplication codes for CPU and GPU.

### *5.1. Modular multiplication*

Figure 1 and Figure 2 are experimental results for modular multiplication on CPU and GPU respectively. In both cases, each slot of an input array of length $n = 2^k$ consisting of machine word size integers is multiplied by a given machine word size integer $\omega$. This type of calculation is typical for FFT algorithms. For both CPU and GPU we compare our implementations of the Montgomery reduction and `double_mul_mod`. For the GPU kernel, we could choose to have a single thread or multiple threads. In our experimentation, we use the latter and in this case each multiprocessor can only process a double-precision floating-point operation at a time which downgrades the performance.

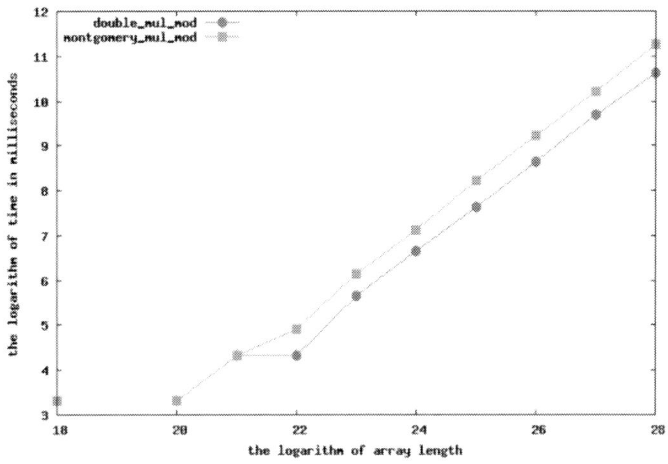


Figure 1: Modular multiplication on CPU. When we increase the number of modular multiplications, the one relying on double-precision floating point computations outperforms the one relying on the Montgomery reduction. Both the array length and the time are scaled by the base 2 logarithm.

Figure 1 shows that `double_mul_mod` is about 1.5 faster than the method based on the Montgomery reduction, when running serial C code on the CPU. On the GPU, Figure 2 shows that `double_mul_mod` is still slightly better than the method based on the Montgomery reduction, which is a surprise to us.

### *5.2. Cooley-Tukey and Stockham FFT over finite fields on a GPU*

It is challenging to figure out what are the best implementation techniques for each of the two formulas. During our experiments, we realized that the pre-computation of the powers $1, \omega, \omega^2, \ldots, \omega^{n/2-1}$ is a necessity, for an $n$-point FFT. This step is rather time-consuming if we implement it naively, that is, first compute those powers in the host sequentially, and then transfer them to the GPU device. Fortunately, this preprocessing is a special form of the exclusive prefix sum and the pre-computation can be achieved by a

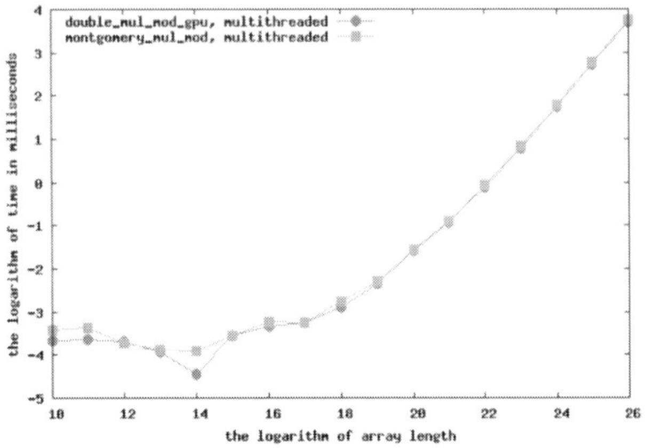


Figure 2: Modular multiplication on GPU. We do the same test as in Figure 1 on GPU, and run the kernels with massive threads. Both the array length and the time are scaled by the base 2 logarithm.

sequence of GPU kernel calls to a subroutine `double_expand`, which takes an array $\{1, \omega, \ldots, \omega^{s-1}\}$ of length $s$ as input and returns $\{\omega^s, \ldots, \omega^{2s-1}\}$ by multiplying each element of the input array with $\omega^s$.

We have noticed that step $\mathbf{S}_3$ of the Cooley-Tukey FFT (as described in Section 3.2) was accessing the powers of $\omega$ by performing larger and larger jumps. For example, while the following formula $(\mathrm{DFT}_2 \otimes I_{2^{k-i}})(I_{2^{i-1}} \otimes D_{2,2^{k-i}})$ operates on a subarray of length $2^{k-i+1}$, the powers $\{1, \omega^{2^i}, (\omega^{2^i})^2, \ldots, (\omega^{2^i})^{2^{k-i}-1}\}$ get accessed. We call *jumped powers at level* $i$ these latter powers. Therefore, we considered pre-computing not only the powers $\{1, \omega, \ldots, \omega^{n/2-1}\}$ but also all jumped powers at level $i$ for each $i$.

To visualize the performance of our implementation, we use the NVIDIA's visual profiler `cudaprof` to analyze CUDA kernel calls. It is very helpful to find out the bottlenecks of an implementation. For instance, Figure 3 shows the kernel statistics where the pre-computation of jumped powers has been done on the host. In this figure, the x-axis shows the CUDA kernel call indices in chronological order and the y-axis is proportional to the GPU time for each kernel. We notice that it spends a fairly large amount of time to move the extra $n/2$ pre-computed powers to the GPU.

If the jumped powers were not computed in advance, the accesses to those powers harm the performance heavily as shown by Figure 4, since those memory accesses to the global memory are non-coalesced, in the step $\mathbf{S}_3$ of the Cooley-Tukey FFT. Up to our knowledge, it is hard to achieve coalesced accesses without pre-computing jumped powers while implementing the Cooley-Tukey FFT.

However, the Stockham FFT avoids such a problem. Indeed, all the accesses to a power of $\omega$ are packed together, resulting in a broadcasting inside a thread block. Figure 5 shows the kernel statistics of the Stockham FFT of size $2^{26}$, which is our best GPU FFT implementation. Our Stockham FFT implementation pre-computes all powers in an extremely fast manner without computing jumped powers. The first and last kernel are for the input and output data transfer, and all the other kernels are run efficiently (the occupancy is 1 for each kernel).

For completeness, in Figure 6 we compare our two GPU implementations for FFT against our C code from `modpn` [14]. This latter library is shipped with the computer algebra system MAPLE and is considered as a reference code for FFT computations over finite fields. Without considering the time spent in host-device data transfer, the speedup we achieve is about 37 for the FFT size $2^{26}$ (this speedup is about 21 if the data movement time is counted). As shown by Figure 7, our Stockham FFT code is about 2 times faster than our Cooley-Tukey FFT code, mainly due to the jumped powers pre-computation.

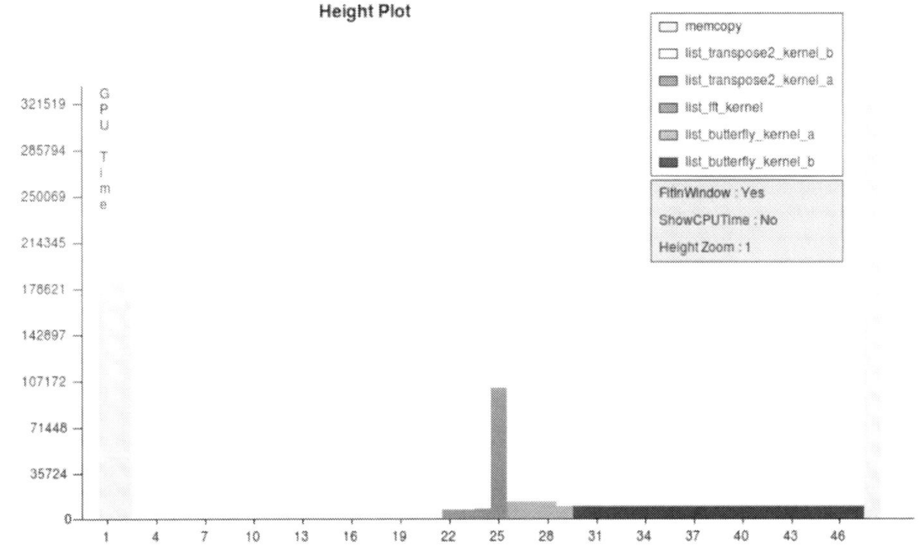


Figure 3: Kernel statistics for the Cooley-Tukey FFT with pre-computed jumped powers. The second kernel moves the extra $n/2$ powers of the primitive root of unity, which affects the overall performance. Note that those method names _a or _b as a suffix, since we implement the same algorithm for handling input data in different ranges.

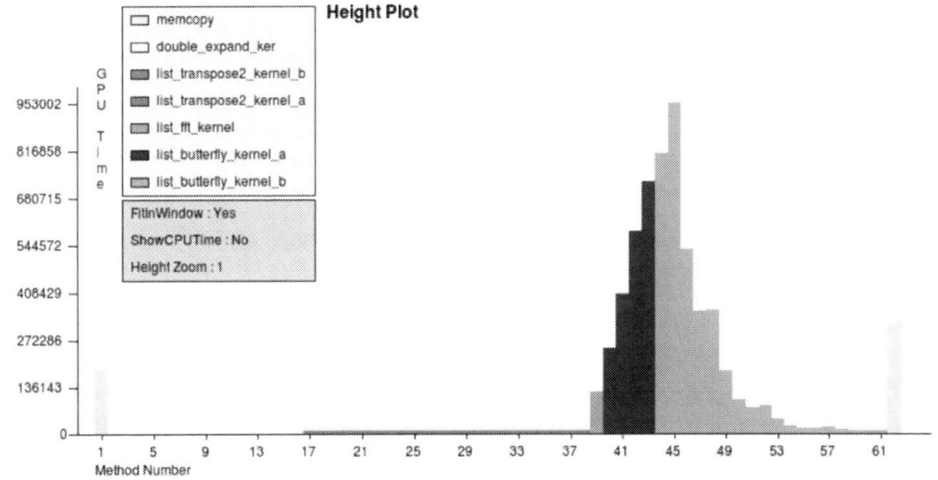


Figure 4: Kernel statistics for the Cooley-Tukey FFT on GPU without pre-computed jumped powers. Time to call the kernel *list_butterfly_kernel* increases significantly, which greatly downgrades the overall performance.

### *5.3. Univariate polynomial multiplication over finite fields*

As a direct application of fast Fourier transforms, we have implemented FFT based univariate polynomial multiplications over finite fields. Figure 8 compares the modpn FFT based polynomial multiplication against our GPU Stockham FFT-based one. The input two polynomials are randomly generated with the same given degree. When the degree is relatively large, the speedup we achieved is about 21 - 37, comparing to the modpn polynomial multiplication.

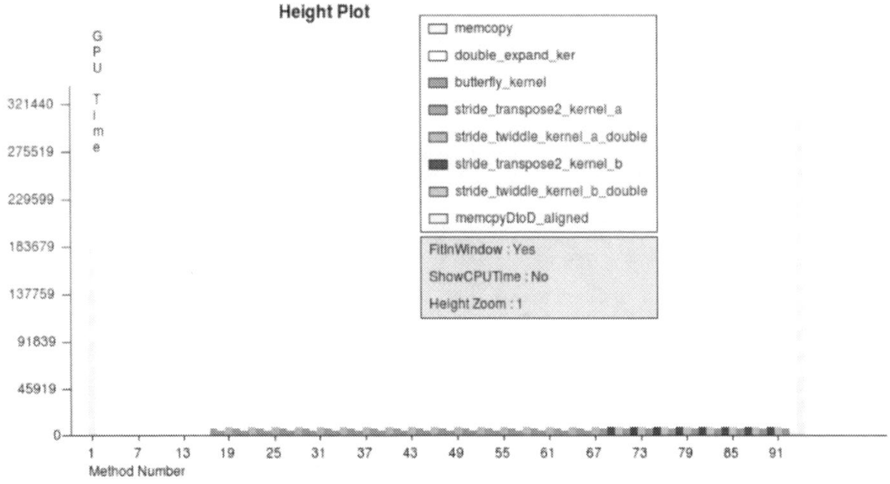


Figure 5: Kernel statistics for the Stockham FFT on GPU. This is our best implementation for 1D FFT. The pre-computation is achieved by the kernel call *double_expand_ker*. The steps $\mathbf{A}_1$, $\mathbf{A}_2$ and $\mathbf{A}_3$ are realized by *stride_transpose2_kernel*, *stride_twiddle_kernel* and *butterfly_kernel*, respectively. All of them are running very efficiently.

| | CT FFT | CT FFT + transfer | Stockham FFT | Stockham FFT + transfer | modpn FFT |
|---|---|---|---|---|---|
| 12 | 1 | 1 | 2 | 2 | 1 |
| 13 | 2 | 2 | 2 | 3 | 1 |
| 14 | 1 | 2 | 2 | 3 | 3 |
| 15 | 2 | 2 | 3 | 3 | 4 |
| 16 | 3 | 3 | 3 | 4 | 10 |
| 17 | 4 | 5 | 3 | 5 | 16 |
| 18 | 6 | 9 | 4 | 7 | 37 |
| 19 | 11 | 15 | 6 | 10 | 71 |
| 20 | 22 | 28 | 9 | 16 | 174 |
| 21 | 44 | 56 | 16 | 28 | 470 |
| 22 | 83 | 105 | 29 | 52 | 997 |
| 23 | 165 | 210 | 56 | 101 | 2070 |
| 24 | 330 | 418 | 113 | 201 | 4194 |
| 25 | 667 | 842 | 230 | 405 | 8611 |
| 26 | 1338 | 1686 | 473 | 822 | 17617 |

Figure 6: Timing of FFT codes on CPU and GPU in milliseconds. The first column is the logarithm of FFT size in base 2. The second and the fourth column show the timing of our Cooley-Tukey and Stockham FFT implementations, without counting the data transfer between GPU and CPU, respectively. The third and the fifth column shows these FFT implementations with the data transfer. The last column shows the `modpn` FFT timing.

## 6. Conclusion and future work

We have presented in detail various issues in implementing efficient fast Fourier transforms over finite fields on the GPU. Our experimental results show that the Stockham formula is well-suited for massively-threaded architectures. In particular, it avoids pre-computing extra powers of primitive roots of unity in a natural way, without downgrading the performance. Our implementation exhibits a significant performance improvement over a reference C implementation. For multiplying two dense univariate polynomials, we have achieved about 30x speedup with respect to the best code available to us. As future work, we would like to implement multidimensional FFTs, and to revisit various modular algorithms in symbolic computation, like evaluation/interpolation based subresultant chain construction

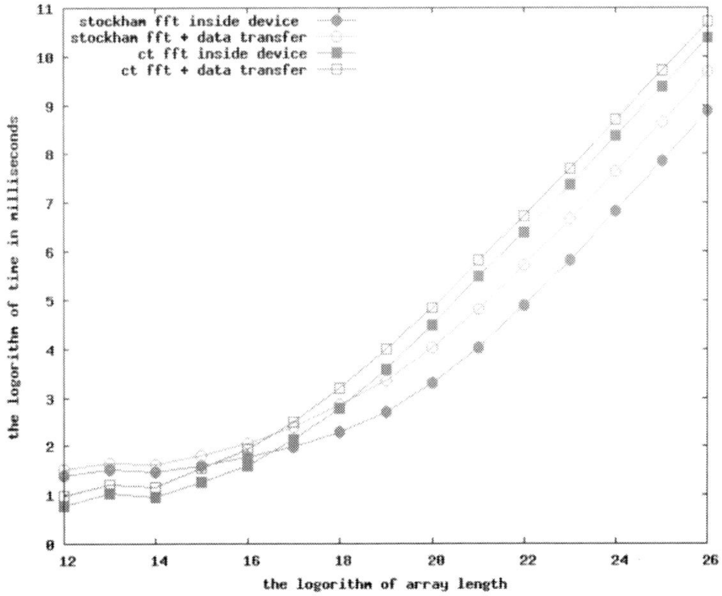


Figure 7: We visualize the comparison among GPU FFT implementations according to the timing from Figure 6. Both the FFT size and the time are scaled by the base 2 logarithm.

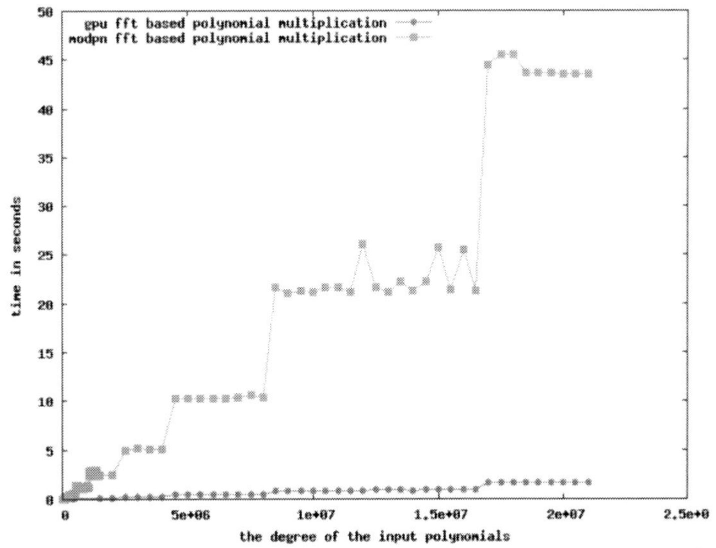


Figure 8: FFT-based dense polynomial multiplication on GPU and CPU. The data transfer has been counted for the GPU code.

of polynomials.

[1] *The MAGMA Computational Algebra System*. http://magma.maths.usyd.edu.au/magma/.
[2] *Nvidia CUDA*. http://www.nvidia.com/object/cuda_home_new.html.
[3] *Nvidia CUDA programming guide 2.3*. 2009.
[4] D. Cantor and E. Kaltofen. On fast multiplication of polynomials over arbitrary algebras. *Acta Informatica*, 28:693–701, 1991.

[5] S. Chellappa, F. Franchetti, and M. Püschel. How to write fast numerical code. In *Summer School on Generative and Transformational Techniques in Software Engineering*, volume 5235 of *LNCS*, pages 196–259, 2008.

[6] J. Cooley and J. Tukey. An algorithm for the machine calculation of complex Fourier series. *Math. Comp.*, 19:297–301, 1965.

[7] J.-G. Dumas, T. Gautier, and C. Pernet. Finite field linear algebra subroutines. In *ISSAC '02: Proceedings of the 2002 international symposium on symbolic and algebraic computation*, pages 63–74, New York, NY, USA, 2002. ACM.

[8] A. Filatei, X. Li, M. Moreno Maza, and E. Schost. Implementation techniques for fast polynomial arithmetic in a high-level programming environment. In *Proc. ISSAC'06*, pages 93–100, New York, NY, USA, 2006. ACM Press.

[9] F. Franchetti and M. Püschel. *Encyclopedia of Parallel Computing*. Springer, 2011.

[10] F. Franchetti, M. Püschel, Y. Voronenko, S. Chellappa, and J. M. F. Moura. Discrete Fourier transform on multicore. *IEEE Signal Processing Magazine, special issue on "Signal Processing on Platforms with Multiple Cores"*, 26(6):90–102, 2009.

[11] J. Gathen and J. Gerhard. *Modern Computer Algebra*. Cambridge University Press, 1999.

[12] N. K. Govindaraju, B. Lloyd, Y. Dotsenko, B. Smith, and J. Manferdelli. High performance discrete fourier transforms on graphics processors. In *SC '08: Proceedings of the 2008 ACM/IEEE conference on Supercomputing*, pages 1–12, Piscataway, NJ, USA, 2008. IEEE Press.

[13] GPUmat. *GPU toolbox for MATLAB*.

[14] X. Li, M. Moreno Maza, R. Rasheed, and É. Schost. The modpn library: Bringing fast polynomial arithmetic into maple. In *MICA'08*, pages 73–80, 2008.

[15] X. Li, M. Moreno Maza, and E. Schost. Fast arithmetic for triangular sets: From theory to practice. *J. Symb. Comp.*, 44(7):891–907, 2008.

[16] P. LinBox. *Exact computational linear algebra*. http://www.linalg.org/.

[17] MathWorks. *Parallel Computing Toolbox*.

[18] M. Monagan and R. Pearce. Parallel sparse polynomial multiplication using heaps. In *ISSAC '09: Proceedings of the international symposium on Symbolic and algebraic computation*, pages 263–270, New York, NY, USA, 2009. ACM.

[19] P. L. Montgomery. Modular multiplication without trial division. *Mathematics of Computation*, 44(170):519–521, 1985.

[20] K. Moreland and E. Angel. The FFT on a gpu. In *HWWS '03: Proceedings of the ACM SIGGRAPH/EUROGRAPHICS conference on Graphics hardware*, pages 112–119, Aire-la-Ville, Switzerland, 2003. Eurographics Association.

[21] M. Moreno Maza and Y. Xie. Balanced dense polynomial multiplication on multicores. In *Proc. of PDCAT'09*. IEEE Computer Society, 2009.

[22] M. Moreno Maza and Y. Xie. FFT-based dense polynomial arithmetic on multi-cores. In D. Mewhort, editor, *Proc. HPCS 2009*, volume 5976 of *LNCS*, Heidelberg, 2010. Springer-Verlag Berlin.

[23] M. Püschel, J. M. F. Moura, J. Johnson, D. Padua, M. Veloso, B. Singer, J. Xiong, F. Franchetti, A. Gacic, Y. Voronenko, K. Chen, R. W. Johnson, and N. Rizzolo. SPIRAL: Code generation for DSP transforms. *Proceedings of the IEEE, special issue on "Program Generation, Optimization, and Adaptation"*, 93(2):232– 275, 2005.

[24] G. Ruetsch and P. Micikevicius. Optimizing matrix transpose in cuda. 2009.

[25] V. Shoup. *NTL: The Number Theory Library*.

[26] T. G. Stockham, Jr. High-speed convolution and correlation. In *AFIPS '66 (Spring): Proceedings of the April 26-28, 1966, Spring joint computer conference*, pages 229–233, New York, NY, USA, 1966. ACM.

[27] P. N. Swarztrauber. FFT algorithms for vector computers. *Parallel Comput.*, 1(1):45–63, 1984.

[28] C. Van Loan. *Computational frameworks for the fast Fourier transform*. Society for Industrial and Applied Mathematics, Philadelphia, PA, USA, 1992.

# A Parallel Algorithm for Connected Component Labelling of Gray-scale Images on Homogeneous Multicore Architectures

**Mehdi Niknam[1,2], Parimala Thulasiraman[1,2], Sergio Camorlinga[2,1]**

[1]Department of Computer Science - University of Manitoba, [2]Telecommunication Research Labs

mhniknam@cs.umanitoba.ca, thulasir@cs.umanitoba.ca, scamorlinga@trlabs.ca

**Abstract**. Connected component labelling is an essential step in image processing. We provide a parallel version of Suzuki's sequential connected component algorithm in order to speed up the labelling process. Also, we modify the algorithm to enable labelling gray-scale images. Due to the data dependencies in the algorithm we used a method similar to pipeline to exploit parallelism. The parallel algorithm method achieved a speedup of 2.5 for image size of 256 X 256 pixels using 4 processing threads.

## 1. Introduction

Computer-aided diagnosis (CAD) systems are examples of diagnostics, screening and detection tools for medical purposes. CAD provides clinicians (e.g. radiologists) a computerized analysis of medical images as a 'second opinion' in detecting lesions, assessing extent and progression of disease, and supporting diagnostic decisions among other things. CAD systems utilize image processing techniques in order to detect and diagnose different diseases from medical images.

The pre-processing techniques such as image filtering and image registration play an important role in enhancing the accuracy of image analysis and later steps in CAD systems. Connected component labeling is a useful tool used in pre-processing stages as well as in image analysis and in post processing stages [1]. In connected component labeling of an image, every set of connected pixels having same gray-scale values are assigned the same unique region label. This region later will be used to identify the suspicious lesions. The fact that the connected component labeling is a fundamental module in medical image processing, optimizing the existing algorithm will result in an improvement of many medical diagnoses and procedures [2, 3, 4].

Medical applications are computationally and data intensive problems. With the recent advancement in multicore architectures, these problems are gaining insight from a whole new perspective [5]. The focus of this paper is to parallelize the sequential connected component labeling algorithm presented by Suzuki et al. [6] on a homogenous multicore architecture in order to study the speedup of the parallel algorithm.

We use OpenMP to parallelize the algorithm. We also apply a similar method mentioned in [1] to enhance Suzuki's algorithm to label the gray-scale images since the original algorithm labels only binary images and the medical images are normally gray-scale. Section 2 provides a brief discussion regarding the existing connected component labeling algorithms. Section 3 presents Suzuki's algorithm in detail. Our parallel algorithm is explained in the section 4. Section 5 outlines the experimental framework we used and the section 6 provide a discussion of the experimental results. Finally, section 7 describes conclusion and future work.

## 2. Background and Related Work

Suzuki et al. [6] classified the connected component labeling methods into 4 different categories: A) Algorithms that repeat passes through an image in forward and backward directions alternately to propagate the label equivalences until no labels change. B) Algorithms using two passes in a way that they assign provisional labels to the connected components and store the label equivalences in an array. Then they resolve the label equivalences using a search algorithm such as union-find algorithm [7] and store the resolved result in a one-dimensional table. In the second pass, they replace the provisional labels with the smallest equivalent label. C) Algorithms that represent the image using hierarchical tree structures and resolve the label equivalences using union-find algorithm. D) And parallel algorithms developed for specific parallel architectures such as mesh and hypercube parallel processors.

Suzuki et al. [6] introduce an algorithm which optimizes the algorithms in category A. Suzuki's algorithm promises a linear time complexity. Suzuki et al. show that maximum of four scans is sufficient for the images with complex geometrical shapes. Wu et al. [8, 9] provide an optimization on Suzuki algorithm that reduces the number of neighboring pixels needed to be examined during the scans. He et al. [10], present a similar algorithm which scans the whole image once and then resolves the label equivalences using recorded run data.

We chose the Suzuki algorithm to parallelize due to its linear time complexity. There are several sequential algorithms to enhance the performance of the Suzuki algorithm; however, due to our best knowledge there is no parallel version for this algorithm. We decided to investigate the performance improvement of the parallel version of the algorithm.

## 3. Suzuki Algorithm

The Suzuki algorithm scans through the image in forward and backward directions using masks shown in Figure 1, in order to assign a provisional label to each pixel. Also, it stores the label equivalences in an additional one dimensional array called label connection table. The provisional labels propagate through the image as well as the label connection table that reduces the number of scans needed to complete the labeling.

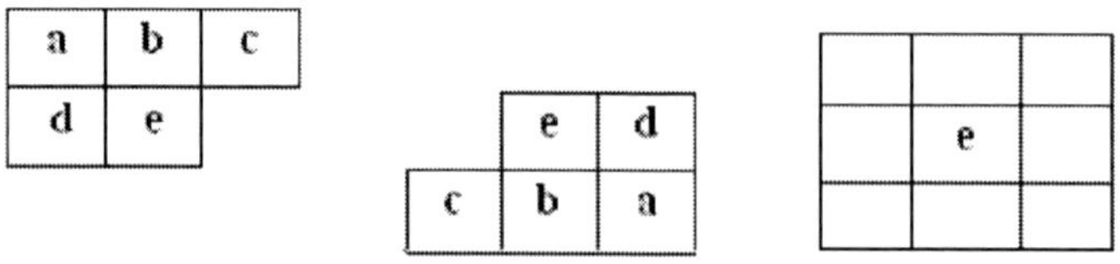


a b c

Figure 1- a) Forward scan mask b) backward scan mask c) 8-connected neighborhood

Suppose that a binary image *b(x,y)* consists of pixel values $F_O$ (indicating objects), and $F_B$ (indicating the background); and a provisional label *m* is initialized to 1. In the first scan, the Suzuki algorithm assigns a provisional label to each pixel at position "e" according to the following equation:

$$g(x,y)=\begin{cases} F_B & \text{if } b(x,y)=F_{B,} \\ m,(m=m+1) & \text{if } \forall\{i,j\in M_s\}g(x-i,y-j)=F_B, \\ T_{\min}(x,y) & \text{Otherwise} \end{cases}$$

$$T_{\min}(x,y)=\min[\{T[g(x-i,y-j)] \mid i,j\in M_S\}]$$

Where $g(x,y)$ stores the provisional labels, $T[m]$ is the label connection table, $(m=m+1)$ indicates an increment of $m$, min(.) an operator calculating the minimum value, and $M_S$ the region of the mask except the object pixel, i.e. $b(x-1,y-1)$, $b(x,y-1)$, $b(x+1,y-1)$, and $b(x-1,y)$.

Also, the label connection table $T[m]$ is updated at the same time as $g(x,y)$ according to the following equation:

$$\begin{cases} \text{non-operation} & \text{if } b(x,y) = F_B, \\ T[m] = m & \text{if } \forall\{i,j \in M_s\} g(x-i,y-j) = F_B, \\ T[g(x-i,y-j)] = T_{\min}(x,y) & \text{if } g(x-i,y-i) \neq F_B. \end{cases}$$

In the next scans, the forward and backward scans are performed alternately and the tables are updated using the following equations:

$$g(x,y) = \begin{cases} F_B & \text{if } g(x,y) = F_B, \\ T_{\min}(x,y) & \text{otherwise,} \end{cases}$$

$$T_{\min}(x,y) = \min[\{T[g(x-i,y-j)] \mid i,j \in M\}].$$

$$\begin{cases} \text{non-operation} & \text{if } g(x,y) = F_B, \\ T[g(x-i,y-j)] = T_{\min}(x,y) & \text{if } g(x-i,y-j) \neq F_B. \end{cases}$$

The forward and backward scans will be performed repeatedly while the following condition is not satisfied:

$$g(x-i,y-j) \neq T_{\min}(x,y) \quad \text{if } g(x-i,y-j) \neq F_B,$$

$$i,j \in M_S.$$

The end result is the image with the final label for each pixel. Figure 2 shows an example of labeling according to the Suzuki algorithm.

| | | | | | |
|---|---|---|---|---|---|
| 0 | 0 | 0 | 0 | 1 | 1 |
| 0 | 0 | 1 | 1 | 0 | 0 |
| 1 | 1 | 0 | 0 | 0 | 0 |

a) A binary image to be scanned

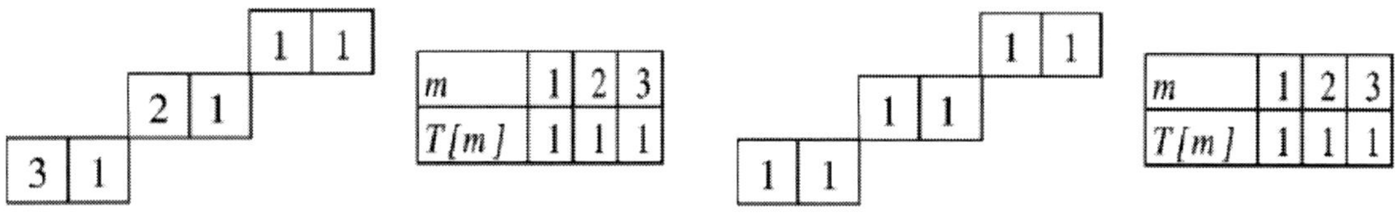


b) After first scan

c) After second scan

Figure 2- Example of labeling using Suzuki algorithm

The following is a pseudo code of the Suzuki algorithm:

```
algorithm Suzuki(data)
    First pass
    for row in data
        for column in row
            assign a label to data[row][col] using Forward Mask
    Next passes
    while there is a change in labels do
        for row in data
            for column in row
                update the labels[row][column] using Backward Mask
        for row in data
            for column in row
                update the labels[row][column] using Forward Mask
return labels
```

## 4. Solution Strategy

In this section we describe our proposed method to parallelize the Suzuki algorithm and its implementation in OpenMP.

### 4.1. Proposed Method

The Suzuki algorithm is sequential in nature. That is, the result of the previous iteration will be used for the next iteration. Two data structures storing the provisional label and the label connection table (see section 3) are updated after labeling a pixel and their updated versions are used to label the next pixel. Due to these data dependencies, it is not possible to perform the labeling using multiple threads without specifying the order of execution. We used a method similar to pipeline in order to exploit parallelism. The following is a detail description of the proposed method to parallelize the algorithm.

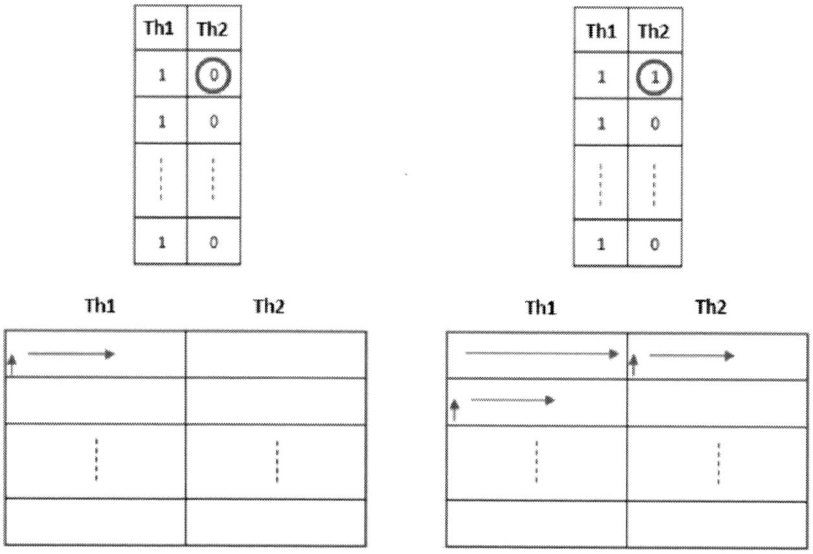


a) Th1 labeling its portion and Th2 is waiting b) Th1 and Th2 are labeling concurrently

Figure 3- Example of the algorithm with 2 threads

As indicated in the pseudo code, the algorithm examines each pixel's neighbors and assigns a provisional label to it accordingly. In our algorithm, the image pixels are distributed between threads row-wise. Assuming there are 2 threads (Th1 and Th2) and the image has 128 rows and 128 columns. The first 64 pixels in the first row will be labeled by Th1 and the next 64 pixels by Th2. However, Th2 cannot start labeling in the same row as Th1, until Th1 is done with labeling its portion due to the sequential nature of the Suzuki algorithm. The Th2 can start labeling when Th1 is done, also Th1 can start labeling the first 64 pixels of the second row at the same time. After Th1 is done with labeling the first portion of the second row, Th2 can label its portion of the second row, and so forth. With this method, we maintain the order of execution of the algorithm as well as exploiting parallelism.

In order to synchronize the thread in the manner mentioned above, we utilized a two dimensional array of size: "number of thread" by "number of rows" called "condition". The thread cannot label its portion of a row and need to wait, if the corresponding value in the array is 0. As soon as Th1 finished the labeling of its portion, it will change the value for Th2 in the condition array to 1. Then Th2 labels its portion and Th1 labels its portion in the next row. Figure 3 shows the image and the condition array for two iterations.

### 4.2. OpenMP Implementation

As can be seen in the iterations example in section 4.1, after the first iteration Th1 and Th2 label their portions concurrently. However, there is some overhead for the synchronization. Each scan in the sequential algorithm includes two loops. In order to parallelize the scan portion of the algorithm in OpenMP as described in the example, we can add *pragma omp parallel for* on the inside loop as shown below:

```
for row in data
    #pragma omp parallel for
    for column in row
        while the condition is not 1 for this thread wait;
        label the pixel;
        if it is end of the row set the condition of the next thread to 1;
```

In the above code, the data will be distributed properly; however the overhead is very high. The number of forking and joining in this code is equal to the number of columns in the image which causes high overhead. In order to deal with this overhead, we use *pragma omp parallel* on the top of the first loop. The cost for the *pragma omp parallel* is only one time fork and joins. We also provide a suitable code to perform the data distribution in the program. Following is the parallel pseudo code that is implemented for the scan portions of the algorithm:

```
chunk = row / number of threads;
#pragma omp parallel
for row in data
    k = calculate the start index for this thread;
    while the condition is not 1 for this thread wait;
    for k in row to k + chunk
        label the pixel;
    set the condition of the next thread to 1;
```

In addition, we used *pragma omp critical* in order to synchronize the access to the label connection table (T[m]). The *pragma omp critical* provides synchronization in a way that only one thread can update the T[m] at a time. There are three scan loops in the Suzuki algorithm as shown in section 3. All three scan loops are replaced with above parallel code. We made some modifications to the Suzuki algorithm to enable labeling of gray-scale images. To distinguish between region of interest (ROI) and background in the images we defined a threshold value. We also modified the algorithm to compare gray values instead of binary values.

## 5. Experiment

We performed a set of experiments to measure the speedup of our proposed parallel version of Suzuki's algorithm. We used 30 gray-scale images in 3 sizes: 128 X 128, 256 X 256, and 512 X 512 pixels from University of Massachusetts Vision Image Archive [11]. The images were labeled using the sequential and parallel algorithm. We examine the sequential algorithm against parallel version using four different numbers of threads: 2, 4,8,16.

The algorithms were implemented in C and OpenMP. The algorithms were tested on one of the five computing nodes of a cluster called helium. The helium cluster has one head node (Sun Fire X4200) and five computing nodes. Each computing node is a Sun Fire X4600 machine which has eight Dual-Core (16 Cores) AMD Opteron 885 2.6 GHZ.

## 6. Results

The Figure 4 shows the speedup of the parallel version of the algorithm. For the small size images (128 X 128), the sequential version is faster than the parallel version. The reason of no speedup for small size images is due to the overhead caused by synchronization of the threads.

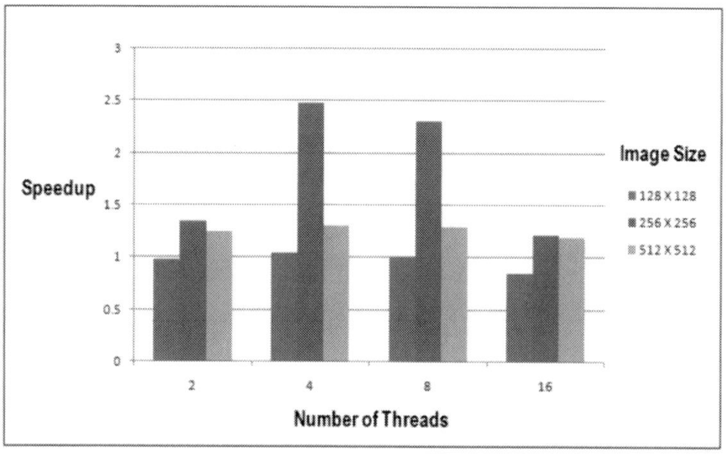


Figure 4- The speedup for different image sizes and different numbers of threads

For the medium size images (256 X 256), the algorithm achieved the best speedup in comparison to other image sizes. The best speedup of 2.5 belongs to the 4 threads while labeling the medium size images. However, in medium size images using 2 and 16 threads the speed up is not significant. While using 16 threads the size of data allocated to each thread reduces. Consequently, the overhead is higher than the amount of exploited parallelism.

For large size images (512 X 512), the speedup is almost similar in all different number of threads. As can be seen there is a trade-off between the data size and the number of threads. The synchronization cost will not be covered by parallelism exploitation in small size images. Similarly, the synchronization overhead is high when using too many threads since the data size allocated to each thread reduces.

According to the results, the overhead of our approach is not decreased linearly with the problem size. One possible reason could be an additional overhead caused by cache miss for image size of 512 X 512 since the image cannot be transferred completely to the cache. The other possible reason could be an extra overhead caused by Non-Uniform Memory Access (NUMA) effect related to the specific architecture used to test the algorithm.

Considering the sequential nature of the Suzuki algorithm, the speedup achieved for medium size (256 X 256) is reasonable and it can be used to label specific medical images. However, to achieve a higher performance as a future work we can investigate using a method which divides the image to number of processes and run the Suzuki algorithm locally, then provide a way to merge the portions in order to achieve a consistent labeling for the image.

## 7. Conclusion

Connected component labeling is a fundamental module in image processing. We provide a parallel version of Suzuki's sequential connected component algorithm in order to speed up the labeling process. We also modified the algorithm to enable labeling of gray-scale images. Due to the data dependencies in the algorithm, we used a method similar to pipeline in order to exploit parallelism. The parallel algorithm achieved the best speedup when labeling medium sized images. Explicitly, the speedup of 2.5 is achieved while labeling image size of 256 X 256 using 4 threads.

As a future work, we can investigate a method to provide a parallel connected component algorithm which divides the image between several threads and each thread utilizes the Suzuki algorithm enhancement ideas proposed in this paper for its local data. Then we integrate the local results to provide a consistent label for the whole image. Also, examining a connected component algorithm on other parallel architectures such as Graphics Processing Unit (GPU) can be performed.

## References

[1] R. D. Yapa, K. Harada, " Connected Component Labeling Algorithms for Gray-Scale Images and Evaluation of Performance using Digital Mammograms," *IJCSNS International Journal of Computer Science and Network Security*, **8**(6), pp. 33–41, 2008.

[2] J. Freixenet, X. Mu˜noz, D. Raba, M. Marti, and X. Cuf´ı, "Yet another survey on image segmentation: Region and boundary information integration," in *Proceedings of the European Conference on Computer Vision (ECCV 2002)*, pp. 408–422, 2002.

[3] B. van Ginneken, B. M. ter Haar Romeny, and M. A. Viegever, "Computer-aided diagnosis in chest radiography: A survey," *IEEE Transcations on Medical Imaging* **20**(12), pp. 1228–1241, 2001.

[4] T. W. Nattkemper, "Automatic segmentation of digital micrographs: a survey," in *Proceedings of MEDINFO 2004, San Francisco, Americal Medical Informatics Association*, 2004.

[5] M. Knaup, S. Steckmann, O. Bockenbac and M. Kachelriess, " Tomographic Image Reconstruction using the Cell Broadband Engine (CBE) General Purpose Hardware", *Proceedings Electronic Imaging, Computational Imaging V, SPIE* Vol 6498, 64980P, pp. 1-10, January 2007.

[6] K. Suzuki, I. Horiba, and N. Sugie, "Linear-time connected-component labeling based on sequential local operations," *Comput. Vis. Image Underst.* **89**(1), pp. 1–23, 2003.

[7] C. Fiorio, J. Gustedt, "Two linear time union-find sterategies for image processing," Theoretical Computer Science, **154**(2), pp. 165–181, Feb. 1996.

[8] K. Wu, E. Otoo, and A. Shoshani, Optimizing Connected Component Labeling Algorithms, In Proceedings of SPIE Medical Imaging Conference, pp. 1965-1976, Apr. 2005.

[9] K. Wu, E. Otoo, and K Suzuki.: Optimizing two-pass connected-component labelling algorithms. *Pattern Analysis and Applications* 12, pp.117-135, 2009

[10] L. He, Y. Chao, K. Suzuki, T. Nakamura, and H. Itoh: A label-equivalence-based one-scan labeling algorithm. Journal of Information Processing Society of Japan 50, pp. 1660-1667, 2009.

[11] Computer Vision Research Laboratory at UMass. University of Massachusetts Vision Image Archive, http://vis-www.cs.umass.edu/~vislib/, Retrieved May 2010.

[12] OpenMP. The OpenMP API Speccification for Parallel Programming, http://openmp.org/wp, Retrieved May 2010.

[13] OpenMP. OpenMP tutorial, https://computing.llnl.gov/tutorials/openMP/, Retrieved May 2010.

# Simulated tempering distributed replica sampling: A practical guide to enhanced conformational sampling

**Sarah Rauscher**[1,2] **and Régis Pomès**[1,2]

[1] Molecular Structure and Function, Hospital for Sick Children, 555 University Avenue, Toronto, ON, Canada
[2] Department of Biochemistry, 1 King's College Circle, University of Toronto, Toronto, ON, Canada

Email: pomes@sickkids.ca

**Abstract**. Simulated tempering distributed replica sampling (STDR) is a generalized-ensemble method designed specifically for simulations of large molecular systems on shared and heterogeneous computing platforms [Rauscher, Neale and Pomès (2009) J. Chem. Theor. Comput. 5, 2640]. The STDR algorithm consists of an alternation of two steps: (1) a short molecular dynamics (MD) simulation; and (2) a stochastic temperature jump. Repeating these steps thousands of times results in a random walk in temperature, which allows the system to overcome energetic barriers, thereby enhancing conformational sampling. The aim of the present paper is to provide a practical guide to applying STDR to complex biomolecular systems. We discuss the details of our STDR implementation, which is a highly-parallel algorithm designed to maximize computational efficiency while simultaneously minimizing network communication and data storage requirements. Using a 35-residue disordered peptide in explicit water as a test system, we characterize the efficiency of the STDR algorithm with respect to both diffusion in temperature space and statistical convergence of structural properties. Importantly, we show that STDR provides a dramatic enhancement of conformational sampling compared to a canonical MD simulation.

## 1. Introduction

Biomolecular systems are challenging to study using molecular dynamics (MD) simulations because of their rugged energy landscapes [1]. The energetic barriers that separate conformational states are often larger than the available thermal energy, and therefore transitions between conformations are statistically rare events. As a consequence, achieving complete Boltzmann sampling of all important conformational states often requires prohibitively long simulation times. Without reaching statistical convergence, it is not possible to draw meaningful, quantitative conclusions regarding the relative populations of different conformational states [2].

Using state-of-the-art supercomputing, atomistic MD simulations with an explicit representation of solvent are currently limited to the nanosecond to microsecond timescale. Therefore, the timescales of many important biomolecular processes are still inaccessible to conventional MD. For example, β-hairpin folding occurs on the microsecond timescale [1], while the folding of two-state proteins occurs over a range of timescales spanning six orders of magnitude, from microseconds to seconds [3]. Sampling the conformational space of folded proteins is computationally demanding. Yet it is perhaps even more challenging to adequately sample the complex conformational landscape of disordered proteins, which lack a well-defined folded state. Molecular simulations of disordered proteins are complicated by the fact that not one but possibly many thousands of states must be sampled in order to characterize a structurally heterogeneous ensemble of conformations [4].

It has recently become clear that the disordered states of proteins have significant biological relevance. While the majority of proteins are folded in their native state, a growing body of evidence from spectroscopy and bioinformatics indicates that a significant fraction of proteins are natively disordered [5]. Disordered proteins are abundant in all kingdoms of life: more than one-third of mammalian proteins [6] and more than 75% of cancer-associated proteins are predicted to contain disordered regions [5]. Despite their biological importance, disordered proteins are poorly understood: they are notoriously difficult to study using experimental approaches. Molecular simulations therefore have the potential to offer tremendous insight into the structural and physicochemical properties of disordered proteins [4]. However, conventional "brute force" MD simulations are insufficient for investigations of these proteins because they often remain trapped in a single conformational basin [7]. In order to provide meaningful insight into the structural properties of disordered proteins, molecular simulation approaches that enhance conformational sampling are required.

Error due to insufficient sampling can be alleviated by utilizing generalized-ensemble algorithms. These methods use a generalized Hamiltonian for the purpose of achieving uniform sampling along a reaction coordinate of interest. Generalized-ensemble algorithms in which replicas of the system execute a random walk in temperature rely on the fact that the free energy surface becomes less rugged at high temperature, resulting in an increased frequency of interconversion between conformational states [8]. Two commonly-used generalized-ensemble algorithms that utilize a random walk in temperature are simulated tempering (ST) [9, 10] and replica exchange (RE) [11-14]. An RE simulation consists of identical copies of the system (replicas) sampling canonical ensembles at different temperatures [12-14]. Exchanges are performed between replicas at neighbouring temperatures according to a Metropolis criterion. Each temperature exchange requires synchronization, which means that RE requires a large, dedicated, and homogeneous computing cluster to function efficiently when applied to complex systems. While ST has no such requirement for synchronization, it does require extensive initial simulations to accurately compute the Helmholtz free energy of the system at each temperature [8, 15]. We have shown previously that ST is the preferred generalized-ensemble algorithm for simple systems [7]. This is because ST exhibits the fastest temperature diffusion, which is correlated with faster convergence of structural properties. However, the computational cost of the initial simulations becomes prohibitive for all but the simplest biomolecular systems [7]. In this paper, we present our implementation of simulated tempering distributed replica sampling (STDR), which is a generalized-ensemble algorithm that combines ST with distributed replica sampling (DR) [7, 16]. STDR offers numerous practical advantages compared to both ST and RE that become important for simulations of complex biomolecular systems [7]. Unlike ST, STDR requires minimal initial simulation; unlike RE, STDR does not require synchronization of replicas or a fixed number of replicas. It accommodates fluctuations in CPU availability and requires minimal communication between replicas. STDR is therefore well-suited for shared computing platforms [7, 16].

This paper provides a guide to applying STDR to complex biomolecular systems and demonstrates the efficiency of the method using a challenging model system. We begin by briefly discussing the details of the algorithm, which has previously been derived [7, 16]. Next, we provide a description of our STDR implementation. We quantify the computational overhead for performing STDR compared to conventional MD simulations and we determine the amount of processor time required for each step of the algorithm. For a specific test case (a disordered peptide in explicit water), we evaluate the random walk in temperature space, the convergence of structural properties, and the enhancement of conformational sampling compared to a canonical MD simulation.

## 2. The STDR Algorithm

In contrast to RE, in which pairwise exchanges of replicas are attempted, STDR considers stochastic moves of individual replicas one at a time [7, 16]. The stochastic move of one replica is coupled to the distribution of all other replicas through a pseudo-energy, called the distributed replica potential energy (DRPE) [16]. In STDR, replicas perform temperature jumps based on the current temperature location of all other replicas. The probability of accepting a move from a temperature $T_i$ to a neighbouring temperature, $T_j$ is:

$$p\left(T_i \rightarrow T_j\right) = \min\begin{cases} 1 \\ e^{-\left(\beta_j - \beta_i\right)E + \left(a_j - a_i\right) - \left(DRPE_j - DRPE_i\right)} \end{cases} \quad (1)$$

where $E$ is the potential energy of the system at the end of the previous MD step, $\beta_i$ and $\beta_j$ are the inverse temperatures, and $a_i$ and $a_j$ are dimensionless Helmholtz free energies corresponding to temperatures $T_i$ and $T_j$, respectively . The first two terms in the exponent are identical to the exchange probability from ST, with the addition of the difference in DRPE between the states for which the replica is at temperature $T_i$ ($DRPE_i$) and temperature $T_j$ ($DRPE_j$) [7]. The DRPE depends upon the current temperatures of all replicas as follows [16, 17]:

$$DRPE = c_1 \sum_{m=1}^{M} \sum_{n=1}^{M} \left[\left(\lambda_{m,linear} - \lambda_{n,linear}\right) - \left(m - n\right)\right] + c_2 \left[\sum_{m=1}^{M} \lambda_{m,linear} - \sum_{m=1}^{M} m\right]^2 \quad (2)$$

Replicas are labeled by indices $m$ and $n$, where $M$ is the number of replicas. The values of $\lambda_{m,linear}$ refer to a linearly-spaced temperature coordinate. In this coordinate, the lowest temperature has $\lambda_{m,linear}=1$, and the highest temperature has $\lambda_{m,linear}$ equal to the number of temperatures. This procedure transforms exponentially-spaced temperatures into a uniformly-spaced coordinate. In STDR, the DRPE enforces approximately homogeneous sampling of a set of temperatures. The effect of the first term is to impose a pseudo-energetic penalty for two replicas sampling the same temperature simultaneously, while the second term imposes a penalty for a concerted drift of replicas towards high or low temperature. The two constants in the equation, $c_1$ and $c_2$, control the magnitude of the DRPE and the relative weight of the two terms [16, 17].

## 3. Implementation Details

### *3.1. Test System: A Disordered Elastin-like Peptide in Explicit Water*

Because one of the main goals of this study is to develop and test a methodology that will provide structural insight into disordered states of proteins, we use a disordered peptide as our test system. Specifically, we use a 35-residue peptide with a sequence derived from elastin, $(GVPGV)_7$ [18]. Elastin is an intrinsically disordered protein that is responsible for the elasticity of the lungs, arteries, and skin. Its structural disorder is essential to its biological role: the increased entropy of the relaxed state relative to the stretched state makes elastin an entropic elastomer [19]. This elastin-like peptide has a complex conformational landscape, which makes it ideally suited to evaluate STDR's ability to enhance conformational sampling. This paper does not focus on the detailed structural properties of this peptide, which will form the basis of future studies.

### *3.2. Computational Platform*

The general purpose cluster (GPC) at the University of Toronto's SciNet consortium was used to perform large-scale STDR simulations of the $(GVPGV)_7$ peptide in water. Each replica was run on an individual node of the GPC. Each Nehalam node has eight 2.5 GHz CPUs and 16 GB of memory. Because the STDR algorithm requires only minimal communication between nodes, the portion of the GPC connected with gigabit Ethernet was utilized. In total, 106 nodes were used for the simulations described here: 105 nodes were used to simulate individual replicas, with one additional node designated as a server. In fact, the number of nodes running concurrently occasionally is slightly less than 106 due to rare node failure and fluctuations in node availability (since the SciNet GPC is a shared computing resource). This is not problematic, however, since STDR is well-suited for such a shared computing platform [7, 17]. In total, these simulations required 1.5 million CPU hours (approximately 106 nodes running for 75 days).

### *3.3. MD Simulation Details*

The simulation system consists of the $(GVPGV)_7$ peptide capped with an acetyl group at the N-terminus and an $NH_2$ group at the C-terminus in a rhombic dodecahedral box with 9663 water molecules. Simulations were performed with an accurate leap-frog stochastic dynamics integrator [20] using GROMACS version 4.0.5 [21]. The OPLS-AA/L force field [22] and the TIP3P model [23] were used for the peptide and water, respectively. Periodic boundary conditions were applied. Calculation of electrostatic forces utilized the particle mesh Ewald summation method [24] with a Fourier spacing of 1.2 Å and a fourth-order interpolation. The real-space Coulombic cutoff was 11 Å. Lennard Jones interactions were cut off at 14 Å. The LINCS algorithm [25] was used to constrain covalent bonds involving hydrogen atoms, and the SETTLE algorithm [26] was used for water, allowing the use of a 2 fs integration time step. The solvated system was initially equilibrated in the isothermal-isobaric ensemble at 300 K and 1 atm for 0.5 ns using the Parrinello-Rahman barostat [27, 28]. All STDR simulations were performed in the canonical ensemble using stochastic dynamics as a thermostat. In addition, a long-time canonical MD simulation was performed for 700 ns, starting with a randomly-selected configuration from the STDR simulation at 300 K.

For the STDR simulation, 105 temperatures exponentially-spaced between 266 K and 749 K were used. At each of these temperatures, a preliminary canonical MD simulation was performed for 15 ns. The lists of potential energies obtained from these simulations were used to compute the dimensionless Helmholtz free energies in equation (1) using the method of Park and Pande [15]. A set of 105 random starting configurations (one for each replica) was obtained from ten 10 ns simulations at 1000 K. At the beginning of the STDR simulation, each replica was started at one of the 105 temperatures. The DRPE constants $c_1$ and $c_2$ were set to 0.004 and 0.002, respectively. Each replica performed $\sim 10^5$ 8 ps MD runs and attempted temperature jumps according to equations (1) and (2), which corresponds to a total of ~800 ns/replica. In total, this resulted in an accumulated simulation time of 84 μs for the entire STDR simulation. This is an order of magnitude more simulation than our previous study [7]. The implementation of STDR makes use of several GROMACS programs including g_energy, trjconv, trjcat, eneconv, and trjorder. These programs are used for analysis and data management, as described in detail in section 3.4.

### *3.4. Implementation of the STDR Algorithm*

#### *3.4.1 Parallelization*

Our STDR implementation incorporates two levels of parallelization:

(1) First, the STDR algorithm is inherently highly parallel: the system is simulated as a set of replicas, each occupying a single node. Each replica executes alternating MD simulations and stochastic temperature jumps. Minimal information sharing between replicas is required: only the current temperature location of all replicas is required to compute the DRPE in equation 2. In the present implementation, a server node collects the temperature information and trajectories of all replicas.

(2) Parallelization also occurs at the level of the individual replica node. Each 8 ps MD simulation is run in parallel using all 8 cores of the node. The time required for 1 ns of simulation is 17.98 hours and 2.12 hours for one core and eight cores, respectively. Thus, the parallel efficiency is 106%. Superlinear scaling is achieved using GROMACS version 4.0, which has been highly optimized to scale well on parallel machines [21].

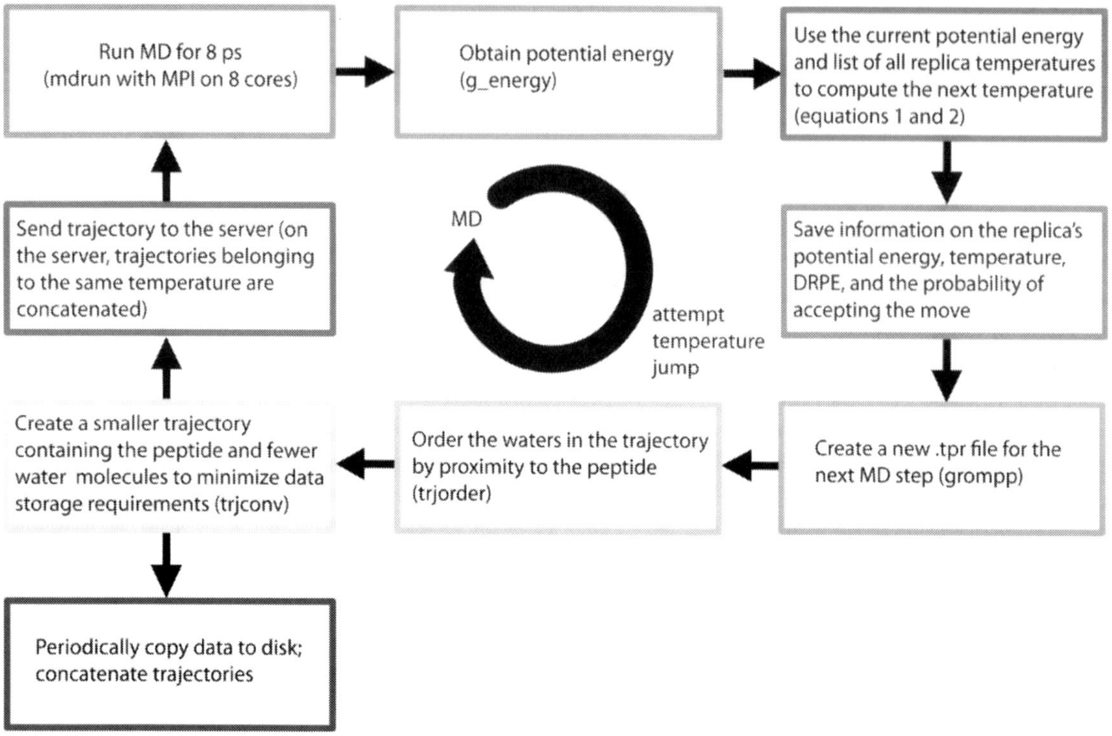


**Figure 1.** Implementation of the STDR Algorithm. The cycle of steps involved in STDR is illustrated as a flow chart. Each cycle consists of two parts: (1) a brief MD simulation (8 ps in length) and (2) an attempted temperature jump. Associated with this temperature jump are several essential data processing steps. During the cycle, all data is stored in shared memory and no disk access is required. The cycle is repeated thousands of times, with occasional interruptions to back up essential data to disk (every few hours) and concatenate trajectories (every two days).

### *3.4.2 The replica nodes*

The specific steps executed by a single STDR replica are illustrated in the flow chart in figure 1. The cycle of steps in a single temperature move begins with a short MD simulation (8 ps in length). The final potential energy from the last step of this MD simulation is obtained using the GROMACS program g_energy. This potential energy value, along with an up-to-date list of all replicas' temperatures (received from the server node), is used to compute the probability of accepting a proposed temperature jump (equations 1 and 2). Information about the temperature jump is stored, including the potential energy, DRPE, probability, and temperature. In order to begin the next MD step, grompp generates a run input file. To minimize the amount of hard disk space required for storage, the water molecules in the MD trajectory are ordered by proximity to the protein using trjorder: only the nearest water molecules are saved in the final trajectory using trjconv. The replica's trajectories are concatenated locally (on the node on which the replica is running). Trajectories corresponding to the same temperature must also be concatenated; this occurs on the server node. The trajectory corresponding to each STDR step must therefore be sent to the server node over the network. The replica periodically backs up the data stored in memory to disk once every 4 hours. This checkpointing limits the amount of data lost in case of a node failure to a maximum of a few hours, which corresponds to a few hundred STDR steps. This approach was designed to minimize disk I/O, minimize the amount of long-term data storage required for trajectories, and maximize the fraction of compute time spent on MD (see figure 2). Because GPFS, the high performance file system of the SciNet GPC, is not designed for accessing many small files, it is important that all trajectory files be concatenated while still in memory. Data locality is maximized by passing the smallest possible amount of data between the server node and the replica nodes. All of the processes on the nodes are controlled by bash scripts that call GROMACS programs, as well as an in-house program that computes the STDR temperature jump criterion (equations 1 and 2). In figure 2, we assess the efficiency of the STDR implementation for the $(GVPGV)_7$ test system by determining the amount of processor time required for each step of the algorithm. The majority of the STDR overhead results from managing the trajectory files (trjorder, trjconv, and concatenation). These steps are not part of the core STDR algorithm. However, the trajectories must be the minimum size possible and concatenated to facilitate structural analysis and long term data storage.

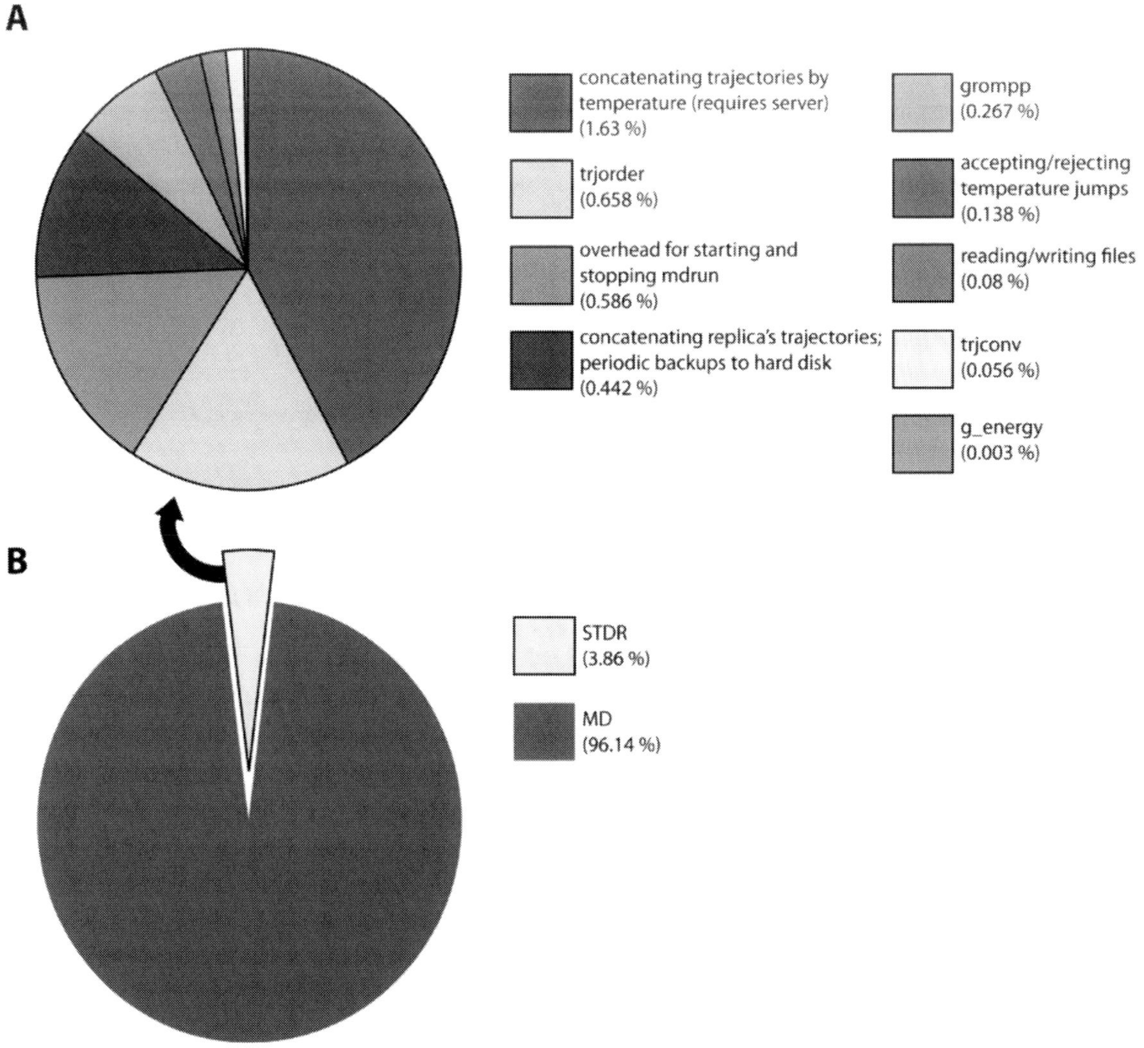


**Figure 2.** Node time required for each step in STDR . (A) The percent of replica node time spent on each step in the STDR algorithm is represented as a pie chart (colours correspond to the same steps illustrated in figure 1). (B) Overall, 96.14 % of node time is spent performing MD, while 3.86 % is spent on all tasks associated with STDR, including management of trajectory files. Note that the mdrun step in figure 1 is coloured blue to corresponding to the overhead for starting and stopping mdrun indicated in figure 2A.

### *3.4.3 The server node*

The server node fulfills two key roles in our implementation of STDR:

(1) It stores the current temperature of each replica. The server maintains a current list of the temperatures of all replicas. This temperature list is frequently communicated to each replica

node, ensuring that an up-to-date temperature list is used in the calculation of the DRPE (equation 2). After every temperature jump, the replica communicates its current temperature to the server.

(2) The server node also collects the short MD trajectories of all replicas. Trajectories must be stored according to temperature to facilitate further structural analysis (described in detail in section 5 below). In order to concatenate trajectories by temperature, the trajectories of individual replicas must be collected in a central location (the shared memory of the server node). When the memory is almost full, the trajectories are concatenated by temperature and moved to disk. By performing all trajectory concatenation in memory on the nodes, disk I/O is kept to an absolute minimum.

## 4. Evaluating Temperature Diffusion

The quality of the random walk in temperature is assessed using several different metrics, which we have defined in our previous work [7]. These metrics are reported in Table 1. Importantly, the sampling of temperatures is very close to perfect homogeneity: the average deviation from homogeneity is only 1.3%. This indicates that the constants $c_1$ and $c_2$ in equation 2 have been appropriately selected to enforce homogeneous temperature sampling. Almost 30% of temperature jumps are accepted, which indicates good overlap of potential energy distributions of neighbouring temperatures, and therefore adequate temperature spacing. The majority of replicas have visited all temperatures at least once, and on average each replica has visited 94% of temperatures.

**Table 1.** Evaluating the diffusion of replicas in temperature space.

| Temperature Diffusion Properties | |
|---|---|
| average deviation from homogeneity[#] $= \frac{1}{M}\sum_{m=1}^{M}\frac{\lvert N_m - \langle N_m\rangle\rvert}{\langle N_m\rangle}$ | 0.013 |
| acceptance ratio $= \frac{\text{number of accepted T jumps}}{\text{total number of attempted jumps}}$ | 0.292 |
| mean free path = average distance travelled between rejected temperature jumps | 0.280 |
| replica speed = (average distance travelled in 50 steps)/50 | 0.051 |
| diffusion coefficient = 0.5 x slope in the linear region of the mean square displacement vs. steps plot | 0.085 |
| average fraction of temperatures visited per replica | 0.94 |

[#] $N_m$ is the number of samples at temperature $m$, and $M$ is the number of temperatures.

In figure 3, the random walk in temperature is shown for one of the replicas. By varying the temperature, the STDR algorithm leads to exploration of conformational states with a wide range of radius of gyration, which is a measure of the spatial extension of the peptide. At low temperature, the peptide samples more compact conformations; increasing temperature allows the peptide to sample more expanded conformations. The random walk in temperature facilitates the peptide's expansion and collapse, preventing it from becoming trapped in any specific conformational state.

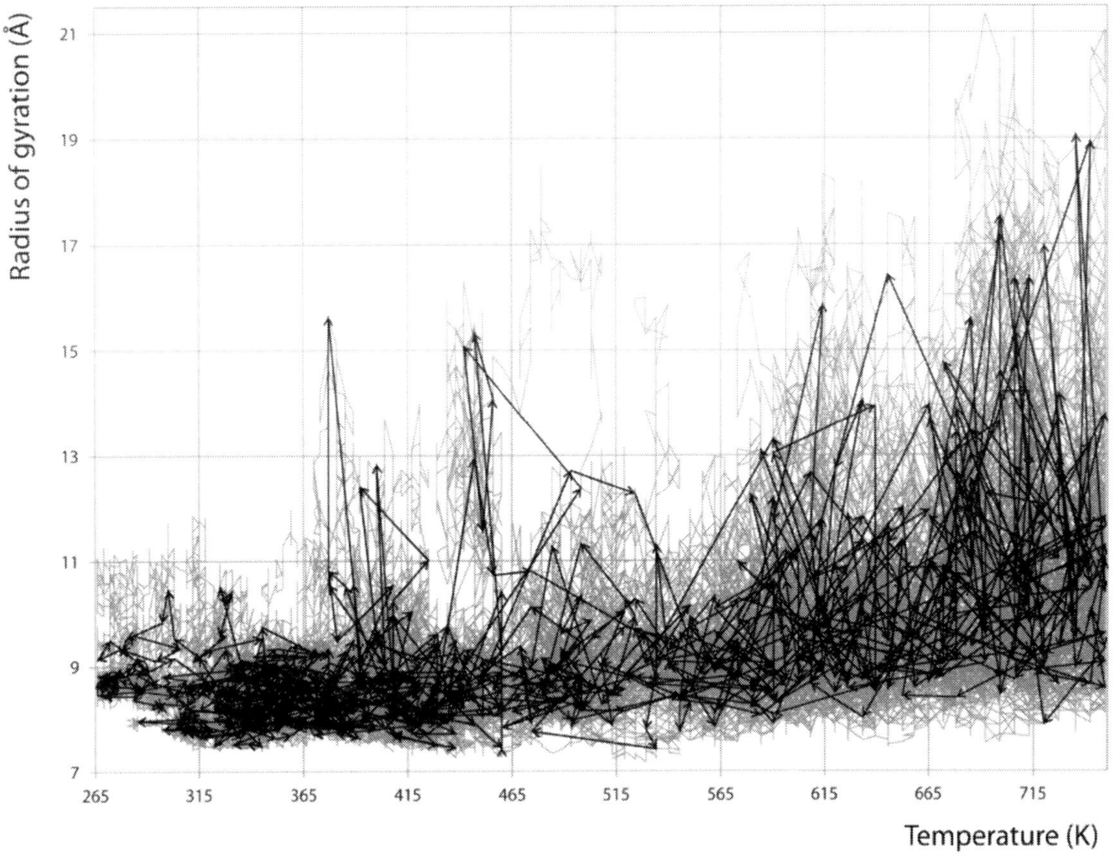


**Figure 3**. A two-dimensional plot of the radius of gyration of the peptide vs. temperature. The complete STDR simulation of a single replica (750 ns in total) is shown in gray. Superimposed in black is the same trajectory including only every hundredth point with arrows for clarity.

## 5. Evaluating Convergence of the Structural Ensemble

### *5.1. Removal of unequilibrated structures*

The STDR simulation of each replica began with a random configuration generated at high temperature. Because these configurations are not representative of the true conformational ensembles, an initial equilibration phase occurs during which these random configurations collapse to more realistic, compact structures. In order to delineate this equilibration phase of the STDR simulation, we look at the running average of radius of gyration, $R_g$, as a function of time (figure 4). Note that in this case, time refers to the accumulated simulation time at 300 K, and not to a continuous MD time trajectory. The average and standard deviation of $R_g$ from the second half of the STDR simulation are used for comparison. The point at which the $R_g$ falls within one standard deviation of the average (in this case, 64.9 ns) is taken to be the point at which the equilibration phase is complete. Configurations accumulated during the equilibration phase are discarded, and all other configurations are used for further structural analysis.

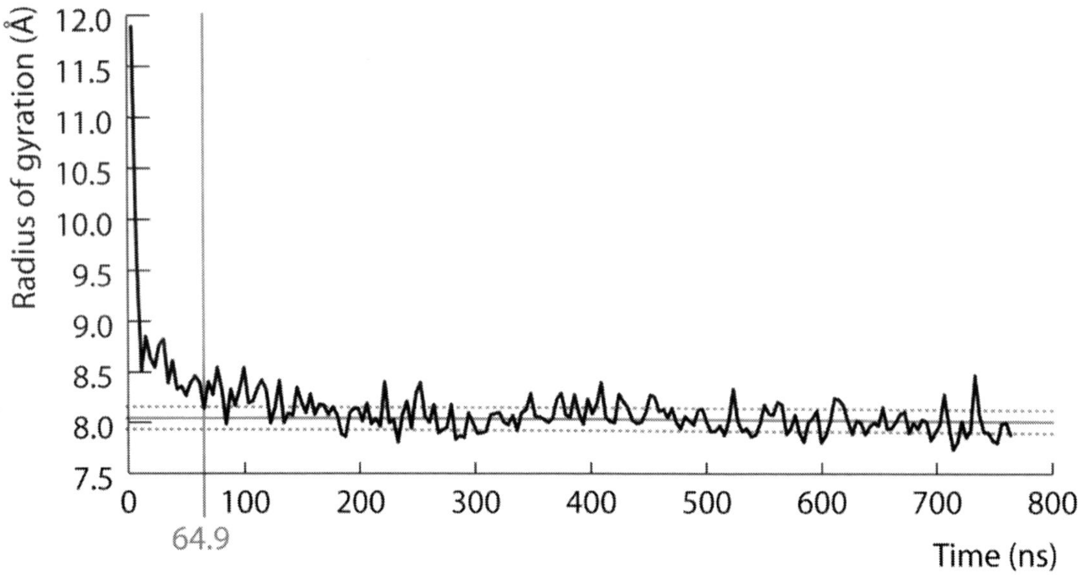


**Figure 4**. Running average of radius of gyration versus time. The radius of gyration is computed for the backbone Cα atoms. The red solid line indicates the average radius of gyration over the second half of the simulation, and the red dotted lines indicate one standard deviation above and below this average. The first point at which the average radius of gyration falls within this range is taken to be the end of the equilibration phase.

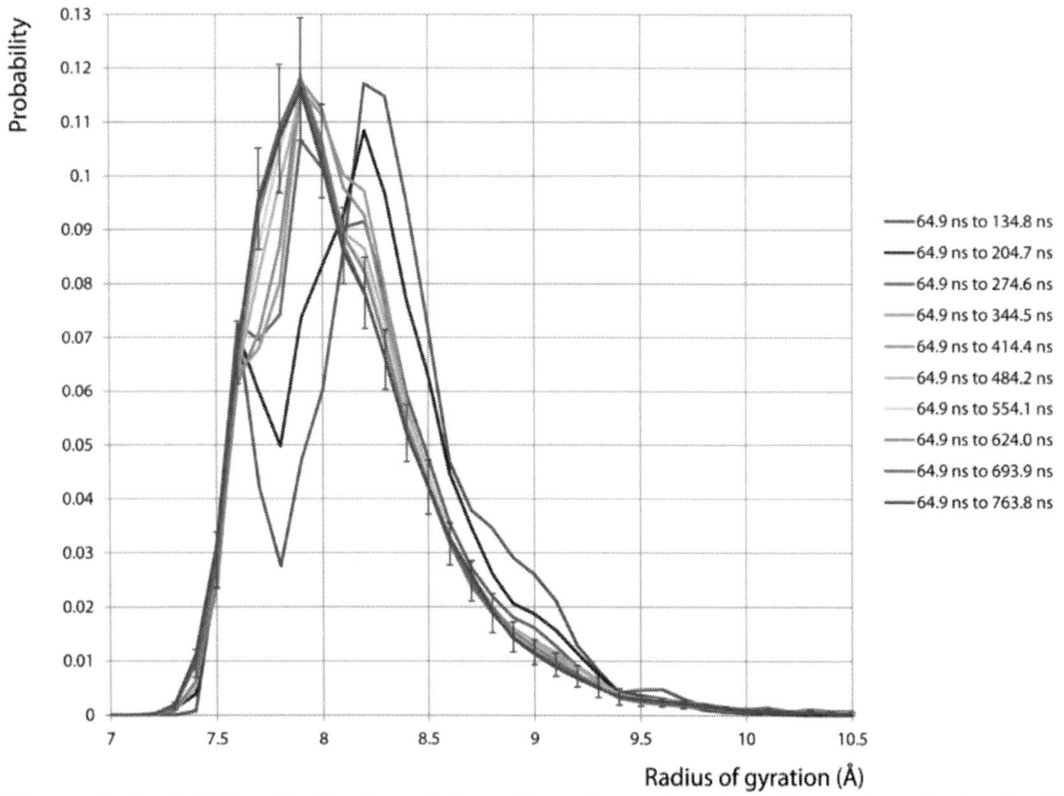


**Figure 5**. Probability distribution of the radius of gyration, $R_g$. Shown here is the $R_g$ distribution computed using incrementally increasing amounts of data from the STDR trajectory at 300 K from 64.9 ns to 763.8 ns (the data from the equilibration phase is not included). Error bars for the $R_g$ distribution computed using the complete trajectory (in red) are the standard error based on block averaging.

*5.2 Convergence of the conformational ensemble*

The probability distribution of the radius of gyration (figure 5) appears to be statistically converged, since it does not change with additional simulation time in the second half of the trajectory at 300 K (after 414.4 ns). It is always possible that longer simulations will reveal longer timescale phenomenon. Thus, it is not possible to be certain in declaring that a simulation or a structural property is converged [2]. Similarly, we note that it is important to investigate the convergence of several structural properties, since they may reach statistical convergence with differing amounts of conformational sampling [7].

## 6. Comparison of Canonical MD and STDR

In order to quantify the sampling enhancement provided by STDR, we compare the conformational ensemble at 300 K with that of an MD simulation in the canonical ensemble at 300 K (figures 6 and 7). First, we monitored the number of intramolecular peptide hydrogen bonds observed with increasing simulation time, shown in figure 6 (a hydrogen bond was counted as observed if it was formed in at least one of the configurations in the trajectory). With the complete STDR trajectory at 300 K, more than 92% of all possible hydrogen bonds were observed in at least one configuration. In contrast, the canonical MD simulation at the same temperature only sampled 11.9% of all possible peptide-peptide hydrogen bonds. We also show the number of hydrogen bonds observed for the replica trajectory depicted in figure 3. This replica has sampled 95% of all possible hydrogen bonds. The STDR simulation has not yet sampled all possible hydrogen bonds, which indicates that unexplored regions of conformational space remain to be sampled. It is also possible that some of the intramolecular hydrogen bonds that have not yet been sampled cannot be made due to structural restraints. Nevertheless, the vast majority of intramolecular peptide hydrogen bonds have been sampled at least once.

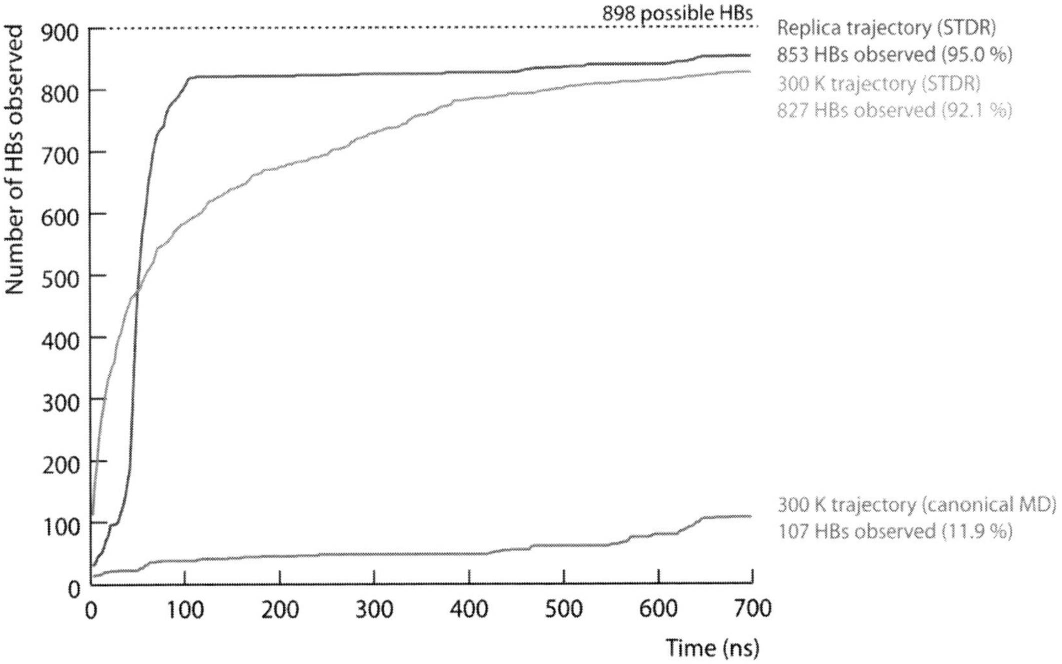


**Figure 6.** Number of intramolecular peptide hydrogen bonds (HBs) observed with increasing simulation time for STDR (green and purple) and canonical MD (red). In purple, the number of HBs observed for the replica trajectory corresponding to figure 3 is shown. The number of possible HBs is computed based on the number of possible contacts in the HB contact plot, figure 7A and 7B (35 x 35 = 1225). From this value, we subtract the diagonal and the two off diagonals (35+34+34); these HBs cannot be formed due to conformational restrictions. Contacts that would involve the N-H of proline residues are also subtracted. Thus, there are 898 possible HBs.

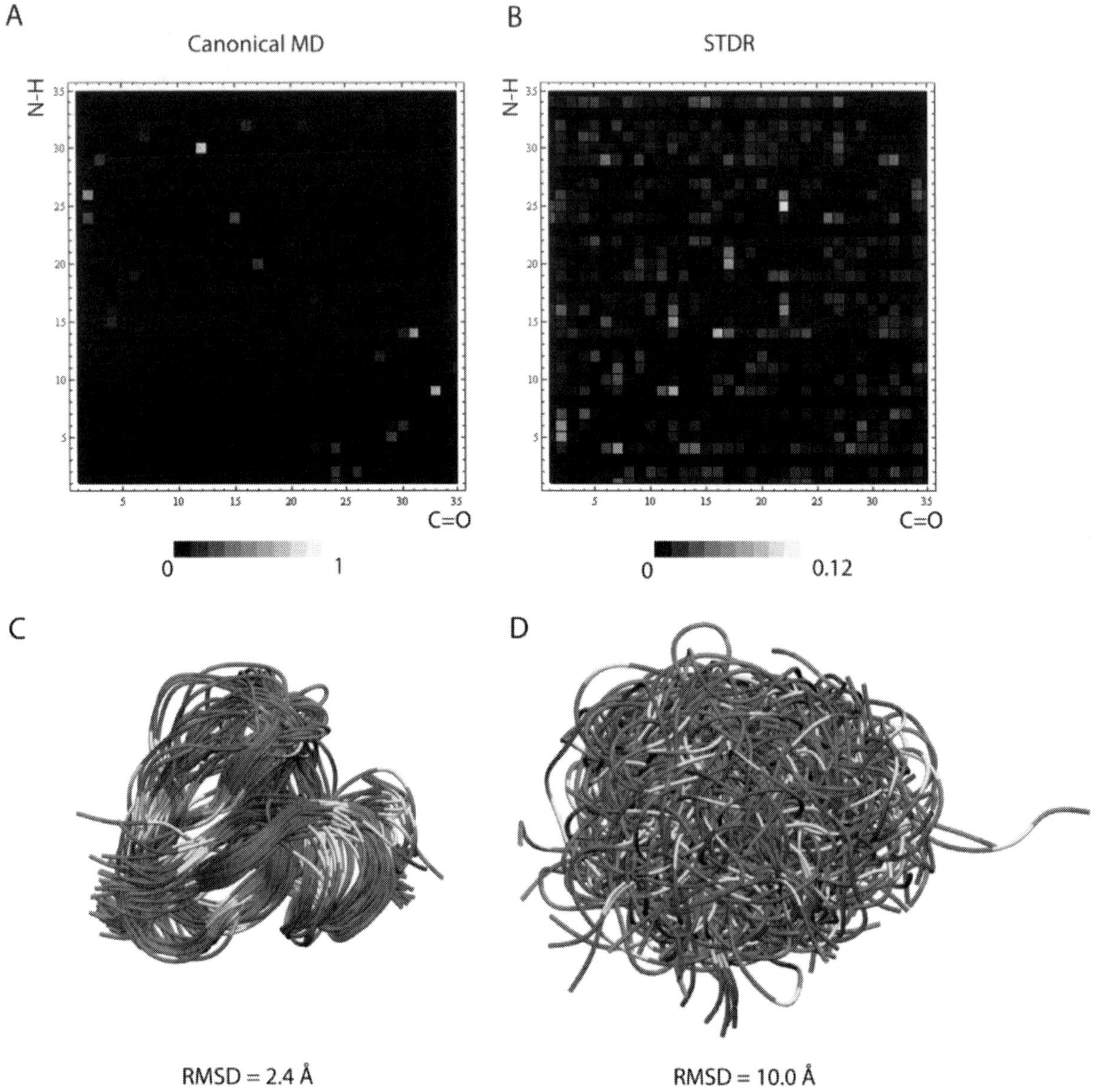


**Figure 7.** Configurations of the disordered $(GVPGV)_7$ peptide from canonical MD and STDR simulations at 300 K. Intramolecular hydrogen-bonding contact maps for the ensembles from canonical MD and STDR are shown in (A) and (B), respectively. The probability of a particular hydrogen bond between backbone C=O and N-H groups is indicated by the colour of the corresponding square. Note that the contact maps have different colour scales; some of the contacts in the canonical MD simulation are highly populated, whereas the STDR ensemble has no contacts populated in more than 12% of structures. In (C) and (D), we show 100 randomly selected configurations from each ensemble and the average root mean-square deviation (RMSD) of backbone atoms compared to the starting configuration. Configurations are shown with a cartoon representation of the backbone using VMD [29] (glycine in orange, valine in purple, and proline in yellow).

The canonical MD simulation was 700 ns in length, which is the same amount of accumulated simulation time at 300 K after removing the initial unequilibrated configurations (refer to section 4). Representative sets of configurations from both the MD simulation and the STDR simulation are shown in figure 7, along with hydrogen-bonding contact maps for these ensembles. The canonical MD simulation exhibits quasi-nonergodicity [30]: even though the system is ergodic, it appears non-ergodic because it samples essentially a single conformational basin throughout the entire simulation (over hundreds of nanoseconds). The configurations from the canonical MD simulation do not differ significantly from each other based on RMSD. In contrast, the configurations from STDR represent a small subset of the thousands of configurations generated by the simulations, and illustrate the heterogeneous set of structures that this peptide can adopt. By comparing the hydrogen-bonding contact maps obtained using canonical MD and STDR (figure 7C and 7D), it is evident that canonical MD severely underestimates the structural heterogeneity of this disordered peptide, while STDR provides a dramatic enhancement in conformational sampling.

## 7. Conclusions

We have presented our implementation of STDR on a large-scale computing cluster, demonstrating the computational efficiency of the approach with a detailed account of the overhead of each step in the algorithm. An atomic-level description of the disordered conformational ensemble of an elastin-like peptide was obtained. Our results push the boundaries of modern MD simulations with nearly 0.1 ms of sampling. By comparing the conformational ensembles obtained using STDR and canonical MD, the sampling enhancement provided by the random walk in temperature is clearly demonstrated. The detailed structural properties of the ensemble of conformations for this peptide will be the subject of future publications. Significantly, this study establishes a methodological framework for future applications of STDR to other disordered states of proteins.

## Acknowledgements

We are grateful to Dr. Tom Rodinger, Chris Neale, and Grace Li for discussions on distributed replica sampling, and to Chris Madill for administration of our in-house data storage servers. We also thank the SciNet support team for helpful advice in tailoring our implementation of STDR to the SciNet GPC architecture. Computations were performed on the GPC supercomputer at the SciNet HPC Consortium. SciNet is funded by: the Canada Foundation for Innovation under the auspices of Compute Canada; the Government of Ontario; Ontario Research Fund - Research Excellence; and the University of Toronto. The Canadian Institutes of Health Research (Grant No. MOP84496) is gratefully acknowledged for support. S.R. is funded by a Canada Graduate Scholarship from the Natural Sciences and Engineering Research Council (NSERC) and the Research Training Center at the Hospital for Sick Children. R.P. is a CRCP chairholder.

## References

[1] S. Gnanakaran *et al.*, Curr. Opin. Struct. Biol. **13**, 168 (2003).
[2] E. Lyman, and D. M. Zuckerman, Biophys. J. **91**, 164 (2006).
[3] B. Gillespie, and K. W. Plaxco, Annu. Rev. Biochem. **73**, 837 (2004).
[4] S. Rauscher, and R. Pomès, Biochemistry and Cell Biology **88**, 269 (2010).
[5] A. K. Dunker *et al.*, BMC Genomics **9**, 25 (2008).
[6] J. J. Ward *et al.*, J. Mol. Biol. **337**, 635 (2004).
[7] S. Rauscher, C. Neale, and R. Pomès, J. Chem. Theory Comput. **5**, 2640 (2009).
[8] X. Huang, G. R. Bowman, and V. S. Pande, J. Chem. Phys. **128**, 205106 (2008).
[9] E. Marinari, and G. Parisi, Europhys. Lett. **19**, 451 (1992).
[10] A. P. Lyubartsev *et al.*, J. Chem. Phys. **96**, 1776 (1992).
[11] A. Mitsutake, Y. Sugita, and Y. Okamoto, Biopolymers **60**, 96 (2001).
[12] Y. Sugita, and Y. Okamoto, Chem. Phys. Lett. **314**, 141 (1999).

[13] M. C. Tesi *et al.*, J. Stat. Phys. **82**, 155 (1996).
[14] A. M. Ferrenberg, and R. H. Swendsen, Phys. Rev. Lett. **61**, 2635 (1988).
[15] S. Park, and V. S. Pande, Phys. Rev. E **76**, 5 (2007).
[16] T. Rodinger, P. L. Howell, and R. Pomès, J. Chem. Theory Comput. **2**, 725 (2006).
[17] T. Rodinger, P. L. Howell, and R. Pomès, J. Chem. Phys. **129**, 155102 (2008).
[18] S. Rauscher *et al.*, Structure **14**, 1667 (2006).
[19] L. D. Muiznieks, A. S. Weiss, and F. W. Keeley, Biochem. Cell Biol. **88**, 239 (2010).
[20] W. F. van Gunsteren, and H. J. C. Berendsen, Molecular Simulation **1**, 173 (1988).
[21] B. Hess *et al.*, J. Chem. Theory Comput. **4**, 435 (2008).
[22] G. A. Kaminski *et al.*, J. Phys. Chem. B **105**, 6474 (2001).
[23] W. L. Jorgensen *et al.*, J. Chem. Phys. **79**, 926 (1983).
[24] T. Darden, D. York, and L. Pedersen, J. Chem. Phys. **98**, 10089 (1993).
[25] B. Hess *et al.*, J. Comput. Chem. **18**, 1463 (1997).
[26] S. Miyamoto, and P. A. Kollman, J. Comput. Chem. **13**, 952 (1992).
[27] M. Parrinello, and A. Rahman, J. Appl. Phys. **52**, 7182 (1981).
[28] S. Nose, and M. L. Klein, Mol. Phys. **50**, 1055 (1983).
[29] W. Humphrey, A. Dalke, and K. Schulten, J. Mol. Graph. **14**, 33 (1996).
[30] C. Chipot, and A. Pohorille, *Free Energy Calculations: Theory and Applications in Chemistry and Biology* (Springer, Berlin, 2007).

Journal of Physics: Conference Series **256** (2010) 012012 doi:10.1088/1742-6596/256/1/012012

# A Parallel Non-Alignment Based Approach to Efficient Sequence Comparison using Longest Common Subsequences

**S. Bhowmick [1], M. Shafiullah[1], H. Rai[1] and D. Bastola[2]**

1. Department of Computer Science, University of Nebraska at Omaha

2. School of Interdisciplinary Informatics, University of Nebraska at Omaha

Email dkbastola@unomaha.edu

**Abstract**. Biological sequence comparison programs have revolutionized the practice of biochemistry, and molecular and evolutionary biology. Pairwise comparison of genomic sequences is a popular method of choice for analyzing genetic sequence data. However the quality of results from most sequence comparison methods are significantly affected by small perturbations in the data and furthermore, there is a dearth of computational tools to compare sequences beyond a certain length. In this paper, we describe a parallel algorithm for comparing genetic sequences using an alignment free-method based on computing the Longest Common Subsequence (LCS) between genetic sequences. We validate the quality of our results by comparing the phylogenetic tress obtained from ClustalW and LCS. We also show through complexity analysis of the isoefficiency and by empirical measurement of the running time that our algorithm is very scalable.

## 1. Introduction

A fundamental operation in bioinformatics involves the comparison of genetic (DNA) sequences. Similarity between genetic sequences is a strong indicator of evolutionarily preserved characteristics. This property has been successfully used in determining pathologically important bacteria, viruses and fungi [1-3].

Among the many sequence comparison tools for mining genetic information, one extremely common technique includes the alignment-based methods. These involve aligning the entire (global alignment, Needleman-Wunsch[4]) or smaller sections (local alignment, Smith-Waterman [5]) of the genetic sequences. The choice of global or local alignment is based on the type of analysis desired. However, both these methods are heavily dependent on the quality of sequence data. Even slight discrepancies resulting from experimental or technical limitations, can significantly affect the comparison results.

Although the fine granularity of comparative analysis is desirable when analyzing specific biological properties such as the single nucleotide polymorphism (SNP), alternative approaches of sequence analysis are becoming increasingly important in dealing with the exponential growth of genetic sequence data, and the classification and the grouping of organisms based on these sequences. Such alternative approaches include the alignment-free methods, which match the relative (as opposed to the exact) order of the base pairs in the sequence [6-9].

Advancements in sequencing technology have provided a deluge of genetic data. The Genbank, a public repository of genetic sequence data, reported 120604423 sequence records in its 178th release in June 15, 2010. Analyzing such large datasets, including the 3 billion bases of the human reference genome, on uniprocessor machines is an extremely time consuming process. It is imperative therefore, to harness the power of high performance computing to facilitate our understanding of this high throughput data.

In this paper, we present an efficient parallel non-alignment method for sequence comparison, based on finding the longest common subsequence (LCS), across genetic sequences. Our preliminary results demonstrate that this technique can be used to identify "signature-sequences" across multiple strains within different species of mycobacterium.

## 2. Sequence Comparison Using LCS

The longest common subsequence algorithm finds the longest subsequence between two strings. In contrast to the substring, the subsequence denotes a series of letters from the string which while being in order, need not be consecutive. For example, between ATCG and CTCAG, the longest common substring is TC, while the longest common subsequence is TCG.

LCS can help identify the key nucleotides across genetic sequences and is considerably less affected by the occasional sequencing error. This method is also useful for identifying potential regions of small mutations by analyzing the portions of the string not present in the LCS.

### *2.1. Computing the LCS*

Our algorithm for computing LCS is inspired by the FAST_LCS method described in [10]. The algorithm involves creating position pairs of identical letters in the sequence and combining them to potential LCS strings as long as their relative order is maintained across the pairs. Given two position pairs (i,j) and (l,m), the corresponding letters would be a potential candidate for LCS only if $i<l$ and $j<m$.

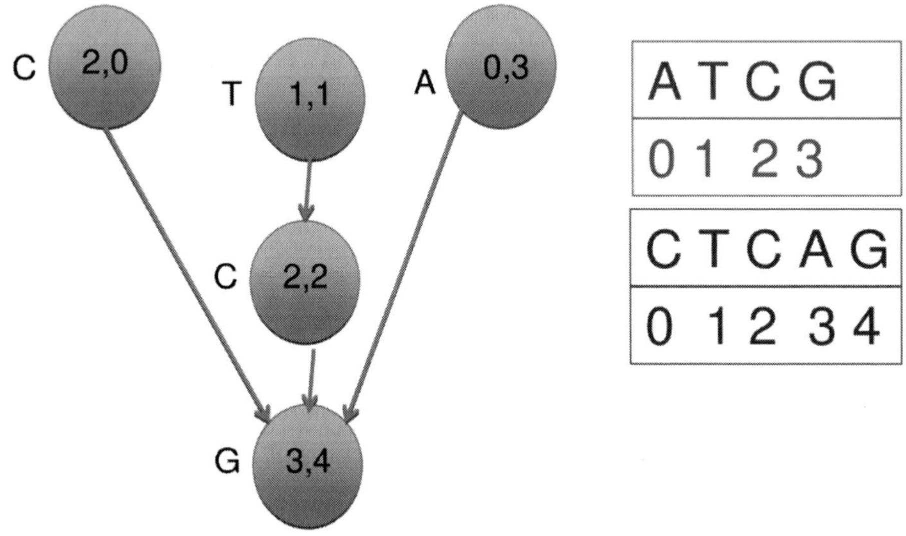


**Figure 1.** An example of computing LCS using position pairs. The position pairs of the two strings ATCG and CTCAG are represented as vertices of the graph. An edge connects vertex (i,j) to (l,k) if ($i<l$) and ($j<k$). The diameter of the tree gives the longest common subsequence.

In the sequences given above; the position pairs are; A:(0,3); C:(2,0) and (2,2); T:(1,1) and G: (3,4). Some strings preserving the relative order are: (i) AG :(0,3)->(3,4); (ii) TG: (1,1)->(3,4); and (iii) TCG: (1,1)->(2,2)->(3,4)

The algorithm consists of obtaining these position pairs and then creating a tree where the vertices represent the position pairs and the edges, their order in the subsequences. The diameter of the tree provides the LCS, as shown in the Figure1.

In order to improve the time to traverse the tree, [10] suggested pruning techniques, whereby some vertices can be deleted from their original positions and later added at a new position. This helps remove redundant edges that do not contribute to the LCS. For example though both TG and TCG are subsequences, the vertex representing the position pair of G can be later reconnected to the position pair of C, instead of being connected to T.

However, pruning is a computationally expensive algorithm since (i) it does not eliminate the initial redundant additions, (ii) deletion of edges requires more operations than additions and (iii) identifying the position of the redundant vertices requires periodic tree traversals.

To avoid these excess operations, we order the pairs such that a vertex appearing later in a string is not visited earlier. In the above example the ordered position pairs will be A:(0,3), T:(1,1), C:(2,0), C:(2,2), and G:(3,4). Therefore, TG will not be connected before TC. Furthermore, we ensure that a new letter is connected only to the longest candidate string, thereby reducing the number of edges to traverse.

Let the length of a string with L different symbols be given by;

$$M = \sum_{i=1:L} m_i ,$$

where $m_i$ is the number of occurrences of the symbol *i*. The complexity of computing LCS, for two strings of length

$$M = \sum_{i=1:l} m_i$$

and

$$N = \sum_{i=1:L} n_i ,$$

is proportional to the number of vertices and is given by

$$O(\sum_{i=1:L} (m_i \times n_i) .$$

### 2.2. *Parallel Implementation of LCS*

One of the simplest methods to utilize multiple processors for sequence comparison is to execute different sets of comparisons in each processor in parallel. Though such embarrassingly parallel algorithm, as described in [10] can improve the performance, the method is not very scalable. That is, the effectiveness of the parallel algorithm is dependent on having a large number of sequence comparisons rather than being adaptable as the number of comparisons and processors change. Furthermore, the lengths of the sequences to be compared are limited by the memory in one processor.

We present a domain decomposition based parallel algorithm, which divides each sequence across the processors and *performs per sequence comparison in parallel co-ordination*. The algorithm, outlined in Figure 2, is as follow; Each sequence is divided across P processors. Then the partial LCS between the sequences in each processor is computed in parallel (Step 1 of Figure 2). Once this is completed, we compute the "unused" portions, which are portions of the sequences before (after) the first (last) positions included in the LCS. These areas are marked as grey in Figure 2.

These "unused" portions can potentially contain parts of the LCS that fall across processors. Therefore the appropriate unused portions are exchanged between processors that contain consecutive portions of the sequences. The pattern of exchange between two processors is given in Step 2 of Figure 2. Note that for each sequence comparison a processor communicates with at most two other processors, therefore the communication pattern is scalable and independent of the total number of processors used.

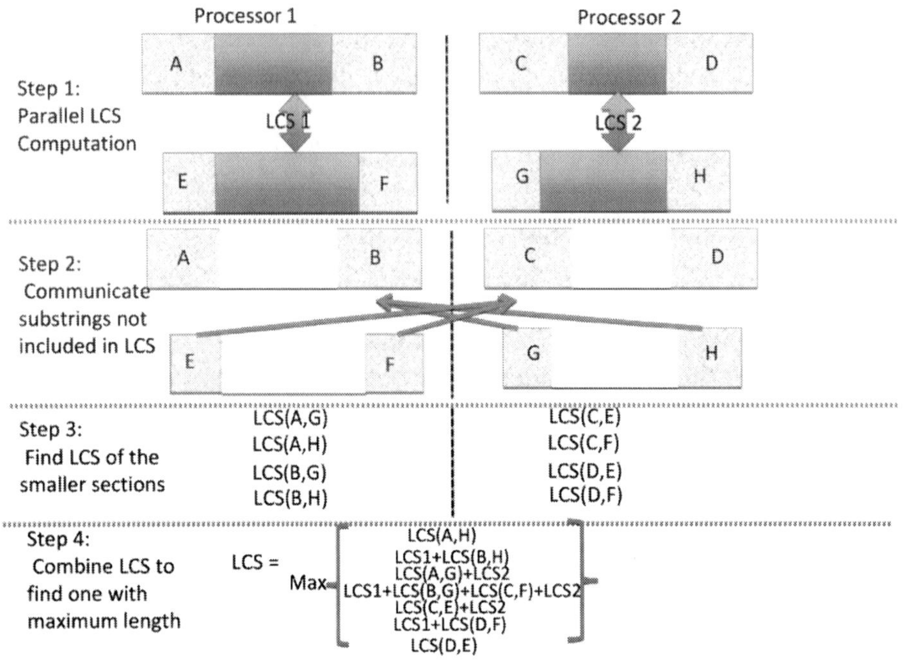


**Figure 2.** A Schematic Diagram of the Parallel LCS Algorithm

After the exchange, the LCS of the appropriate unused sequences are computed, Step 3 of Figure 2 and then finally the partial LCS portions are combined using a reduction operation. Based on the partial sequences that are compared, several overlapping LCS can be observed. We combine these portions to find the non-overlapping LCS strings of maximum length (Step 4 of Figure 2).

### 2.3. *Complexity of the Parallel Algorithm*

Given two sequences with L symbols of length

$$M = \sum_{i=1:L} m_i$$

and

$$N = \sum_{i=1:L} n_i \, ,$$

the complexity of the sequential algorithm is

$$O\left(\sum_{i=1:L} (m_i \times n_i)\right).$$

In the parallel version each sequence is divided across P processors. Therefore the length of the sequences in each processor is approximately M/P and N/P. Based on this partition, the time to compute the partial LCS in each processor is

$$O\left(\sum_{i=1:L} ((m_i / P) \times (n_i / P))\right).$$

We assume that the unused portions form approximately x% of the total length of the strings. The communication per processor is therefore for the exchange step,

$$O\left(\frac{x}{100}(m_i + n_i)\right)$$

and for the combination step,

$$O(\log P).$$

The total communication volume across all processors, when x represents a very small percentage, is therefore

$$O\left(\frac{Px}{100}(m_i + n_i) + P\log P\right) \approx O(P\log P).$$

Without loss of generality, we assume that M> N. Therefore the maximum execution time of the sequential algorithm is $O(M^2)$and the maximum memory required to store the sequences is $O(2M)$. Therefore the memory required per processor to maintain isoefficiency should be at least

$$O(\sqrt{P} \times 1/2(\log P)/P) = O\left(\log P/\sqrt{P}\right).$$

Therefore the scalability of the algorithm is

$$O\left(\log P/\sqrt{P}\right).$$

Thus our parallel implementation is highly scalable.

## 3. Experimental Results

We conducted two sets of experiments. In the first experiments we focused on clustering different species of mycobacteria based on their genetic sequences and using multiple sequence alignment. The results showed that LCS clustering (sequential) indeed mimics the clustering obtained by multiple sequence alignment. In our second set of experiments we applied the parallel LCS algorithm to compare long sequences of *Mycobacterium tuberculosis* genome and compared the execution and change in results as more processors were used.

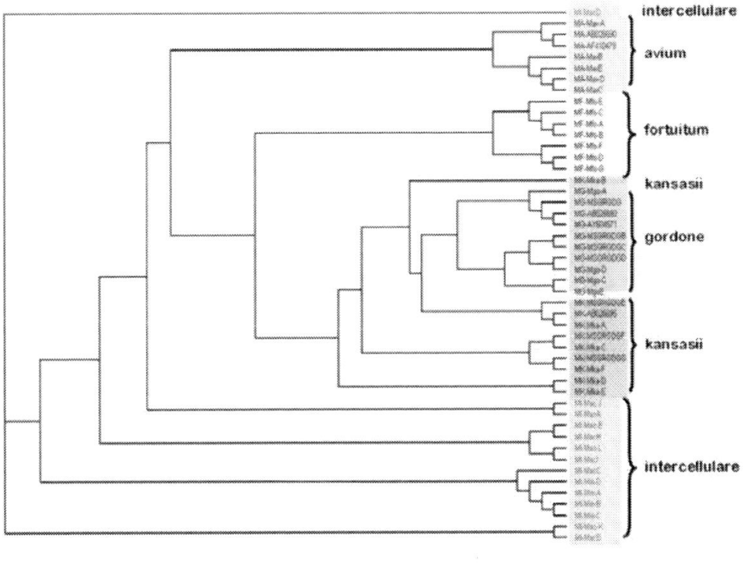


**Figure 3a**. Clustering of sequences using multiple sequence alignment. The different colors indicate the different species of mycobacterium

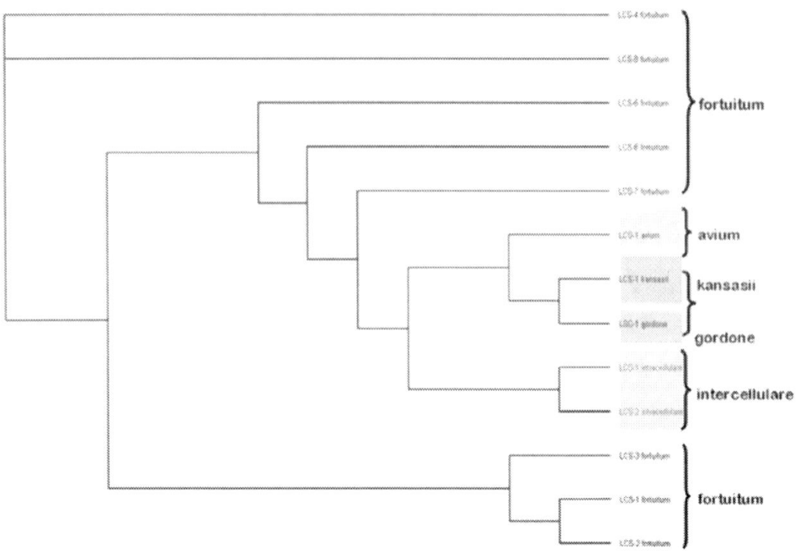


**Figure 3b.** Clustering of sequences using longest common subsequences across species. The different colors indicate the different species of mycobacterium. The species intercellulare and fouruitum have multiple LCS due to internal mutations.

### *3.1. Comparison of LCS and MSA Clustering*

In our first set of experiments we computed the LCS of a portion of the genomic sequences (average length 200 base pairs) of five species of *Mycobacterium* on a sequential program. We observed that the number of distinct LCS produced per sequence pair is proportional to their variance in their base pair composition. It is observed that the sequence with low variance in composition (avium, kansasii, gordone) show a single LCS while divergent sequences (fortuitum, intercellulare) show multiple-LCS fragments.

We compared the phylogenetic clustering of the Mycobacterium using only the LCS sequence (Figure 3b) from each group of species with that obtained through multiple sequence alignment (MSA), using ClustalW[11] (Figure 3a) of all sequences across all the species. Our results indicate that the LCS clustering follows the MSA clustering very closely.

**Table 1.** The length of the LCS (over 16 to 128) for pairwise comparison of Genome Sequences from *Mycobacterium tuberculosis*.

| Processors | LCS of Sequence 1 Sequence 2 | LCS of Sequence 1 Sequence 3 | LCS of Sequence 1 Sequence 4 | LCS of Sequence 2 Sequence 3 | LCS of Sequence 2 Sequence 4 | LCS of Sequence 3 Sequence 4 |
|---|---|---|---|---|---|---|
| 16 | 1161 | 1118 | 926 | 1161 | 850 | 833 |
| 32 | 1128 | 1100 | 914 | 1087 | 843 | 830 |
| 64 | 1103 | 1098 | 914 | 1025 | 846 | 817 |
| 128 | 1111 | 1105 | 971 | 1008 | 894 | 857 |

### 3.2. *Parallel Implementation of LCS*

We implemented LCS based on the parallel algorithm sketched in Figure 2. Our parallel experiments were conducted on the Firefly a 1,151 node linux cluster with distributed memory architecture connected through inifiniband [12]. We used MPI to implement the parallel program.

We performed a pairwise comparison over portions of four genome sequences of *Mycobacterium tubercolosis*, with an average length of 4000 base pairs. The genome sequences used for this study were from Mycobacterium_tuberculosis_CDC1551, Mycobacterium_tuberculosis_F11, Mycobacterium_tuberculosis_H37Ra and Mycobacterium_tuberculosis_H37Rv.

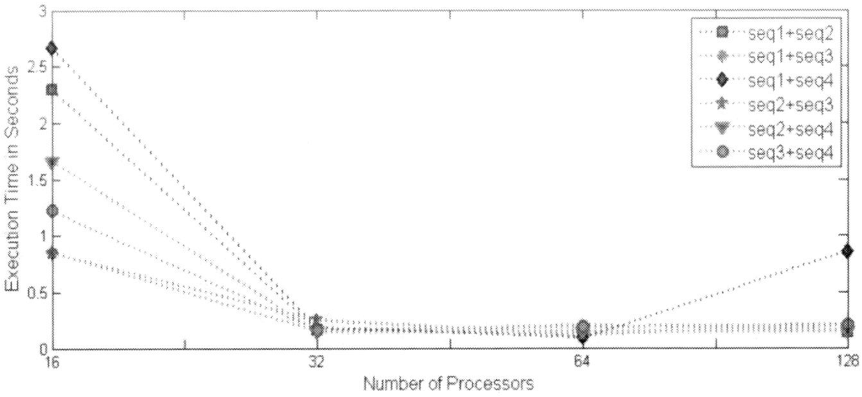


**Figure 4.** The execution time of parallel LCS algorithm as the number of processors are increased.

Each pair of sequences (total six pairs) was divided into a set of processors ranging from 16 to 128. In the first step we compared portions of the sequence that were allocated to the same processor (Step 1 in Figure 2). During Step 2 we observed that the LCS for the 16 and 32 processors started from at most the third position in the partial sequences, that is we miss only 3-4 base pairs from the left end of each sequence due to the domain decomposition. This indicates that there are only miniscule unused portions in each string that have not been compared. The total fraction of base pairs not computed by forgoing the exchange step per processor is at most 4XP/4000=.001xP. Therefore for 16 and 32 processors the maximum loss is at most 1.6% and 3.2% respectively. Given that this value is less than the percentage of errors in generating the genetic sequences [13], we decided that the comparison results would not be significantly affected by eliminating the exchange step. The cutoff portions were even smaller for the 64 and 128 processors (1-2 sequences from the left) resulting in similar percentage of potential errors.

The total length of the LCS was obtained by adding the partial LCS (which are of course non-overlapping) obtained from each processor. Table 1 shows the length of the LCS across the six sequence pairs. Since we are not computing the LCS across processors (Step 2 in Figure 2) the values differ as the number of processors change. However note that there is only a slight variation in the values for each sequence pair, indicating that they can be reliably used for clustering. This is in accordance to the spirit of using alignment-free methods, where the focus is on obtaining the relative rather than the absolute values.

These results lead us to conjecture that, if the unused regions are small, as in this case, we can obtain an accurate clustering of genome sequences by just obtaining this partial value of the LCS. Figure 4, gives the lower bound on the execution time of the parallel LCS. As can be seen from the results, the current version of the algorithm (implementing only Step 1) is highly scalable.

Note that due to the availability of over a thousand nodes and the mutual independence between the LCS computations of any two pairs of sequences, we could potentially execute the parallel experiments for pair-wise comparison simultaneously. That is, when the LCS of one pair of sequences is being computed over n processors, the LCS of another pair can be computed at the same time over other *n* processors. Thus, we are employing a two level parallelism. In the first level we use the embarrassingly parallel nature of pair wise comparison to allocate each pair over a group of processors. In the second level we use a more tightly coupled distributed memory paradigm to compute the actual LCS.

## 4. Conclusions

We have developed a scalable parallel algorithm for computing LCS between strings. Our results demonstrate that longest common subsequence can be used as an effective non-alignment based technique for genetic sequence comparisons. Our parallel experiments show that LCS is indeed very scalable and can be used to long genomic sequences. In future, we intend to investigate between occurrences of multiple LCS across entire genome sequences of the same species and their relation to intra-species mutations.

## 5. References

[1] Rispail N, Soanes D.M., Ant C, Czajkowski R, Grünler A, Huguet R, Perez-Nadales E, Poli A, Sartorel E, Valiante V, Yang M, Beffa R, Brakhage AA, Gow NA, Kahmann R, Lebrun MH, Lenasi H, Perez-Martin J, Talbot NJ, Wendland J and Di Pietro A 2009 Comparative genomics of MAP kinase and calcium-calcineurin signalling components in plant and human pathogenic fungi. *Fungal Genet Biol.* **46** 287-98

[2] Dong QJ, Wang Q, Xin YN, Li N and Xuan SY 2009 Comparative genomics of Helicobacter pylori. *World J Gastroenterol* **15** 3984-91

[3] Ma LJ, van der Does HC, Borkovich KA, Coleman JJ, Daboussi MJ, Di Pietro A, Dufresne M, Freitag M, Grabherr M, Henrissat B, Houterman PM, Kang S, Shim WB, Woloshuk C, Xie X, Xu JR, Antoniw J, Baker SE, Bluhm BH, Breakspear A, Brown DW, Butchko RA, Chapman S, Coulson R, Coutinho PM, Danchin EG, Diener A, Gale LR, Gardiner DM, Goff S, Hammond-Kosack KE, Hilburn K, Hua-Van A, Jonkers W, Kazan K, Kodira CD, Koehrsen M, Kumar L, Lee YH, Li L, Manners JM, Miranda-Saavedra D, Mukherjee M, Park G, Park J, Park SY, Proctor RH, Regev A, Ruiz-Roldan MC, Sain D, Sakthikumar S, Sykes S, Schwartz DC, Turgeon BG, Wapinski I, Yoder O, Young S, Zeng Q, Zhou S, Galagan J, Cuomo CA, Kistler HC and Rep M 2010 Comparative genomics reveals mobile pathogenicity chromosomes in Fusarium. *Nature.* **464** 367-73

[4] Needleman SB and Wunsch CD 1970 A general method applicable to the search for similarities in the amino acid sequence of two proteins. *J Mol Biol* **48** 443-53

[5] Smith TF and Waterman M S 1981 Comparison of biosequences. *Adv. Appl. Math.* **2** 482–89

[6] Domazet-Loso M and Haubold B 2009 Efficient estimation of pairwise distances between genomes. *Bioinformatics.* **25** 3221-7

[7] Elemento O and Tavazoie S 2007 Fast and systematic genome-wide discovery of conserved regulatory elements using a non-alignment based approach. *Genome Biology* **6** R18

[8] Quest D, Tapprich W and Ali H 2007 A grammar based methodology for structural motif finding in ncRNA database search. *Comput Syst Bioinformatics Conf.* **6** 215-25

[9] Jia C, Liu T, Zhang X, Fu H and Yang Q 2009 Alignment-free comparison of protein sequences based on reduced amino acid alphabets. *J Biomol Struct Dyn.* **26** 763-9

[10] Chen Y, Wan A and Liu W 2006 A fast parallel algorithm for finding the longest common sequence of multiple biosequences. *BMC Bioinformatics* **7** S4

[11] Larkin MA, Blackshields G, Brown NP, Chenna R, McGettigan PA, McWilliam H,Valentin F, Wallace IM, Wilm A, Lopez R, Thompson JD, Gibson TJ, Higgins DG 2007 Clustal W and Clustal X version 2.0. *Bioinformatics.* **23** 2947-8

[12] Holland Computing Center. http://hcc.unl.edu/firefly

[13] Sundquist A, Ronaghi M, Tang H, Pevzner P and Batzoglou S 2007 Whole-Genome Sequencing and Assembly with High-Throughput, Short-Read Technologies. *PLoS ONE* e484

**Acknowledgement**

This project was supported by the NIH grant number P20 RR016469 from the INBRE Program of the National Center for Research Resources.

Journal of Physics: Conference Series **256** (2010) 012013 doi:10.1088/1742-6596/256/1/012013

# Self-Organizing Maps on the Cell Broadband Engine Architecture

**Sabine M. McConnell**

Department of Computing and Information Systems, Trent University, 1600 West Bank Drive, Peterborough, ON, Canada, K9J 7B8

E-mail: `sabinemcconnell@trentu.ca`

**Abstract.** We present and evaluate novel parallel implementations of Self-Organizing Maps for the Cell Broadband Engine Architecture. Motivated by the interactive nature of the data-mining process, we evaluate the scalability of the implementations on two clusters using different network characteristics and incarnations (PS3$^{TM}$console and PowerXCell 8i) of the architecture. Our implementations use varying combinations of the Power Processing Elements (PPEs) and Synergistic Processing Elements (SPEs) found in the Cell architecture. For a single processor, our implementation scaled well with the number of SPEs regardless of the incarnation. When combining multiple PS3$^{TM}$consoles, the synchronization over the slower network resulted in poor speedups and demonstrated that the use of such a low-cost cluster may be severely restricted, even without the use of SPEs. When using multiple SPEs for the PowerXCell 8i cluster, the speedup grew linearly with increasing number of SPEs for a given number of processors, and linear up to a maximum with the number of processors for a given number of SPEs. Our implementation achieved a worst-case efficiency of 67% for the maximum number of processing elements involved in the computation, but consistently higher values for smaller numbers of processing elements with speedups of up to 70.

## 1. Introduction and Motivation

Jointly developed by Sony, Toshiba, and IBM, the CBEA has recently developed a foothold in Scientific and High Performance computing using both of its incarnations: the PowerXCell 8i and the Cell processor used in the PS3$^{TM}$gaming console. The latter in particular has provided a low-cost environment capable of 150 GFlops (single-precision) at a cost of just under $300, which has led to the construction of clusters containing this architecture. Perhaps most notably, the US army has recently purchased 2 200 additional PS3$^{TM}$units to extend a smaller existing cluster. On the other hand, the PowerXCell is used in IBM's RoadRunner, which was the first general-purpose computer to break the Petaflop barrier. Example applications using the Cell processor for scientific computing include matrix optimization by Kurzak et *al.* [14], biomolecular simulation by Fabritiis [5], Fast Fourier Transform by Chow et *al.* [4], and Ray Tracing by Benthin et *al.* [2], with a major focus on the use of single units.

Despite the powerful potential of the CBEA that was demonstrated by a flurry of activities over the past five years, programming for this architecture has proven difficult because of its hybrid nature and the requirement for explicit memory management. One application area that sparked a substantial amount of interest in the Cell Broadband Engine is *data mining*, which is defined as the extraction of novel and useful information from large databases. One particular learning technique that is frequently used for data mining is the concept of a Self-Organizing Map, originally developed by Kohonen [13]. Recent applications of SOMs include the construction of Parton Distribution Functions parameterizations [10], an analysis of spectra [3], clustering of spatial-temporal precipitation data [11], and cluster analysis of spectrum sources [17].

With the rapid growth of the data sets in recent years, the iterative nature of data mining (combined with a multitude of techniques) has introduced a challenge that is unmatched by typical batch-processing environments; there is a current need for platforms that allow interactive data mining on the order of minutes or less, ideally at a low cost. The goal of this paper is to provide and evaluate a set of implementations of one particular learning algorithm for such an interactive and, for one of the incarnations of the architecture, low-cost computing environment. In particular, we are interested in the scalability and processor utilization of our approaches for two different clusters, using both incarnations of the CBEA in combination with two different networks.

The remainder of this paper is organized as follows. In Section 2 we give an overview of the relevant background material, including the architecture, the particular learning algorithm whose implementations we discuss here, parallel versions of said algorithm, and previous work connecting data-mining techniques to this particular architecture. In Section 3 we introduce the parallel implementations of the Self-Organizing Maps. We present the results and discussions in Section 4, followed by conclusions and future work in Section 5.

## 2. Background

### *2.1. Data Mining on the CBEA Architecture*

The Cell Broadband Engine Architecture connects a 3.2 GHz hyper-threaded simplified PowerPC (Power Processing Element, or PPE) with eight Synergistic Processing Elements (SPEs) via a ring-like Element Interconnect Bus with a maximum theoretical bandwidth of 25.2 GB/s. Able to achieve 25.6 GFlops (single precision), each of the single-threaded SPEs requires only 4 Watt, which initially made the Cell architecture an attractive alternative to contemporary architectures. Each SPE contains a vector processor, a 256KB private memory area called the *local store*, 128 128-bit registers, and a Memory Flow Controller. The architectural design focuses on performance and bandwidth rather than latency and ease of programming (severely restricting the portability of any implementation using the SPEs), and memory has to be managed explicitly by the developer.

A number of data-mining techniques have been parallelized to date on a single Cell processor, using up to six of the SPEs that are available to the application developer via a Software Development Kit. For example $k$-means clustering and the $k$ Nearest Neighbors algorithm are implemented by Buehrer et *al.* [1] and $k$-means clustering, APriori for Association Mining, and a Radial Basis Function Network by Duan and Strey [7]. Implementation of SOMs for a single Cell processor are limited to a student project available for download at the IBM developerWorks web site. However, this project did not implement double buffering, and was restricted to a single SPE only. Extensions to clusters for an Artificial Neural Network and Support Vector Machine are presented by Wyganowski [25] and McConnell et *al.* [18] for Decision Trees and Artificial Neural Networks.

### *2.2. Self-Organizing Maps*

The SOM, originally proposed by Kohonen [13], is a variant of an Artificial Neural Network which shares characteristics with Vector Quantization algorithms. The map consists of a set of neurons represented by weights and arranged in a typically 2-d structure. The neurons are repeatedly presented with randomly drawn input samples, for which the closest neuron is determined. The closest neuron, and its neighborhood as determined by a neighborhood function, are then adjusted to move closer to the input sample, resulting in a non-linear mapping. Kohonen originally proposed two versions of the algorithm: sequential training and batch training. Both typically use the Euclidean distance to determine the closest neuron for a given input sample, and are trained over multiple iterations through the input data.

In the sequential training approach, the neurons in the map are updated after each input sample. After determination of the winning node, the updates are calculated based on the difference between the values of the winning node and that of the input sample according to the rule $\mathbf{m}_i(t+1) = \mathbf{m}_i(t) + \alpha h_{ci}(t)[\mathbf{x}(t) - \mathbf{m}_i(t)]$, where $\mathbf{m}_i(t)$ denotes the $m_{th}$ neuron at iteration $t$, $\alpha$ the learning rate, $h_{ci}$ the neighborhood function, and $\mathbf{x}$ the input vector. Both the learning rate and the neighborhood function, which is typically a Gaussian function, decrease over time. The runtime of the SOM depends on the sizes of the map and the data set, but also on the characteristics of the data. Often, convergence of the algorithm occurs over a large number of iterations through the input data.

The batch training algorithm implemented here omits the learning rate and updates the neurons only once for each iteration, after all input data is presented. The new values of the neurons are calculated as weighted averages of the data assigned to the neurons. Furthermore, the neighborhood function is often reduced to $h_{ic} = \delta(i, c)$, so that only the winning neuron itself is updated.

### *2.3. Parallel implementations of SOMs*

Several variants of SOMs have been parallelized, an introduction to and evaluation of several parallelization techniques is, for example, given by Kwiatowski et *al.* [15]. Parallelizations of the two original versions of the algorithm can be grouped by the level of parallelism they exploit. *Network-partitioning* techniques distribute parts of the feature map but require copies of the input data at each of the processing elements. Such partitioning results in limited scalability, not only because the number of nodes in a given map is typically small compared to the size of the data itself, but also because of the increased memory requirements. Furthermore, if the online algorithm is used, synchronization between elements is required after presentation of each input element in order to select the best matching unit.

In contrast, *data-partitioning* approaches distribute the input data but require copies of the weight vectors at each processing element. Strupl and Neruda [21] show the performance advantages of partitioning the input data. If the batch algorithm is used, then global synchronization is necessary only after each input sample is presented, but still required at each iteration.

Using Kohonen's SOM as a starting point for four new algorithms, Huntsberger and Ajjimarangsee [12] present implementations on a NCUBE/10 hypercube, while Filippi and Lawson discuss the implementation of the SOM on the SMART neurocomputer, which was built as a distributed memory systems using a linear processor array [8]. Filippi and Lawson investigated both a round-robin assignment of the input data and map, and network partitioning across the elements in the processor array, for 10 000 iterations and maps containing 100 and 900 nodes. Openshaw and Turton demonstrate superlinear speedup on a Cray T3D [19] for an 8x8 map and a data set containing 145 791 samples. Finally, Schikuta and Weidman achieve near-linear speedup for four different implementations (distributed or replicated weights and data) on a 16-node Intel iPSC/860 hypercube [20]. Motivated in part by the advantages of larger map sizes, Silva and Marques [22] implement a combination of both partitioning approaches for a 3-cluster machine built on Fast Ethernet, achieving a speedup of 1.27 for 30 162 records with 15 attributes. Yang and Ahuja [23] report a speedup of 3.14 for a cluster of four processors. Lawrence et *al.* [16] evaluate the performance of a parallel batch-partitioning SOM (using different implementations of network and data partitioning) on a 16-node SP2, for map sizes of up to 64 nodes and three data sets. Finally, in the context of preserving data privacy, Gorgonio and Costa[9] implement map partitions used with local data, combined with fusion of the partial results at a central site. To our best knowledge, no implementations exist for Self-Organizing Maps on a CBEA cluster.

## 3. Algorithms and Experimental Setup

In order to evaluate the opportunities and challenges for the performance of the Self-Organizing Maps on the CBEA, we implemented the following SOM variants. All versions use a data-partitioning approach which requires copies of the weights to be distributed to each of the processing elements, rather than copies of the data which would limit scalability severely for larger data sets, particularly in view of the small local store for the SPEs. In all algorithms, updates to the weights are computed and stored for the winning node for each input datum, and used to update the map once per iteration after complete passes through the input data.

The algorithms below differ with respect to whether single or multiple Power Processing Elements and Synergistic Processing Elements are used. Beginning with a sequential version for comparison, we then included the Synergistic Processing Elements of a single Cell processor, followed by a parallel version using only the PPEs. Finally, we combined multiple PPEs and SPEs, which required two levels of synchronization. Where appropriate, we used double buffering to overlap computation and communication.

### *3.1. Algorithm 1: Sequential version using the PPE only*

This version is implemented as baseline for comparison. Algorithm 1 shows a high-level view of this algorithm, which is identical to Kohonen's batch learning approach.

**Algorithm 1** Batch-Learning Self-Organizing Map

Initialize the map
**while** Termination criteria not satisfied **do**
  **for all** Input data **do**
    Determine the winning node for the input datum
    Calculate and store update to weights representing winning node
  **end for**
  Update map
**end while**

### *3.2. Algorithm 2: Data parallel version using multiple PPEs*

The data set is scattered over the different PPEs which communicate with each other via MPI. Initially, the root scatters the input data to the other processor and broadcasts the initial weights. At each iteration, processors calculate updates which are then reduced at the root. The root then updates and broadcasts the weight matrix at each iteration. Disregarding the initial scatter operation that only occurs once and therefore is amortized over an increasing number of iterations, the need to compute the final updates at each iteration through the root represents the main serial component in this implementation. Again, this component will increase with the number of processing elements and the size of the weight map, but not with the size of the input data, and synchronization occurs via reductions using MPI.

### *3.3. Algorithm 3: Data-parallel version using multiple SPEs of a single Cell processor*

We distribute the distance calculation to the SPEs, while the PPE is responsible for calculation of the overall weight updates at each iteration. Each SPE receives a copy of the weights, but is responsible for calculating winners and updates for parts of the input data. At each iteration

**Algorithm 2** Distributing the distance calculation over multiple PPEs

Initialize the map at the root
Broadcast map to the PPEs
Scatter data to PPEs
**while** Termination criteria not satisfied **do**
  **for all** PPEs **do**
    **for all** Assigned data **do**
      Determine the winning node for the input datum
      Calculate and store updates to weights representing winning node
    **end for**
  **end for**
  Reduce map updates at PPE
  Broadcast updated map to each PPE
**end while**

through the input data, the PPE has to wait for all SPEs to complete the pass through the assigned data, followed by an idle period for the SPEs while they wait for the PPE to compute the overall updates for the weights. The need to compute the final updates at each iteration through the PPE constitutes the main serial component in this implementation, which will increase with the number of SPEs and the size of the weight map, but not with the size of the input data. In this case, synchronization between PPE and SPEs is implemented using mailboxes.

**Algorithm 3** Distributing the distance calculation over multiple SPEs

Initialize the map at the PPE
Send copies of the map to each SPE
Send starting address of assigned data to each SPEs
**while** Termination criteria not satisfied **do**
  **for all** SPEs **do**
    **for all** Assigned data **do**
      Determine the winning node for the input datum
      Calculate and store updates to weights representing winning node
    **end for**
    Send updates for map to PPE
  **end for**
  Calculate map updates at PPE
  Send copies of the map to each SPE
**end while**

### *3.4. Algorithm 4: Data parallel version using multiple PPEs and SPEs*

With the goal of exploring the limitations to using the full computational power of the Synergistic Processing Elements in each of the processors, we combine the previous two variants. Synchronization now requires two levels: communication between SPEs and PPE to communicate partial updates calculated at the SPEs to the PPE which will then calculate final updates for the data allocated to a single CBEA. This is then followed by global synchronization using MPI in order to communicate the updates to the root, at which the updates to the map will be reduced. After the updates are applied to the map by the root, the new weights are broadcast to each processor. This process is repeated for each iteration through the input data.

**Algorithm 4** Distributing the distance calculation over multiple PPEs and SPEs

```
Initialize the map at the root
Broadcast map to the PPEs
Send copies of the map to each SPE
Scatter data to PPEs
Send starting address of assigned data to each SPEs
while Termination criteria not satisfied do
  for all SPEs do
    for all Assigned data do
      Determine the winning node for the input datum
      Calculate and store updates to weights representing winning node
    end for
    Send updates for map to PPE
  end for
  Calculate updates at the PPEs
  Reduce map updates at root and calculate new weights
  Broadcast updated map to each PPE
  Send copies of the map to each SPE
end while
```

*3.5. Experimental Setup*

We investigated the scalability of our approaches on two fundamentally different CBEA clusters: a small local cluster, and a 44-node SHARCNET system. The local cluster consists of five PS3$^{TM}$consoles, one of which is used as a login node only. The consoles are connected via a Fast Ethernet Switch (EE2400-SS), resulting in a dedicated 100Mb/s Ethernet. The PS3$^{TM}$cluster uses OpenMPI as the message-passing interface, combined with Torque as the resource manager, and Fedora 9. The SHARCNET cluster *prickly* consists of a set of varying nodes, which includes 8 nodes of 2 PowerXCell 8i processors each, in addition to one dual-socket dual-core Xeon processor, and three dual-socket quad-core Xeon processors. Running RedHat Enterprise Linux 5.2, these nodes are connected via Gigabit Ethernet.

We developed the applications using the IBM CELL Broadband Engine Software Development Kit (SDK) for Multicore Acceleration, Version 3.0. For both clusters, we evaluate our implementations with respect to the size of both the (numerical-valued) data set and the map, the number of iterations (which is generally on the order of a few hundred for real datasets) through the data, the number of Cell processors, and the number of Synergistic Processing Elements. Typically, the data used for data-mining is a by-product of some other collection process, and therefore often quite noisy and biased. For most data-mining algorithms, single-precision computation, as implemented here, is therefore sufficient. All results shown below are averaged over five runs.

## 4. Results and Discussion

We evaluated each of the algorithms introduced above on both clusters. The results are organized as follows. First, we present the evaluation for the use of the PPEs only in Section 4.1, while disregarding the computational power of the SPEs in order to determine the effects of the network for both clusters. Section 4.2 then introduces the evaluation of the implementation

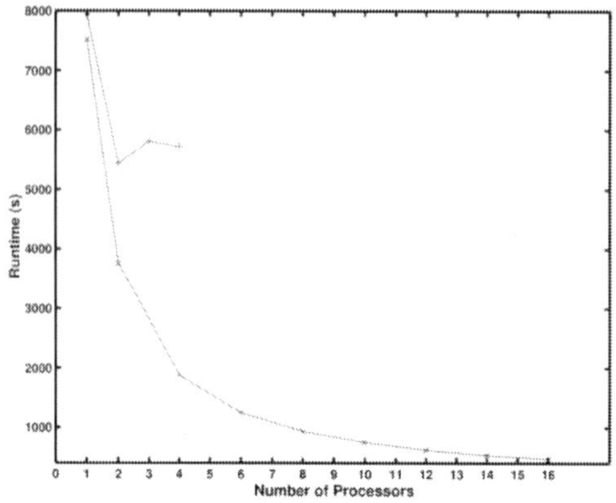


**Figure 1.** Run times (in seconds) for the parallel algorithm using varying numbers of PPEs, no SPEs. Shown are the results for the PS3$^{TM}$cluster for 1 to 4 processors (+), and 1-16 PPEs on the PowerXCell 8i cluster (∗), for 500 iterations, 163 840 input values, 4 attributes, and a map of size 16x16.

for a single Cell processor but multiple SPEs. Finally, we discuss the results achieved when combining multiple Cell processors with multiple SPEs in Section 4.3.

### *4.1. Scalability without Synergistic Processing Elements*

In order to evaluate the effects of the networks, we implemented a version of the SOM that carried out all computation on the PPEs themselves, without involvement of the SPEs (Algorithms 1 and 2). Figure 1 shows the run times for various numbers of processors on both clusters. From the data shown in the figure we see that the speedup (sequential execution divided by the parallel execution time) scaled linearly with the number of processors used for the SHARCNET cluster. For this cluster, the resulting efficiency (speedup divided by the number of processors used) was close to 1 throughout, for example, 0.99 for 16 processors. In contrast, for the smaller cluster the speedup achieved for 2 processors was close to 1.5, but then dropped when the number of processors was increased. In this case, the run time increased and efficiency dropped by half when going from 2 to 4 processors, which clearly shows the restrictions associated with our low-cost cluster.

Reducing the number of iterations or doubling the sample size while holding the map size constant resulted in the same trend for the speedup and efficiency, for both clusters, with almost identical values. For example, our implementations achieved an efficiency of 0.98 for the SHARCNET cluster for 100 iterations, which indicates that the startup cost is amortized early. With respect to a reduction in map size, the efficiency of the implementation on the SHARCNET cluster dropped slightly to .94 for 16 processors. Overall, the implementation scaled well with increasing map sizes and iterations for the SHARCNET cluster, but not for the console cluster. The poor scalability of the implementations on the local cluster was not only rooted in the slower network, but also in the substantial additional overhead to access the network as implemented in the consoles themselves, and shows the limitations of low-cost clusters for applications requiring a fair amount of synchronization.

### *4.2. Scalability with Synergistic Processing Elements for Single Cell Processor*

The performance of both incarnations of the Cell processor was then evaluated using only one PPE but multiple SPEs (Algorithm 3), with the goal of determining whether any performance gains are available from the PowerXCell 8i processor (double-precision performance set aside, which is not an issue for this particular data-mining technique). Figure 2 shows the results for increasing number of SPEs for four different combinations of map sizes and sample sizes, for both incarnations of the Cell processor.

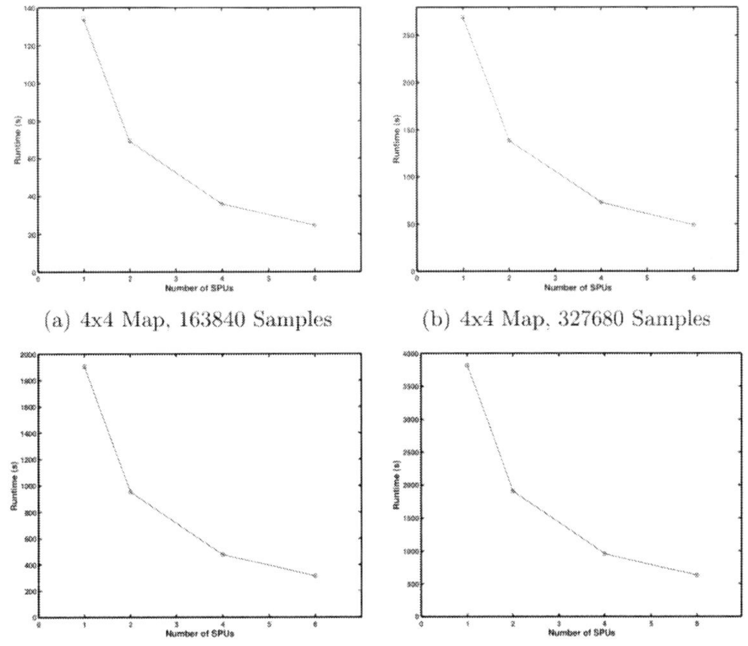


(a) 4x4 Map, 163840 Samples (b) 4x4 Map, 327680 Samples

(c) 16x16 Map, 163840 Samples (d) 16x16 Map, 327680 Samples

**Figure 2.** Run times (in seconds) for both incarnations of a single Cell Processor, using multiple SPEs, for 500 iterations and 4 attributes. Shown are the time for the PS3$^{TM}$(+), and the PowerXCell 8i (□).

The results for both incarnations are almost identical, indicating that the subtle differences in architecture for the two different incarnations do not have an effect for our implementations when adding the SPEs. Overall, the experimentally determined serial fraction is small, and the efficiency is close to 1 with a minimum of 91%. A comparison of the values for single processors depicted in Figures 1 and 2d) shows that the speedup achieved for the use of a single SPE (1911s for console) as opposed to the PPE only (7516s for console) is roughly 4 for both incarnations (sample size 163 840 samples, 4 attributes, 16x16 map, 500 iterations).

Table 1 shows the speedup for the console, for varying map sizes and numbers of SPEs, when increasing the number of SPEs and relative to the use of a single SPE for the Euclidean distance calculation. In general, the achieved speedup tends to be slightly larger for larger map sizes and data sets, for which the initial overhead of mailbox communication with the SPEs (starting addresses of data and weights for DMA transfers, plus parameter settings) is better amortized. Overall, the speedup scales linear with the number of SPEs when using a single Cell processor.

### *4.3. Scalability with Synergistic Processing Elements for Multiple Cell Processors*

Following the implementations utilizing either multiple PPEs without SPEs, or multiple SPEs of a single Cell processor, we then combined both approaches. Our evaluation included the scalability for an increasing number of processors for a given number of SPEs, and the scalability

**Table 1.** Speedup for Increasing Map Size and Numbers of SPEs (4 attributes, 500 iterations, 163 840 Samples)

| *Map* | *2 SPEs* | *3 SPEs* | *4 SPEs* | *5 SPEs* | *6 SPEs* |
|---|---|---|---|---|---|
| 4x4 | 1.96 | 2.95 | 3.75 | 4.56 | 5.52 |
| 16x16 | 2.00 | 3.07 | 3.98 | 4.97 | 6.11 |
| 32x32 | 2.00 | 3.07 | 4.00 | 4.99 | 6.14 |

for an increasing number of SPEs for a given number of processors. Figure 3 shows the run times for both clusters as the numbers of SPEs and processors are increased, initially for up to four processors only for comparison purposes (run times and speedup for larger numbers of processors for the SHARCNET cluster will be discussed below). As can be seen from the data shown in the

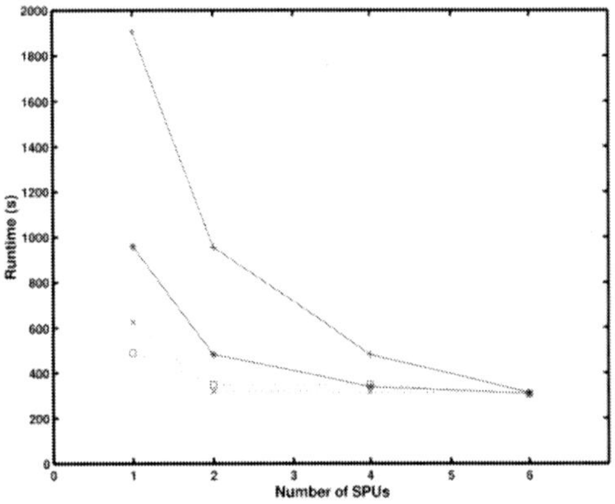


(a) Ps3$^{TM}$Cluster

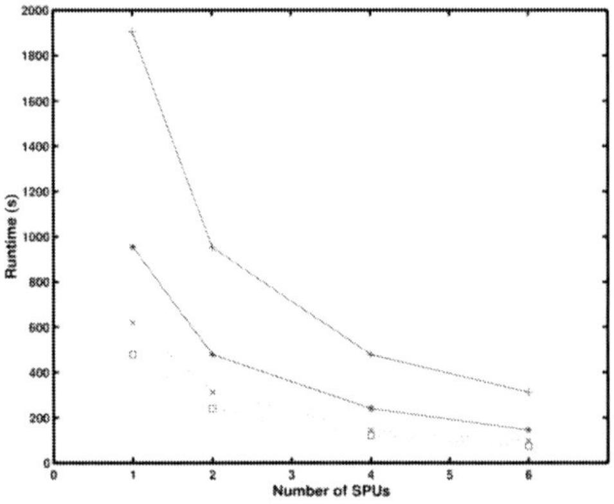


(b) PowerXCell 8i Cluster

**Figure 3.** Run times (seconds) for multiple PPEs and SPEs on the local cluster (left) and for up to 4 processors on the SHARCNET cluster (right). Shown are the results for one (+), two (∗), three (x), and four processors (□), for 163 840 samples, 16x16 map, and 500 iterations.

figure, the run times for the local cluster only level off for more than two processors, regardless of how many SPEs are used, which is caused by the slow network access and slower network. This effect is also illustrated in Figure 4, which shows the speedups for increasing number of processors and various numbers of SPE on the local cluster only.

The speedup for 1 processor scales linearly with the number of SPEs. For two processors, the speedup drops when adding the fourth PPE, but scales linear up to this point. Performance for three and four processors is poor, with no improvement to the run times for 4 processors when adding SPEs.

Figure 5 shows the run times for increasing number of samples on the local cluster, and illustrates the reason for the poor performance of the local cluster. The SPEs tend to spend a large fraction of the code in synchronization or waiting for data: 59%, 55%, and 54% are spend in the vectorized portion of the code, for 163 840, 327 680, and 491 520 samples, respectively.

In conclusion, the local cluster shows severely limited scalability for more than 2 SPEs and processors because of the poor utilization of the SPEs.

In contrast to the effect observed on the local cluster, the fraction of vectorized time compared

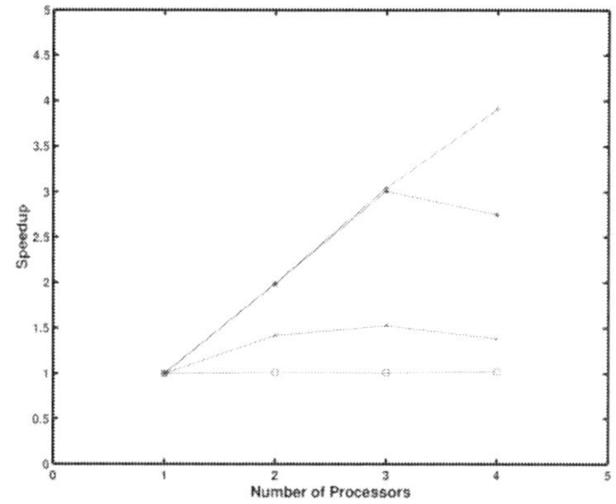


**Figure 4.** Speedups for the PS3$^{TM}$cluster, using multiple PPEs and SPEs (163 840 samples, 16x16 map, 500 iterations). Shown are the results for one (+), two (∗), four (x), and six (□) SPEs.

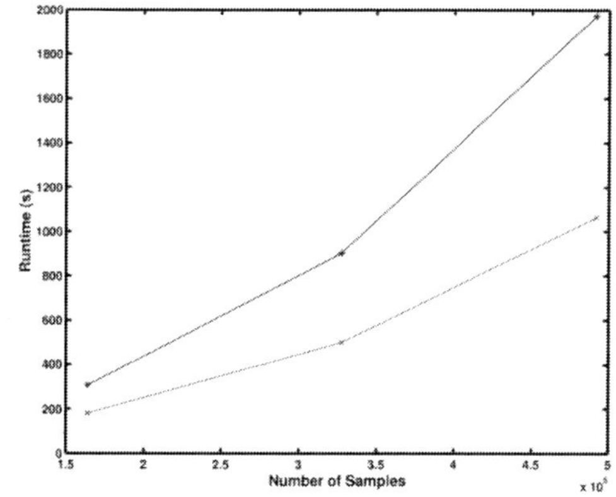


**Figure 5.** Run times on the PS3$^{TM}$cluster (four PPEs, six SPEs, 163 840 to 491 520 samples, four attributes, 16x16 map, and 500 iterations). Shown are the total time for the PPE (+), the total time for the SPE (∗), and the time for the vectorized portion of the code for the SPE (x).

to overall runtime does not drop as dramatically for the SHARCNET cluster: 98% for one processor, which steadily decreases to 82% for 16 processor. Figure 6 and Table 2 present these results from slightly different perspectives.

Figure 6 shows the speedups for increasing number of processors and various numbers of SPEs on the SHARCNET cluster. In this case, the speedup refers to the decrease in runtime as the number of processors is increased and the number of SPEs is held constant. The data shown in the figure indicates that for a for smaller number of processors, the speedup is roughly linear with the number of SPEs. For any given number of utilized SPEs, the speedup levels off for some number of processors. The higher the number of SPEs, the earlier the peak speedup is reached (for example, the maximum speedup occurs for 7 processors for 6 SPEs, but for 13 processors for one SPE). The maximum speedup of 18 occurred for two SPEs and 14 processors.

Table 2 shows the speedups achieved on the SHARCNET cluster, for varying numbers of SPEs and Cell processors, but now with reference to the total number of processing elements. Starting with a single PPE and SPE, linear speedup would result in a value of 6 in the table when

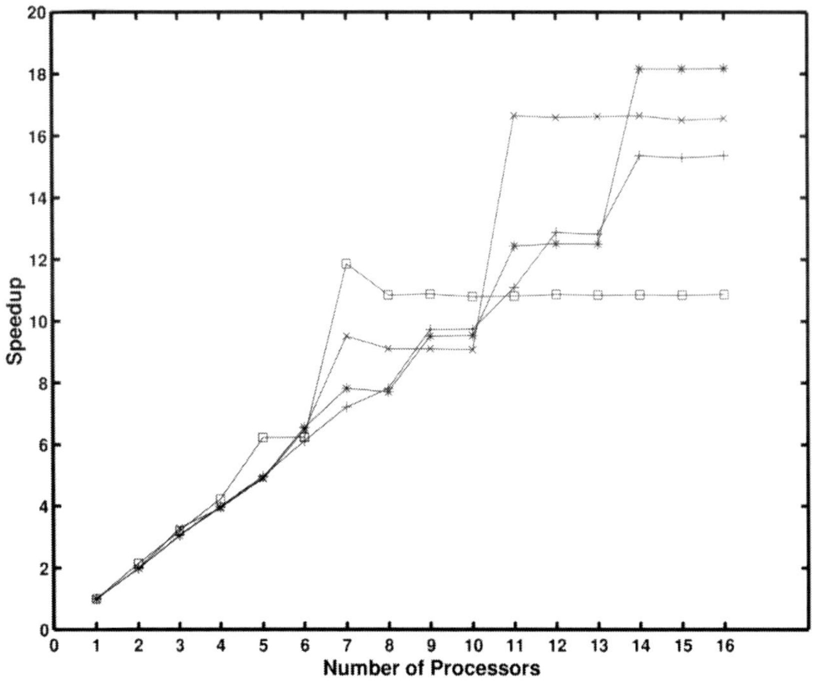


**Figure 6.** Speedup for implementation using multiple PPEs and SPEs on the PowerXCell 8i cluster (163 840 samples, four attributes, 16x16 map, 500 iterations). Shown are the results for one (+), two (∗), four (x), and six (□) SPEs.

fixing the number of PPE (across a row), a value of 16 when fixing the number of SPEs (down a column), and a value of 96 for the bottom right entry using 6 SPEs in each of the 16 PPEs. The data shown in the table indicates that, for a given number of processors, the speedup grows linear with the number of SPEs. The maximum speedup is reached earlier for larger numbers of SPEs, and roughly equal numbers of Cell processors and SPEs achieve the highest efficiencies close to 1. The maximum speedup is 71%, in the case of 7 processors and 6 SPEs, the minimum efficiency of 67%is reached for the maximum number of processing elements (96 SPEs in total). While the slow network characteristics restricted the scalability of our implementation on the PS3$^{\mathrm{TM}}$cluster, we were able to use up to 60 processing elements with an efficiency close to 1 on the SHARCNET cluster.

## 5. Conclusions and Future Work

We evaluated the scalability of various SOM implementations on two clusters employing different networks and incarnations of the Cell processor. The performance of the implementation for a single Cell processor using multiple SPEs was comparable for both incarnations, and scaled well with respect to the size of the data set, the number of iterations and the size of the map. The slower network and access times for the PS3$^{\mathrm{TM}}$cluster resulted in a lower fraction of time spent in vectorized code for the SPEs when multiple Cell processors were used, even without the use of the SPEs. This demonstrates that the use of such a low-cost cluster may be severely restricted. When using multiple Cell processors in combination with multiple SPEs for the PowerXCell 8i cluster, the speedup grows linearly with increasing number of SPEs for a given number of processors. For a given number of SPEs, the speedup initially grows linear with the number of

**Table 2.** Speedup for Increasing Numbers of Processors and SPEs (16x16 map, 4 attributes, 500 iterations, 163 840 Samples)

| *Processors* | *1 SPE* | *2 SPEs* | *4 SPEs* | *6 SPEs* |
|---|---|---|---|---|
| 2 | 1.99 | 3.99 | 7.96 | 12.88 |
| 3 | 3.07 | 6.11 | 13.15 | 19.18 |
| 4 | 3.99 | 7.92 | 15.76 | 25.37 |
| 5 | 4.98 | 9.88 | 19.59 | 37.37 |
| 6 | 6.12 | 13.11 | 25.89 | 37.46 |
| 7 | 7.22 | 15.64 | 38.10 | 71.14 |
| 8 | 7.82 | 15.41 | 36.38 | 65.12 |
| 9 | 9.74 | 19.04 | 36.38 | 65.25 |
| 10 | 9.75 | 19.08 | 36.26 | 64.82 |
| 11 | 11.10 | 24.88 | 66.60 | 64.92 |
| 12 | 12.87 | 25.02 | 66.34 | 65.20 |
| 13 | 12.82 | 24.98 | 66.48 | 65.12 |
| 14 | 15.36 | 36.33 | 66.59 | 65.15 |
| 15 | 15.30 | 36.32 | 66.04 | 65.10 |
| 16 | 15.36 | 36.35 | 66.24 | 65.20 |

processors, but then levels off.

Our implementation achieves an efficiency of 67% for the maximum number of processing elements involved in the computation (16 processors, with 6 SPEs), but consistently close to an efficiency of 1 for smaller numbers of PPEs and roughly equal numbers of PPEs and SPEs. Despite the current uncertainty of the future of the Cell architecture, it is likely to survive in some scaled-down version. Therefore, it might be worthwhile to consider the implementation of additional techniques as long as the clusters are available, given our successful implementation on a single Cell processor, and the scalability of the approach on the SHARCNET cluster.

**Acknowledgments**

This work was made possible by a Fellowship Award from the Shared Hierarchical Academic Research Computing Network (SHARCNET:www.sharcnet.ca).

[1] Buehrer, G., Parthasarathy, S., Goyder, M.: Data Mining on the Cell Broadband Engine. In: Proceedings of the 22nd Annual International Conference on Supercomputing, pp.26–35. ACM, New York, NY, USA (2008)

[2] Benthin, C., Wald, I., Scherbaum, M./ Friedrich, H.: Ray Tracing on the Cell Processor. IEEE Symposium on Interactive Ray Tracing, Salt Lake City, UT (2006)

[3] Bieroza, M., Baker, A., and Bridgeman, J.: Exploratory Analysis of Excitation-Emission Matrix Fluorescence Spectra with Self-Organizing Maps as a Basis for Determination of Organic Matter Removal Efficiency at Water Treatment Works. Journal of Geophysical Research, Vol. 114 (5) (2009).

[4] Chow, A., Fossum, G., Brokenshire, D.: A Programming Example: Large FFT on the Cell Broadband Engine, IBM White Paper (2005)

[5] DeFabritiis, G.: Performance of the Cell Processor for Biomolecular Simulations. Computer Physics Communications, Vol. 176(11-12), pp. 660–664 (2007)

[6] Demian, V. and Mignot, J.-C.: Implementation of the Self-Organizing Feature Map on Parallel Computers. Parallel Processing: CONPAR 92, Vol. 634, pp. 775-776 (1992)

[7] Duan, R., Strey, A.: Data Mining Algorithms on the Cell Broadband Engine. In: Proceedings of the 14th International Euro-Par Conference on Parallel Processing, pp. 653–675. Springer, Berlin/Heidelberg (2008)

[8] Filippi, E. and Lawson, J. C.: A Parallel Implementation of Kohonen's Self-Organizing Maps on the SMART

Journal of Physics: Conference Series **256** (2010) 012013 doi:10.1088/1742-6596/256/1/012013

Neurocomputer. LNCS: New Trends in Neural Computation, Springer Berlin / Heidelberg, pp. 388-393 (1993)
[9] Gorgonio, F. L. and Costa, J.: Parallel Self-Organizing Maps with Application in Clustering Distributed Data. IEEE International Joint Conference Neural Networks (IJCNN) pp. 3276-3283 (2008)
[10] Honkanen, H, Liuti, S.: New Angle on the Parton Distribution Functions: Self-Organizing Maps. 18th International Spin Physics Symposium. AIP Conference Proceedings Vol. 1149, pp. 293-296 (2009)
[11] Hsu, K.-C., Li, S.-T.: Clustering Spatial Temporal Precipitation Data Using Wavelet Transform and Self-Organizing Map Neural Network, Advances in Water Resources, Vol. 33 (2), pp.190-200.
[12] Huntsberger, T. and Ajjimarangsee, P.: Parallel Self-Organizing Feature Maps for Unsupervised Pattern Recognition. International Journal of General Systems, Vol. 16(4), pp. 357-372 (1990)
[13] Kohonen, T.: The Self-Organizing Map. Proceedings of the IEEE, Vol. 78(9), pp. 1551-1558.
[14] Kurzak, J., Alvaro, W., Dongarra, J.: Optimizing Matrix Multiplication for a Short-Vector SIMD Architecture - CELL processor. Parallel Computing 35, pp.138–150 (2009)
[15] Kwiatkowski, J. et al.:Performance Evaluation Of Different Kohonen Network Parallelization Techniques. Proceedings of the International Symposium on Parallel Computing in Electrical Engineering (2006)
[16] Lawrence, R. D., Almasi, G. S., Rushmeier, H. E.: A Scalable Parallel Algorithm for Self-Organizing Maps with Applications to Sparse Data Mining Problems. Data Mining and Knowledge Discovery, Vol. 3, pp. 171-195 (1999).
[17] Torniainen, I. et al.: Cluster Analyses of Gigahertz-Peaked Spectrum Sources with Self-Organising Maps. Astronomy and Astrophysics, Vol. 48(2), pp. 483-498 (2008)
[18] McConnell, S., Patton, D., Hurley, R., Blight, W., Young, G.:Interactive Data Mining on a CBEA Cluster, LNCS: High Performance Computing Systems and Applications, Vol. 976, pp. 269-282 (2010)
[19] Openshaw, S. and Turton, I.: A Parallel Kohonen Algorithm for the Classification of Large Spatial data sets. Computers & Geosciences. Vol. 22(9), pp. 1019-1026 (1996)
[20] Schikuta, E. and Weidmann, C.: Data Parallel Simulation of Self-Organizing Maps on Hypercube Architectures, Springer Studies In Fuzziness And Soft Computing Series, Self-Organizing Neural Networks: Recent Advances and Applications, pp. 245-278 (2001)
[21] Strupl, D. and Neruda, R.: Parallelizing Self-Organizing Maps. LNCS:SOFSEM'97: Theory and Practice of Informatics, Vol. 1338, pp. 563-570 (1997)
[22] Silva. B. and Marques, N.: A Hybrid Parallel SOM Algorithm for Large Maps in Data-Mining. Proceedings of the 13th Portugese Conference on Artificial Intelligence (EPIA07), Workshop on Business Intelligence, Portugal (2007)
[23] Yang, M.-H. and Ahuja, N.: Data Partition Method for Parallel Self-Organizing Map. International Joint Conference on Neural Networks (IJCNN'99), Washington, DC, pp. 1929-1933 (1999)
[24] Wang, H., Takizawa, H., Kobayashi, H.: A Performance Study of Secure Data Mining on the Cell Processor. In: International Symposium on Cluster Computing and the Grid, pp.633-638 (2008)
[25] Wyganowski, M.: Classification Algorithms on the Cell Processor. MSc. Thesis, Department of Computer Engineering, Rochester Institute of Technology, Rochester, NY (2008)

# CUDA-accelerated genetic feedforward-ANN training for data mining

**Catalin Patulea, Robert Peace and James Green**

School of Systems and Computer Engineering, Carleton University, Ottawa, Canada K1S 5B6

E-mail: cpatulea@sce.carleton.ca, rpeace@sce.carleton.ca, jrgreen@sce.carleton.ca

**Abstract.** We present an implementation of genetic algorithm (GA) training of feedforward artificial neural networks (ANNs) targeting commodity graphics cards (GPUs). By carefully mapping the problem onto the unique GPU architecture, we achieve order-of-magnitude speedup over a conventional CPU implementation. Furthermore, we show that the speedup is consistent across a wide range of data set sizes, making this implementation ideal for large data sets. This performance boost enables the genetic algorithm to search a larger subset of the solution space, which results in more accurate pattern classification. Finally, we demonstrate this method in the context of the 2009 UC San Diego Data Mining Contest, achieving a world-class lift on a data set of 94682 e-commerce transactions.

## 1. Introduction

Genetic algorithms (GAs) are a stochastic, evolutionary approach to machine learning for pattern classification. While greedy methods will get stuck at local extrema, GAs are theoretically capable of asymptotically reaching the global optimum [1]. However, because they require the evaluation of approximately 50 to 100 classifiers at each iteration, and the algorithm is repeated for several thousand iterations, they are extremely compute-intensive. The use of artificial neural networks (ANNs) as the classifier exacerbates the computational requirements because ANNs themselves are very compute-intensive.

Compared to a traditional sequential implementation of GA training of ANNs, the use of graphical processing units (GPUs) as parallel processors provides significant performance improvements at a fraction of the price of alternatives such as cluster computing. GPUs, however, use a specialized programming paradigm which must be taken into account to leverage their full processing power. In this report, we show an implementation of GA training of ANNs which achieves an order-of-magnitude speedup over a sequential algorithm.

## 2. Background

### *2.1. Artificial Neural Networks*

ANNs are composed of nodes connected by weighted directed edges. In a feed-forward network architecture, nodes are organized into an input layer, one or more hidden layers, and an output layer. Each layer is fully connected to the next. Figure 1 demonstrates the feedfoward ANN structure. Each of the hidden nodes at hidden layer $n$ perform a transcendental function with inputs equal to the outputs of layer $n-1$ each multiplied by the weight associated with the edge through which they are fed from layer $n-1$ to layer $n$. In a radial basis function (RBF) network, the nodes in each hidden layer calculate the distance between their inputs and a centre vector, and pass a weighted distance to a RBF. Nodes in a sigmoid function network perform a similar operation with a sigmoid function in place of the RBF. The output node is typically linear, computing a weighted sum of its inputs. The output of the ANN is a single real value which is interpreted as a classification confidence. ANN computation is dominated by the number of multiplications at the directed edges, proportional to the number of features in the data set and the number of hidden nodes, and by the transcendental functions at the hidden nodes, which are typically exponentiations. ANN training has been implemented previously on GPU hardware, however these efforts have focused on training through backpropagation [2] as opposed to genetic algorithms.

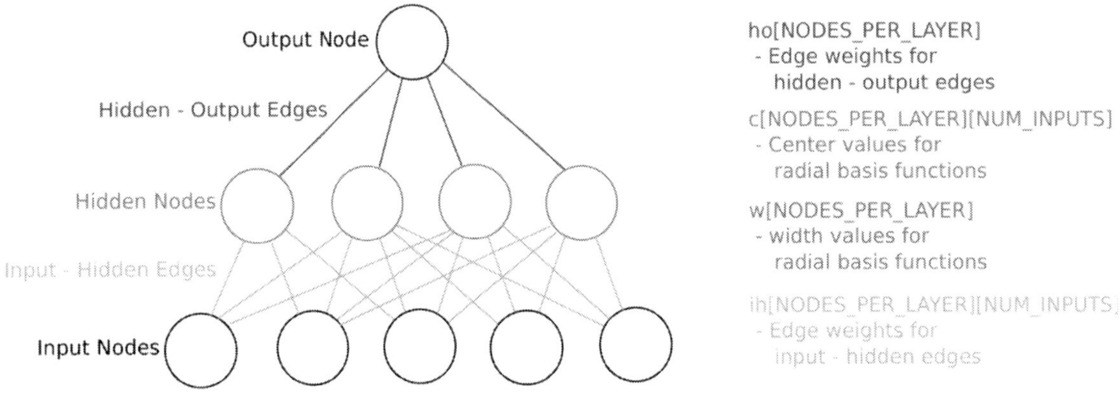


**Figure 1.** ANN structure and related chromosome design. Each element in the chromosome is a real-valued number.

### *2.2. Genetic Algorithms as ANN Training Agents*

Genetic algorithms (GAs) use populations of candidate solutions to simultaneously explore the search space of ANN parameters [1]. The candidate ANNs are randomly mutated, mated and selected during each of several generations. Mutation and mating is done using a problem-specific representation of candidate solutions and "genetic operators." Selection requires calculating the fitness of each candidate ANN then preferentially selecting candidate ANNs based on fitness. The fitness of a classifier is a classifier accuracy metric such as sensitivity. The competitive bias imposed by selection attempts to mimic the "survival of the fittest" principle often seen in nature. Because GAs perform an unbiased search of a solution space and ANNs can be of arbitrary complexity, this technique is suitable for data mining applications using large or complex data sets.

The computation required for GA training of ANNs is proportional to the number of generations ($10^1$-$10^4$), to the size of each generation ($10^1$-$10^2$ candidates) and to the size of the data set ($10^2$-$10^6$ instances).

### *2.3. The CUDA Platform*

NVidia's Compute Unified Device Architecture (CUDA) is a programming platform for massively parallel GPUs found in off-the-shelf graphics cards. GPUs consist of several dozen independent floating-point units, providing significant speedup to data-parallel compute-intensive applications [3]. Each unit is connected to on-board memory using a very wide bus, enabling high memory bandwidth provided certain memory access rules are respected. These features make CUDA an ideal platform for GA-ANN training.

### *2.4. Sample Application*
Our demonstration classifier is designed for the 2009 UC San Diego Data Mining Contest "E-commerce Transaction Anomaly Classification" [4] and with training data thereof. The training data consist of 94682 instances with 19 features of mixed types and a 1:50 binary class imbalance. The evaluation metric is "lift at 20%", which can be understood as the ratio of the true positive rate in the top 20% ranked instances to the overall positive rate of the data set. Lift at 20% is commonly used in marketing research in order to select the subset of a population which is most likely to respond to solicitation.

## 3. Classifier Design
### *3.1. ANN Structure Design*
While our GA-ANN classifier was designed for the 2009 UC San Diego Data Mining Contest data set, it has been designed in order to be capable of adapting to a variety of data sets. We have developed a flexible ANN topology, allowing for either one or two layers of hidden nodes and an arbitrary number of hidden nodes per layer (within the limits of GPU memory). In addition, each layer of the ANN is capable of performing RBF or sigmoid calculations.

The classifier receives as inputs real-valued features which are standardized prior to processing by the ANN. The classifier outputs a single real value, in the range of 0 and 1, which represents the confidence with which the pattern is a member of the positive class. These confidence values can be used for ranking during lift calculations, or for sensitivity or specificity measures with an arbitrary value chosen as a threshold for positive identification of a pattern. Therefore, the fitness function is not limited to lift at 20% as presented in this report.

### *3.2. Genetic Algorithm Design*
For a single-layer ANN, as used in our final design, each candidate ANN is represented by a chromosome which consists of four arrays of real-valued numbers. Edge weights between input nodes and hidden layers, edge weights between hidden and output layers, width values for RBFs at hidden nodes and center values for RBFs at hidden nodes are all encoded, and can take on any real value. Figure 1 demonstrates the chromosome structure and its relation to ANN structure. This structure is simple yet powerful; individual edges are 'removed' by the genetic algorithm when their weight is set to zero and nodes are 'removed' by the genetic algorithm when their RBF width value is set to zero. Thus, while the topology of the ANN is limited to those with $m$ layers and $n$ nodes per layer, the genetic algorithm searches all topologies with $m$ layers and between 0 and $n$ nodes per layer.

The genetic mutation and crossover operations which we have implemented are based on the methods defined by Montana and Lawrence [1]. Of the operations presented in this paper, the MUTATE NODES and CROSSOVER WEIGHTS operations were determined to provide an optimal balance between performance and classifier accuracy, and have been implemented in all of our experiments.

### *3.3. Data Preprocessing Steps*
In order to handle nominal feature values in data sets, we have implemented an algorithm which, given a tab separated values (TSV) file, converts nominal feature values in the data

set into numeric values. The algorithm replaces each nominal value in the data set with the logarithm of the probability ratio between the positive and negative classifications observed given the nominal value. This is inspired from the discriminant function of a Bayesian classifier trained only on that particular feature. As a result, meaningful numeric values can be extracted from nominal values without domain knowledge relating to the values. This improves on the traditional orthogonal coding method [5, slide 17] by avoiding allocation of a disproportionate number of input nodes to nominal features. The logarithm of the ratio of all positive classifications to all negative classifications is assigned to nominal values which do not appear sufficiently in the data set and for values which appear only in blind test data sets; this prevents the classifier from assigning significant values to nominal values for which little or no inference can be done.

Due to the inherent limitations of floating-point (FP) representation of real values, the range of inputs to the ANN plays a crucial part in the numerical stability of the classifier during training and during classification. In particular, our use of exponential functions resulted in an ANN very sensitive to output saturation. In these cases, while the input to the `expf` function is well within the range of representible FP numbers the mathematical output of the exponentiation is too large to be represented in FP. This causes degenerate output values such as infinity or NaN (not a number). These special values propagate through the ANN edges to produce degenerate values at the output of the classifier.

To avoid this issue, we standardize each feature individually before training and before classification. One method of standardization is to use the sample mean and sample standard deviation as shifting and scaling parameters (referred to as "z-score normalization" in [6]). However, sample standard deviation is not robust against outliers, while interquartile range (IQR) does tolerate outliers [7]. We use a variant of IQR: we estimate the range between the $10^{th}$ and $90^{th}$ percentiles and use that as the scaling parameter. After shifting and scaling, 80% of the resulting feature values are in the range $(-1, 1)$. This range of inputs to `expf` is narrow enough to always produce a valid FP value at the output. The standardization does not affect the classification itself because there are scaling and shifting parameters within the ANN (input to hidden edge weights and hidden center parameters, respectively) which are already part of the search space.

## 4. Implementation of GA Training of ANNs

Our GA begins by generating an initial population of random candidate ANNs, where each chromosome in the population is a set of parameters describing one ANN. Feature values for all training instances are loaded into GPU global memory in preparation for ANN computations. The following steps are then executed in sequence for each generation of the GA:

(i) Compute the ANN output corresponding to each training instance, as calculated by each candidate ANN in the population.

(ii) Find the threshold that defines the outputs which fall in the top 20% for each candidate ANN's set of output values.

(iii) Compute the number of top 20% instances which are truly positive, again for each candidate ANN. This allows us to calculate lift, which is used as the fitness value for each candidate ANN.

(iv) Apply genetic operators: mating, mutation and selection. The selected candidate ANNs become the new population, to be used in the next generation.

Items 1, 2 and 3 were all implemented on the CUDA platform and are described in detail below. Performance results are given as the combined time for one generation of these three items but exclude one-time initialization. Item 4 is not very computationally intensive and was not parallelized.

### *4.1. Computation of ANN Output Values*

ANN output values are computed by reading the training instances as real-valued feature vectors, applying them to the input nodes of the network (see Figure 1), and storing the output of the network. Specifically, each feature value is first scaled by the corresponding input-hidden edges. These values are then propagated to all hidden nodes, which apply a radial basis function. Finally, the hidden node outputs are scaled by the hidden-output edges and summed at the output node. A single real value is output from the network and stored in GPU memory. Parameters for edge weights and radial basis functions are given in the chromosome of each candidate ANN.

Feature data are organized to efficiently use memory bandwidth. The requirement for optimal use of CUDA memory bandwidth is that blocks of threads simultaneously access sequential addresses in memory, resulting in a coalesced single-clock cycle access. We store feature data in feature-major order (values for all instances of feature 1, all instances of feature 2, etc.). Therefore, when the thread block reads data for the first feature across different instances, the memory accesses can be coalesced. Similarly, accesses are coalesced for all subsequent features.

ANN output calculation is particularly well-suited to the CUDA architecture because it has a very high ratio of mathematical operations to memory accesses (160 multiplications, 4 exponentiations and 80 bytes of memory accesses for our ANN topology). In addition, each output value depends only on the current ANN parameters and one training instance's feature values. Therefore, parallelization of this step scales particularly well with hardware capabilities.

### *4.2. Calculation of Top 20% Threshold*

Next, we must calculate the $k^{th}$ largest value in each candidate's list of output values, where $k = 0.20 * n$ and $n$ is the number of output values (number of training instances). We can accomplish this by iterating through all $n$ unsorted output values and inserting each into a binary minheap of fixed size $k$. The heap size is kept fixed by removing the root (smallest node) after each insertion. At the end of this process, the root of the heap is the $k^{th}$ largest output value. The insertion and removal can be combined into a single operation which costs $O(\log k)$ time. The heap occupies $4 * k$ bytes of device shared memory.

Because fast on-chip memory (16 KB) is not large enough for our value of $k$ ($4 * 18936 =$ 74 KB), we instead use $p$ (5) passes of $\frac{k}{p}$-sized heaps. Each pass examines only values which are less than the threshold calculated by the preceeding pass. For example, the third pass examines only values below the 8% threshold and finds the top 4% threshold of this subset of values. This threshold therefore is the 12% threshold of the full list. The threshold calculated by the last pass is the desired 20% threshold.

The overall complexity of our algorithm is $O(np \log \frac{k}{p})$. Each top 20% calculation is performed independently for each candidate ANN. Therefore, the parallel performance of this step scales with hardware capabilities when there are enough candidates to occupy the entire device.

### *4.3. Computation of Number of Truly Positive Top 20% Instances*

We wish to compute the number of truly positive instances in the top 20% of scores. Assuming the training instances are unsorted, we must read from memory each instance's true class and corresponding ANN output value and compare the output value to the candidate's threshold. For $n$ training instances, this requires $2 * n$ memory accesses. However, by pre-sorting the training data such that the $n_p$ positive instances appear first in memory, we need only read the first $n_p$ ANN outputs, knowing that they correspond to the positive instances. This reduces the number of memory accesses to $n_p$, which is significant particularly for training data with low positive rate (in our case, $2 * n = 189364$ while $n_p = 2080$). Sorting the training instances by class is a cheap operation which is performed only once during initialization and can be done by the host CPU.

Counting of positive instances above the candidate's threshold is performed independently, in parallel, for each candidate ANN. However, because there may not be enough candidates to fully occupy the GPU, we also split the counting process for each candidate into 256 independent sub-counts: the first sub-counter processes training instances 0, 256, 512, ..., the second sub-counter instances 1, 257, 513, ..., such that the sub-counts can be executed in parallel. The striped assignment of training instance indices to sub-counters is necessary to take advantage of the full memory bandwidth of the GPU. The host CPU then aggregates the sub-counts by summing, which is a very lightweight operation.

Finally, each count is divided by the number of positive instances to give a positive rate in the top 20%. Then, the ratio of this positive rate to the overall positive rate is the lift of each candidate. These operations are also comparatively lightweight and executed on the host CPU.

## 5. Results

We first present the computational performance of our algorithm for classifier training, followed by the estimated classification performance of the classifier designed for the UC San Diego data mining contest.

### 5.1. *Experimental Setup*

To evaluate the performance of our system, we compared execution time of a sequential algorithm on a commodity CPU and an optimized algorithm on commodity GPUs. The goal was to estimate the speedup of our algorithm with respect to sequential without using the CPU's multiprocessing capabilities. Each series represents one type of experimental hardware (Table 1). Each data point is an average over 10 runs of the algorithm with the same parameters.

**Table 1.** Experimental hardware. Free RAM is OS-reported available memory before running each experiment.

| Legend Label | Hardware | Clock | Free RAM | OS |
|---|---|---|---|---|
| x86 | Intel Core 2 Q9450 CPU | 2.66 GHz | 7.5 GB | Linux 2.6 (64-bit) |
| GTX260 | NVidia GTX260 GPU | 1.24 GHz | N/A | Windows 7 (64-bit) |
| GTX275 | NVidia GTX275 GPU | 1.4 GHz | N/A | Linux 2.6 (64-bit) |

### 5.2. *Performance Results*

In most cases, the GPU implementation resulted in approximately an order of magnitude speedup over a serial implementation of equivalent code running on a modern x86 desktop processor (Figure 2). For population sizes below 10, observed speedup decreases and no speedup is observed in the single-candidate case. This is due to GPU kernel invocation overhead and data copying dominating execution time. However, in these cases, both implementations are still quite fast (20 ms per generation).

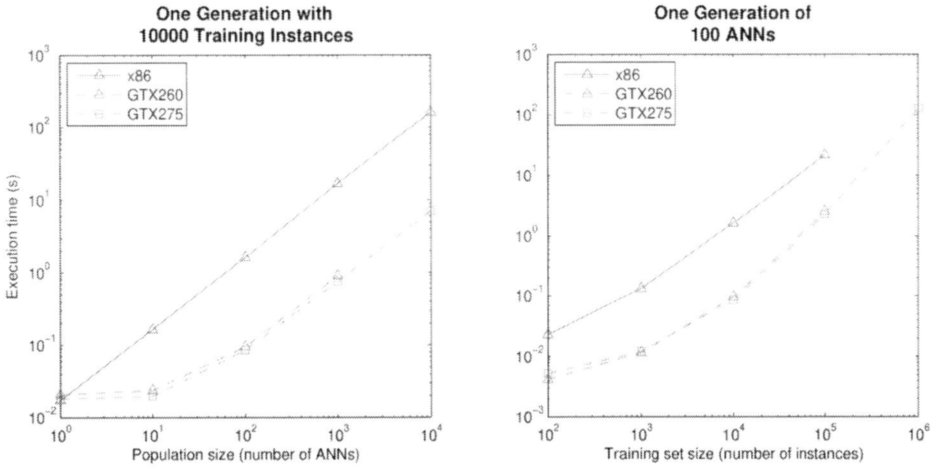


**Figure 2.** Training performance as a function of population size (left) and training set size (right). Note logarithmic axes.

*5.3. Classification Performance*

The classifier was trained over 12000 generations; the fitness metric during this training was lift at 20% calculated on a training set which consisted of 70% of the data set, chosen at random during initialization. After each generation, the candidate ANN with the highest fitness value was used to classify 20 random subsets of the hold-out data set (each consisting of exactly one half of the hold-out data set), and the average of the lift at 20% of these classifications was recorded. Figure 3 demonstrates experimental results across 12,000 generations of population 50. The final estimated lift is 4.51 out of a maximum lift of 5.

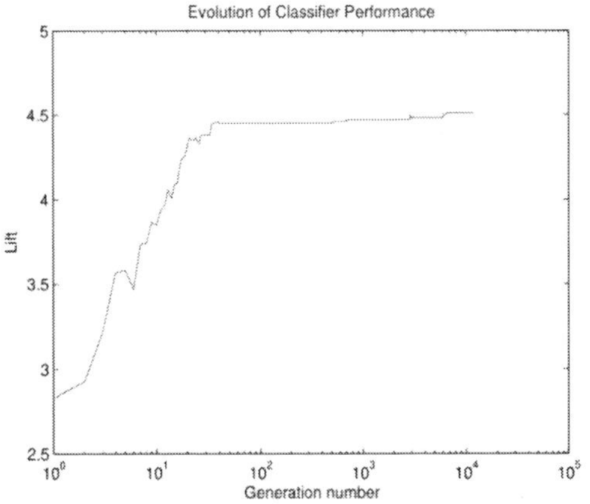


**Figure 3.** Evolution of classifier lift. Note logarithmic generation number axis.

## 6. Conclusion

We have presented an implementation of GA training of feedforward ANN classifiers for the CUDA platform for GPU programming. By carefully designing memory organization, algorithm computational load and memory access patterns, we have obtained a 10-fold speedup compared to a conventional sequential CPU implementation. A multipass approach was required in the selection step of classifier evaluation to compensate for the limited amount of fast on-chip GPU memory. Our method scales across population and training set sizes and is expected to be useful in other data intensive machine learning and data mining tasks.

## 7. Future Work

Montana and Lawrence [1] have demonstrated that a brief period of hill climbing after training with a genetic algorithm may increase the accuracy of classifier results. Thus, we could improve the accuracy of our methods by incorporating hill-climbing methods, such as the parallel backpropagation method described by Oei, Friedland and Janin [2].

In addition, parallelization methods of genetic algorithms involving multiple populations [8] may interact favourably with data mining applications. Seperate populations may be trained on different subsets of the training data, allowing for a more thorough search of the solution space which the training set presents.

[1] Montana D J and Davis L 1989 *Proceedings of the eleventh international joint conference on artificial Intelligence* vol 123 (Citeseer) pp 762–767 URL `http://citeseerx.ist.psu.edu/viewdoc/download?doi=10.1.1.77.3838&rep=rep1&type=pdf`
[2] Oei C, Friedland G and Janin A 2009 URL `http://www.icsi.berkeley.edu/pubs/techreports/TR-09-008.pdf`
[3] Halfhill T 2008 *Microprocessor Report*
[4] Diego U S 2009 UC San Diego Data Mining Contest URL `http://mill.ucsd.edu/`
[5] Nieminen P 2008 Multilayer Perceptron Neural Networks URL `http://users.jyu.fi/~nieminen/dm2008mlp/dm_mlp.pdf`
[6] Han J and Kamber M 2006 *Data mining: concepts and techniques* (Morgan Kaufmann) chap 2.4.2 ISBN 1558609016
[7] Andersen R 2008 *Modern Methods for Robust Regression* (SAGE) chap 2 ISBN 1412940729
[8] Alba E and Troya J M 1999 *Complexity* **4** 31–52 ISSN 10762787

# *Ab initio* Calculations of the Linear and Nonlinear Optical Properties of Amino Acids

**D Tokarz, A Tuer, R Cisek, S Krouglov, and V Barzda**[1]
Department of Chemical and Physical Sciences, Department of Chemistry, Department of Physics, and Institute for Optical Sciences, University of Toronto,
3359 Mississauga Road North, Mississauga, ON L5L 1C6, Canada

E-mail: virgis.barzda@utoronto.ca

**Abstract**. A number of proteins can assemble into chiral structures that display strong nonlinear optical activity. For instance, proteins such as myosin and collagen exhibit intense second harmonic generation (SHG). A large number of experimental studies on the SHG of proteins have been conducted; however few predictive models have been proposed that reliably relate the macroscopic SHG properties to the amino acids present in the peptidic chain. In this study, the linear polarizability (α), first (β) and second hyperpolarizability (γ) of all twenty amino acids was investigated by time-dependent Hartree-Fock calculations under physiological conditions. *Ab initio* calculations were performed using the GAMESSUS computational chemistry package. We have found that the aromatic amino acids give rise to the largest mean α, β and γ values. With this finding, we hope to apply this method to protein structures in order to understand how second harmonic signal is generated from individual amino acids, as well as, recognize how manipulation of the secondary structure of proteins might enhance SHG and third harmonic generation (THG).

## 1. Introduction

Nonlinear optical microscopy based on second harmonic generation (SHG) and third harmonic generation (THG), is currently an emerging technique used in studying biological structures (see review [1]). Harmonic generation microscopy is used to achieve distinct contrast in living tissue [2] [3]. Although both SHG and THG are coherent light processes, SHG only occurs in media with non-central symmetry. THG can be generated in homogeneous media, but when focusing under tight conditions with a microscope, THG is generated at interfaces [4]. For example, THG has been found to originate from multilayer structures such as biological membranes and cell walls [5] [6] while SHG has been previously found to originate from proteins such as myosin and collagen which consist of ordered helical structures [7] [8] [9]. Due to symmetry restrictions, media exhibits SHG signals at well-defined laser polarizations [10]. This information can be used to determine the orientation of proteins in tissues [10]. Therefore, information related to the organization of the secondary structures of proteins can be extracted from SHG data [10].

[1] To whom any correspondence should be addressed.

A large number of experimental studies on the SHG of proteins have been conducted though, few predictive models have been proposed that reliably relate the macroscopic SHG measurements to the structure of proteins. For further understanding of the origin of SHG and THG from various biological structures, modeling of the nonlinear response from molecular structures with various structural organizations is required. In this study, we investigated a relatively simple model for interpreting the nonlinear optical properties of proteins. This was done by examining the influence of the primary structure of proteins to generate both second harmonic and third harmonic signal. *Ab initio* computational studies were performed using the GAMESSUS computational chemistry package (Gordon Group, Ames Laboratory, Iowa State University) [11]. The time-dependent coupled perturbed Hartree-Fock (TDHF) approach was used to calculate the frequency dependent linear and nonlinear optical properties of individual amino acids [12] [13]. Specifically, an average for the linear polarizability, α, was expressed as well as values for the average first hyperpolarizability tensor, β, descriptive of SHG and the average second hyperpolarizability tensor, γ, descriptive of THG.

## 2. Theory

The dipole moment of a molecule when induced by an electric field can be expressed in a Taylor series as [14]

$$\mu = \mu_o + \alpha_{ij} E_j + \frac{1}{2!} \beta_{ijk} E_j E_k + \frac{1}{3!} \gamma_{ijkl} E_j E_k E_l + \ldots \tag{1}$$

where $\mu_o$ is the permanent dipole moment of the molecule in the absence of an electric field, $E$ is the electric field, $\alpha$ is the linear polarizability, and coefficients $\beta$ and $\gamma$ are the first and second hyperpolarizabilities, respectively. In equation (1) the repeated indices imply summation.

In general, $\alpha$, $\beta$, and $\gamma$ are dependent on the frequency, ω, of the incident electric field. The induced dipole yields linear polarization as well as second harmonic generation at 2ω and third harmonic generation at 3ω.

The mean linear polarizability can be defined as

$$\langle \alpha \rangle = \frac{1}{3} \left( \alpha_{xx} + \alpha_{yy} + \alpha_{zz} \right) \tag{2}$$

The mean second hyperpolarizability can be expressed as

$$\langle \gamma \rangle = \frac{1}{5} \left( \gamma_{xxxx} + \gamma_{yyyy} + \gamma_{zzzz} + \gamma_{xxyy} + \gamma_{xxzz} + \gamma_{yyxx} + \gamma_{yyzz} + \gamma_{zzxx} + \gamma_{zzyy} \right) \tag{3}$$

The first hyperpolarizability is zero for all molecules with inversion symmetry. The first hyperpolarizability is a third rank tensor described by a 3 x 3 x 3 array. All 27 components of the three dimensional array can be used to find the mean first hyperpolarizability by calculating an average related to the modulus

$$\langle \beta \rangle = \frac{1}{27} \sqrt{\sum_{ijk} \beta_{ijk}^2} \tag{4}$$

The expressions (2), (4), and (5) were used to express the linear and nonlinear optical properties of all amino acids.

## 3. Method

Calculations of the linear polarizability, the second- and third-order hyperpolarizabilities at the fundamental wavelength (1028nm) of our home-built Yb:KGd$(WO_4)_2$ laser [15] was achieved by applying the *ab initio* time-dependent coupled perturbed Hartree-Fock (TDHF) method at the Restricted Hartree-Fock (RHF) level using GAMESSUS program [12] [13] running parallel on the GPC supercomputer at the SciNet GPC Consortium at the University of Toronto. Calculations were performed

Journal of Physics: Conference Series **256** (2010) 012015 doi:10.1088/1742-6596/256/1/012015

on a single node which included 8 processors. Geometry optimization of each amino acid did not take longer than a single day. Similarly, the corresponding polarizability and hyperpolarizabilities of each amino acid were calculated within one day. Output geometry optimization files were as large as 430 000 kB whereas hyperpolarizability calculations generated files no larger than 2300 kB.

The ground state geometries of the amino acids were fully optimized without any symmetry restrictions using the 6-311++G** basis set at the DFT/B3LYP level at pH 7 [16]. These molecular structures were then used for nonlinear optical property calculations.

As suggested in literature, reliable calculation of the nonlinear optical properties requires increased convergence parameters which included the following GAMESS options: ITOL=30 (the products of primitives whose exponential factor is less than $10^{-30}$ are omitted), ICUT=20 (integrals less than $10^{-20}$ are overlooked), INTTYP=HONDO (HONDO/Rys integrals evaluated), ATOL=1.0D-07 (the tolerance for convergence of first-order results is $1.0^{-7}$), BTOL=1.0D-07 (tolerance for convergence of first-order results is $1.0^{-7}$) and NCONV=10 (when the absolute value of the density change between two consecutive self-consistent field cycles is less than $10^{-10}$ then convergence is achieved) [17]. Nonlinear optical calculations were also done using 6-311++G** basis set. It has been shown previously that a basis set augmented with diffuse functions is necessary when computing the nonlinear optical properties of molecules [18].

## 4. Results

The mean linear polarizability, first hyperpolarizability, and second hyperpolarizability for all twenty amino acids at pH7 were found by applying equations (2), (4), and (3), respectively. Results are shown in table 1, where a few trends are observed.

**Table 1.** The average linear polarizability, first hyperpolarizability, and second hyperpolarizability of amino acids organized by chemical derivative.

| Amino Acid | $\langle\alpha\rangle$ ($10^{-24}$ esu) | $\langle\beta\rangle$ ($10^{-32}$ esu) | $\langle\gamma\rangle$ ($10^{-36}$ esu) |
|---|---|---|---|
| **Hydrocarbon** | | | |
| Alanine | 7.94 | 5.50 | 9.80 |
| Glycine | 6.24 | 6.11 | 8.27 |
| Isoleucine | 12.9 | 6.15 | 13.5 |
| Leucine | 12.9 | 7.30 | 14.6 |
| Proline | 10.8 | 8.54 | 15.8 |
| Valine | 11.2 | 6.83 | 12.7 |
| **Alcohol** | | | |
| Serine | 8.33 | 4.06 | 8.02 |
| Threonine | 10.0 | 4.73 | 9.46 |
| **Sulfur-containing** | | | |
| Cysteine | 10.8 | 14.1 | 16.0 |
| Methionine | 14.3 | 8.74 | 17.0 |
| **Second carboxylic acid** | | | |
| Aspartic acid | 11.1 | 4.26 | 18.5 |
| Glutamic acid | 12.8 | 7.52 | 19.2 |
| **Carboxylic acid derivative** | | | |
| Asparagine | 10.7 | 4.51 | 10.4 |
| Glutamine | 12.3 | 6.84 | 10.6 |

| | | | |
|---|---|---|---|
| **Aromatic** | | | |
| Phenylalanine | 17.6 | 14.7 | 22.7 |
| Tryptophan | 22.0 | 19.8 | 34.8 |
| Tyrosine | 18.0 | 7.22 | 18.5 |
| **Second amino group** | | | |
| Arginine | 16.3 | 9.73 | 14.5 |
| Histidine | 13.6 | 21.5 | 14.2 |
| Lysine | 14.1 | 5.43 | 13.3 |

From table 1, larger amino acid structures give rise to increased linear polarizabilities due to increased electron density and delocalization. Of all the different chemical derivatives, the aromatic amino acids, phenylalanine, tryptophan, and tyrosine have the largest linear polarizability values. Aromatic amino acids consist of conjugated residues which contribute to the increased polarizability due to the delocalization of $\pi$-electrons. Aromatic amino acids, may for this reason, play a significant role in the nonlinear response of proteins [19]. In particular, tryptophan has the largest mean linear polarizability as well as the largest mean first and second hyperpolarizability values. In general, a large linear polarizability is necessary to obtain substantial hyperpolarizabilities, though there are additional factors associated with increased hyperpolarizability values [20].

Due to symmetry restrictions, the SHG response of a molecule is highly directionally dependent where each individual $\beta$ tensor varies. Eighteen of twenty-seven unique components of the first hyperpolarizability tensor were calculated where Kleinman symmetry was assumed for the other nine $\beta$ tensor elements. An overall orientational average $\beta$ was calculated using equation (4). In particular, phenylalanine, tryptophan, and histidine demonstrate the largest $< \beta>$ values. These molecules are polarizable in multiple directions and therefore, the overall average of $\beta$ appears to be large. We express an orientational average for the first hyperpolarizability however, when amino acids are aligned, such as in proteins, $\beta$ is best expressed by its individual tensor components.

From calculations of the second hyperpolarizability, aromatic amino acids demonstrate the largest $\gamma$ values due to increased $\pi$-electron delocalization. However, the second carboxylic acid amino acids including aspartic acid and glutamic acid also give rise to large $\gamma$ values. Although these two amino acids are not conjugated, both of them contain two carboxylic acid groups, which are deprotonated at pH 7. Electron density is delocalized amongst the deprotonated carboxylic acid groups at both ends of the amino acid which therefore likely enhances the polarizability of aspartic acid and glutamic acid.

## 5. Conclusions

We have demonstrated that *ab initio* calculations can help in interpreting the origin of SHG and THG from the primary structure of proteins. Each amino acid alone does not give rise to high nonlinear optical signals however; increased SHG and THG can be achieved by coherently summing the nonlinear response of many amino acids. In particular, the aromatic amino acids are highly polarizable due to $\pi$-electron delocalization and may be desirable candidates for increasing SHG and THG when forming secondary structures. We hope to use the fundamental understanding of the origin of SHG and THG from amino acids to manipulate the secondary structure of proteins in order to enhance the response of SHG and THG.

## Acknowledgements

Computations were performed on the GPC supercomputer at the SciNet HPC Consortium. SciNet is funded by: the Canada Foundation for Innovation under the auspices of Compute Canada; the Government of Ontario; Ontario Research Fund - Research Excellence; and the University of Toronto.

The authors also gratefully acknowledge support in part by the Natural Sciences and Engineering Research Council of Canada and the Ontario Centres of Excellence.

**References**

[1] Cisek R, Spencer L, Prent N, Zigmantas D, Espie G and Barzda V 2009 *Photosynth Res* **102** 111-41
[2] Sheppard C J R, Gannaway J N, Kompfner R and Walsh D 1977 *Ieee Journal of Quantum Electronics* **13** D100-D00
[3] Barad Y, Eisenberg H, Horowitz M and Silberberg Y 1997 *Appl Phys Lett* **70** 922-24
[4] Ward J F and New G H C 1969 *Phys Rev* **185** 57-72
[5] Muller M, Squier J, Wilson K R and Brakenhoff G J 1998 *J Microsc-Oxford* **191** 266-74
[6] Squier J A, Muller M, Brakenhoff G J and Wilson K R 1998 *Opt Express* **3** 315-24
[7] Plotnikov S V, Millard A C, Campagnola P J and Mohler W A 2006 *Biophys J* **90** 693-703
[8] Freund I and Deutsch M 1986 *Opt Lett* **11** 94-96
[9] Stoller P, Reiser K M, Celliers P M and Rubenchik A M 2002 *Biophys J* **82** 3330-42
[10] Campagnola P J, Millard A C, Terasaki M, Hoppe P E, Malone C J and Mohler W A 2002 *Biophysical Journal* **82** 493-508
[11] Schmidt M W, Baldridge K K, Boatz J A, Elbert S T, Gordon M S, Jensen J H, Koseki S, Matsunaga N, Nguyen K A, Su S J, Windus T L, Dupuis M and Montgomery J A 1993 *J Comput Chem* **14** 1347-63
[12] Sekino H and Bartlett R J 1986 *J Chem Phys* **85** 976-89
[13] Karna S P and Dupuis M 1991 *J Comput Chem* **12** 487-504
[14] Buckingham A D and Orr B J 1967 *Q Rev Chem Soc* **21** 195-&
[15] Major A, Cisek R, Sandkuijl D and Barzda V 2009 *Laser Phys Lett* **6** 272-74
[16] Becke A D 1993 *J Chem Phys* **98** 5648-52
[17] Cundari T R, Kurtz H A and Zhou T 2001 *J Chem Inf Comp Sci* **41** 38-42
[18] Shelton D P and Rice J E 1994 *Chem Rev* **94** 3-29
[19] Guthmuller J and Simon D 2006 *J Phys Chem A* **110** 9967-73
[20] Marder S R, Gorman C B, Meyers F, Perry J W, Bourhill G, Bredas J L and Pierce B M 1994 *Science* **265** 632-35

# Numerical Simulations of Thermal Convection in Rapidly Rotating Spherical Shell

**Constantine Nenkov and Richard Peltier**
Department of Physics, University of Toronto
Toronto, Ontario, Canada M5S 1A7
E-mail: nenkov,peltier@atmosp.physics.utoronto.ca

**ABSTRACT.** We present a novel numerical model used to simulate convection in the atmospheres of the Gas Giant planets Jupiter and Saturn. Nonlinear, three-dimensional, time-dependant solutions of the anelastic hydrodynamic equations are presented for a stratified, rotating spherical fluid shell heated from below. This new model is specified in terms of a grid-point based methodology which employs a hierarchy of tessellations of the regular icosahedron onto the sphere through the process of recurrent dyadic refinements of the spherical surface.

We describe discretizations of the governing equations in which all calculations are performed in Cartesian coordinates in the local neighborhoods of the almost uniform icosahedral grid, a methodology which avoids the potential mathematical and numerical difficulties associated with the pole problem in spherical geometry. Using this methodology we have built our model in primitive equations formulation, whereas the three-dimensional vector velocity field and temperature are directly advanced in time.

We show results of thermal convection in rapidly rotating spherical shell which leads to the formation of well pronounced prograde zonal jets at the equator, results which previous experiments with two-dimensional models in the limit of freely evolving turbulence were not able to achieve.

*Key words:* computational fluid dynamics, multigrid methods, icosahedral mesh, convection, anelastic model,zonal jets *PACS:* 02.60.Cb, 02.70.Dh, 92.60.Bh, 92.60.Hk, 96.15.Hy

## 1. Introduction

Understanding how large-scale zonal flows are generated on the Gas Giant planets is a longstanding problem in planetary and geophysical fluid dynamics (Busse [1], Williams [2], Cho and Polvani [3], Peltier and Stuhne [4], Aurnou and Heimpel [5], Glatzmaier et al. [6]). Strong eastward equatorial jets are the dominant dynamical feature for both Jupiter and Saturn, and there is a growing body of evidence that such features are also predominant on the other two Gas Giant planets Uranus and Neptune, although the direction of the equatorial jets are in the opposite direction.

We study the emergence and evolution of large-scale zonal flows, as observed on the Gas Giant planets, using a newly developed 3D GCM in spherical shell geometry. The physical basis of the model is the anelastic approximation of the hydrodynamic equations of motion, continuity, and classical thermodynamics. We describe a representative investigation of the convective response of a layer of Boussinesq fluid in rapidly rotating spherical shell subject to isothermal temperature boundary conditions. The physical scaling is governed by the 3 non-dimensional parameters: Taylor, Prandtl and Rayleigh numbers, while the depth of the shell is a variable parameter.

The model is specified in terms of a grid-point based methodology which employs a hierarchy of tessellations derived from the successive dyadic refinements of the spherical icosahedron. One major advantage of this multi-grid methodology is that it allows for nearly linear growth of complexity in operation count as opposed to the spectral transform models, which by their nature are at least quadratic in computational complexity. Another potential advantage is the absence of pole problems, and therefore the ability of the code to capture important features of the dynamics in the polar regions. An added bonus of this new methodology is the possibility for greater local control over the computational mesh.

The spatial operators of the model are constructed around a multitude of quasi-regular icosahedral grids at each spherical level for the surface part, and second or higher order finite differences in the radial direction. They use new, performance optimized data structures which enhances the stability of the code in long-term integrations. Another novelty is the introduction of multigrid finite element methods for the solution of a fully 3D generalized Elliptic equation that arises as a consequence of the anelastic approximation. These finite element methods are explicitly defined in terms of the icosahedral grid structure and takes into account the curvature of the spherical shell in the form of an asymptotic series expansion.

This work is direct continuation of previous work of Stuhne and Peltier (1996, 1999 [7, 8]), who reported results with this novel grid-point based methodology within the framework of one-layer barotropic and shallow water models.

## 2. The model and its grid structure

For the purposes of this study, our 3D numerical simulations of thermal convection in rotating spherical shell is assumed to be governed by the Boussinesq approximation of the anelastic system. The dynamical equations in this approximation can be written in the following reference independent form:

$$\frac{\partial \mathbf{u}}{\partial t} + \mathbf{u} \cdot \nabla \mathbf{u} = \nabla^2 \mathbf{u} - \left(\frac{Ra}{Pr}\right) \nabla p' + \left(\frac{Ra}{Pr}\right) \bar{g}(r) T \hat{\mathbf{r}} - Ta^{1/2} (\mathbf{e}_z \times \mathbf{u}) \tag{1}$$

$$\nabla \cdot \mathbf{u} = 0 \tag{2}$$

$$\frac{\partial T}{\partial t} + \mathbf{u} \cdot \nabla T = \left(\frac{1}{Pr}\right) \nabla^2 T \tag{3}$$

under the assumption of constant density profile. In the above system $\mathbf{u}$ is the 3D velocity field, $T$ is the temperature field, and $\bar{g}(r)$ is the radial profile of acceleration due to gravity. The principle three non-dimensional scaling parameters are: Rayleigh $Ra$, Taylor $Ta$ and Prandtl $Pr$ numbers.

The hierarchy of grids obtained through the process of recursive dyadic refinement is best illustrated graphically in Figure 1 which shows the mesh structures of grid levels 0 through 5. This sub-devision of the computational domain leads to quasi-regular coverage of the surface of the sphere that has no pole problem.

The connectivity of the nearest and next-to-nearest grid points on the surface of the sphere, as well as the full three-dimensionality of the computational mesh forming the so called *computational molecule* is shown in Figure 2.

The spatial differential operators of the model are constructed using affine basis states on the surface of the sphere and finite differences with the possibility of variable resolution in the vertical.

Other important details concerning the anelastic model are the following:

- dynamical core consists of a dry adiabatic atmosphere in a non-hydrostatic formulation
- non-dimensionalized Cartesian form of the dynamical equations
- 4D Cartesian embedding of spherically constrained 3D velocity field
- vertical component of the Coriolis force and the centrifugal force taken into account
- instantaneous pressure adjustment of velocity field carried out by a generalized 3D elliptic solver
- freely adjustable spatial and temporal resolution
- isothermal shell boundaries with linearly varying gravity with radius

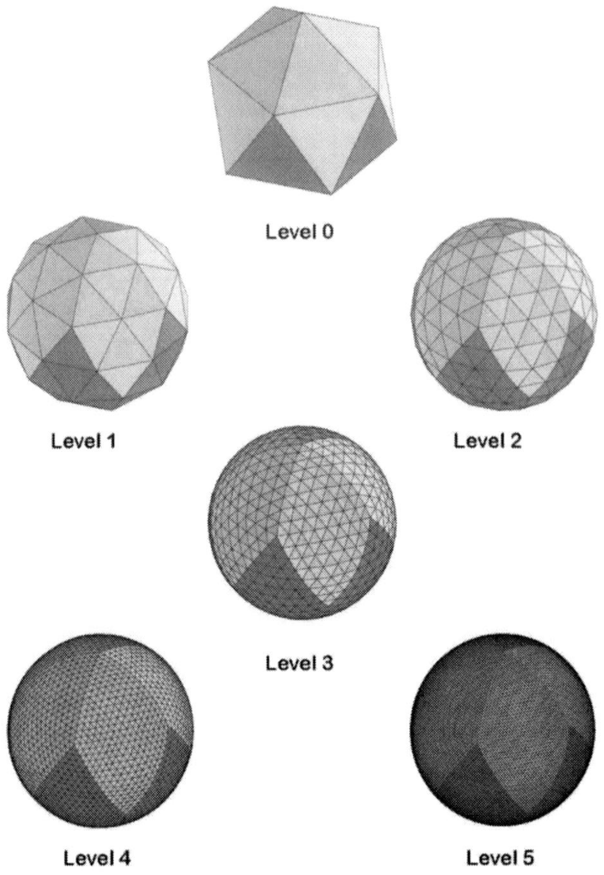


**FIG. 1.** Dyadic refinement of the computational mesh structure on the surface of the sphere for refinement levels $l$ = 0 up to level $l$ = 5 of the basic icosahedron.

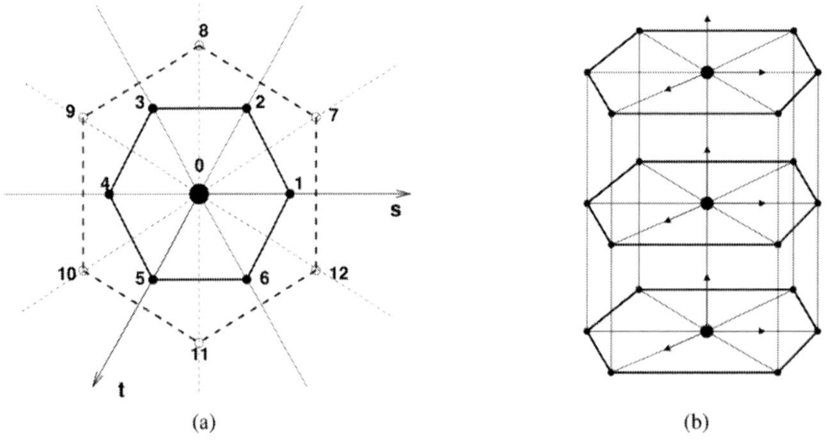


**FIG. 2.** Nearest and next-to-nearest neighbours grid points on the surface of the sphere (a) and nearest neighbours grid points belonging to the 3D computational molecule (b).

- the initial state is a thermally perturbed hydrostatic equilibrium
- the end state is zonal geostrophic flow - invariant along the axis of rotation, with alternating prograde/retrograde surface jets .

## 3. Thick-thin zonal flow formation

We apply our icosahedral model to the problem of zonal jet formation in the atmospheres of the Gas Giant planets. We investigate the formation and evolution of these zonal jets, which are driven by vigorous convection and strong Coriolis force. Such a test case is of particular relevance to the outer layers of the gas giant planets where a number of open questions associated with the formation and evolution of coherent structures await solution.

For this problem, the model parameters we have employed were selected to reflect (within the technical limitations of a fully 3D model) our current understanding of the Jovian dynamics. The Rayleigh and Taylor numbers were chosen to be as high as is numerically feasible although they are many orders of magnitude smaller than those assumed for Jupiter and Saturn. Regardless of this limitation, our ultimate goal is to demonstrate that the anelastic icosahedral model can simulate successfully the formation and evolution of zonal flows in spherical shell geometry. These zonal flows are spontaneously generated out of purely random thermal initial conditions under the joint action of strong rotation and vigorous convection.

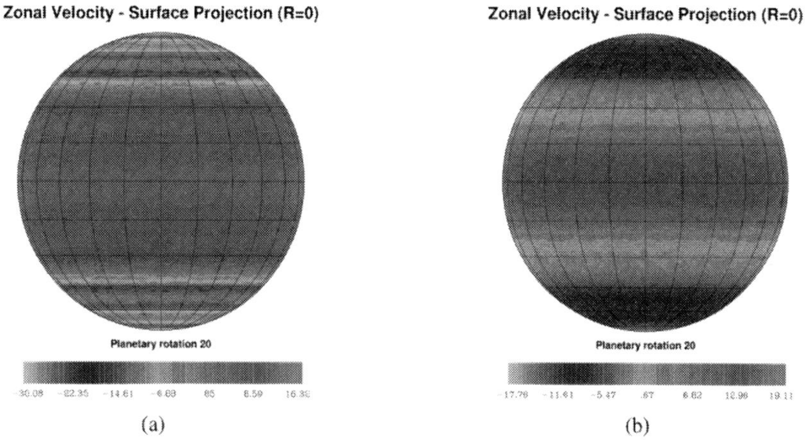


**FIG. 3.** Surface projections of zonal velocity at the outer shell boundary for free-rigid velocity boundary condition: (a) thick shell depth 0.65%; (b) thin shell depth 0.25%.

As one example of successful numerical simulation we present results from a global flow comparison study for thick-thin shell experiment in the Boussinesq approximation. The important point here is that the anelastic model has been run at more extreme values of the control parameters ($Ra = 10^7$, $E = 10^{-5}$ or $Ta = 4 \times 10^{10}$). The intent was to test the code under strong forcing of both convection and rotation where the characteristic dynamical time scale is very short in comparison to the diffusion time scale. The model was run with free-rigid velocity boundary conditions. The effect of the shell-depth on the width of the equatorial zonal flow is represented in Figure 3. The two panels show surface projections of the mean zonal velocity field at the free outer boundary level of the shell. For the thick shell (shell depth 65%) the prograde equatorial jet is much wider in comparison to the equatorial jet for the thin shell (shell depth 25%). Consequently, the strength of the broader zonal flow is presumably diminished in comparison to that for the shell with shallower depth. This figure serves primarily an illustrative purpose since the actual shell-depth at which more realistic numerical experiments for Jupiter and Saturn should be run would be considerably smaller (shell depth $|sim10\%$). It should also be noted that this shell geometry would be computationally much more demanding in a 3D numerical experiment with the full anelastic icosahedral formulation because of the increased complexity of the dynamical system.

## 4. Conclusions and future work

Numerical simulations of global convection in rapidly rotating spherical shells remain one of the outstanding challenges for computational fluid dynamics. To more truthfully simulate the global zonal flows as observed on Jupiter and Saturn, for instance, including the broad prograde equatorial jets and multiple alternating jets at higher latitudes, the new anelastic icosahedral model in spherical geometry should be run at considerably higher spatial resolution (presumably multigrid level $l = 7, 8$ on a spherical surface, and number of radial levels in the vertical $R > 100$) to ensure that the small scale structure of the flow is adequately represented. In addition, the simulations should be performed at shell depths $< 0.1$ in order to capture the development and evolution of multiple jets not only in the equatorial region, but also the complex jet structure in the polar regions as recently observed by the Cassini spacecraft. The hope is that the new code will be very well suited for this because it does not have a "pole problem" and the resolution on the surface of the sphere is almost regular, including the polar regions. At the same time the principle scaling parameters, the Rayleigh number $Ra$ and the Taylor number $Ta$, should be boosted up in accord with the changed shell geometry. A rough estimate could be: $Ra > 10^8 \div 10^{10}$, and $Ta > 10^{10} \div 10^{12}$. These parameter values are still many orders of magnitude below the physical values for Jupiter and Saturn, but will allow to test the performance of the code at more extreme regimes. Even higher Rayleigh numbers will be needed to simulate the Ice Giants - Uranus and Neptune, where it is believed that strong convection dominates over the Coriolis force and vigorous angular momentum mixing forces their zonal winds in the retrograde direction. All this undoubtedly will put a premium on computer resources in terms of memory and speed and ultimately will require more powerful computer architectures.

The newest addition to this model is that the radial direction can be treated also in a fully spectral manner using Chebyshev polynomials which are the polynomials with the smallest error bounds for the radial components of the gradient and Laplacian operators. This new option allows for a hybrid gridpoint-spectral formulation of the model which can considerably enhance the flexibility of the model to operate in different regimes.

The non-tensor nature of the nearly uniform icosahedral grid in 3D (see Doyd, 2001 [9] and references therein) still poses a huge research challenge. Apart from programming complexities, the accuracy of the spatial operators is not optimal since the affine basis states on the surface of the sphere are not orthogonal to each other. Current gridpoint methods however lack explicit conservation of momentum and thermal fluxes, as well as kinetic energy conservation. In long term integrations and strongly forced thermal convection the icosahedral gridpoint methodology can suffer from non-linear aliasing instabilities, in addition to problems with insufficient spatial resolution near the boundaries. To alleviate such shortcomings we started experimenting with non-uniform Chebyshev grids in the radial direction that are able to resolve well thermal boundary layers at the edges of the spherical shell, although this approach can lead to severe constraints on explicit and semi-implicit time stepping schemes. Regardless of these principle limitations, our recent numerical experiments show that the hybrid formulation of the spatial operators of the model with spectral Chebyshev representation on stretched radial mesh is superior to the fully gridpoint formulation.

A natural evolution of the code that can dramatically improve the conservation properties of the model is the conversion of the code into fully 3D finite volume formulation. This will undoubtedly improve flux balancing of the differential operators, thereby allowing long integration runs at more extreme values of the scaling parameters that are needed in order to reach statistically equilibrated final states of the dynamical system.

Currently the model is being run very efficiently on a single node of the IBM Power6 machine at SciNet with OpenMP 3.0 multithread option. Future upgrades of the model require cross node parallelization with the MPI 3.0 library that will improve the efficiency of the code at high resolution that is absolutely necessary in the strong convective regime.

In conclusion, we believe that the general framework for globally convecting systems in spherical geometry based on the icosahedral methodology, is robust and capable of resolving some of the more complicated processes that arise in nonlinear geophysical and astrophysical hydrodynamics. With continuing refinement and improvement of the code, we are encouraged to believe that the methodology we have based this research on may ultimately rival the currently prevalent spectral transform methods in this field. One very significant ingredient in achieving this goal is to parallelize the code using the newest Message Passing Interface (MPI) standard that ultimately will allow the model to be run efficiently on massively parallel computer architectures.

## 5. Acknowledgments

The authors thank the organizers of the HPCS 2010 conference on High Performance Scientific Computing for valuable comments and guidance that significantly improved the manuscript. Support for this research has been made available through grants from the Natural Sciences and Engineering Research Council of Canada. Computing resources were provided by the SciNet Consortium at the University of Toronto.

## References:

[1] Busse, F.H., "A simple model of convection in the Jovian atmosphere", Icarus **20** (1976) 255–260.
[2] Williams, G.P., "Jupiter's atmospheric circulation", Nature **257** (1975) 778.
[3] Cho, J.Y.K., Polvani, L.M., "The morphogenesis of bands and zonal winds in the atmospheres on the giant outer planets", Science **273** (1996) 335–337.
[4] Peltier, W.R., Stuhne, G.R., "The upscale turbulent cascade — shear layers, cyclones and Gas Giant bands", in: Pearce, R.P. (Ed.), Meteorology at the Millennium, Academic Press, New York, 2002, p. **43**.
[5] Heimpel, M.H., Aurnou, J.M., "Turbulent convection in rapidly rotating spherical shells: A model for equatorial and high-latitude jets on Jupiter and Saturn", Icarus **187** (2007) 540–557.
[6] Glatzmaier, G.A., Evonuk, M., Rogers, T.M., "Differential rotation in giant planets maintained by density-stratified turbulent convection", Geophys. Astrophys. Fluid Dynamics **103** (2009) 31–51.
[7] Stuhne, G.R., Peltier, W.R., "Vortex erosion and amalgamation in a new model of large scale flow on the sphere", J. Comp. Phys. **128** (1996) 58–81.
[8] Stuhne, G.R., Peltier, W.R., "New icosahedral grid-point discretizations of the shallow water equations on the sphere", J. Comp. Phys. **148** (1999) 23–58.
[9] Boyd, J.P., "Chebyshev and Fourier Spectral Methods", Dover Publications, 2001.

Journal of Physics: Conference Series **256** (2010) 012017 doi:10.1088/1742-6596/256/1/012017

# A Pipelining Implementation for Parsing X-ray Diffraction Source Data and Removing the Background Noise

**Michael A. Bauer***, **Alain Biem**†, **Stewart McIntyre*** **and Yuzhen Xie***

*University of Western Ontario, London ON, Canada

†IBM Research, Yorktown Heights, United States

E-mail: bauer@uwo.ca; biem@us.ibm.com; smcintyr@uwo.ca; yxie@csd.uwo.ca

**Abstract.** Synchrotrons can be used to generate X-rays in order to probe materials at the atomic level. One approach is to use X-ray diffraction (XRD) to do this. The data from an XRD experiment consists of a sequence of digital image files which for a single scan could consist of hundreds or even thousands of digital images. Existing analysis software processes these images individually sequentially and is usually used after the experiment is completed. The results from an XRD detector can be thought of as a sequence of images, generated during the scan by the X-ray beam. If these images could be analyzed in near real-time, the results could be sent to the researcher running the experiment and used to improve the overall experimental process and results. In this paper, we report on a stream processing application to remove background from XRD images using a pipelining implementation. We describe our implementation techniques of using IBM Infosphere Streams for parsing XRD source data and removing the background. We present experimental results showing the super-linear speedup attained over a purely sequential version of the algorithm on a quad-core machine. These results demonstrate the potential of making good use of multi-cores for high-performance stream processing of XRD images.

## 1. Introduction

Synchrotrons are large electron storage rings that produce X-ray beams of unprecedented brightness. Synchrotron facilities, such as the Canadian Light Source (CLS) in Saskatchewan and the Advanced Light Source (ALS) in the US, allows scientists to use synchrotron beams to study various materials. In order to understand and predict properties of materials, detailed information about the structure of the material at the atomic-scale level is needed. Synchrotrons can be used to generate X-rays in order to probe materials at the atomic level. One approach is to use X-ray diffraction (XRD) to do this. The Laue X-ray diffraction (XRD) technique has been used to determine the atomic-scale structures of materials based on the Bragg's Law [17]. When a material that consists of a periodic atomic structure, such as a crystal, is irradiated with X-rays, it produces a diffraction pattern showing numerous sharp spots, called Bragg diffraction peaks. By measuring and analyzing the positions and intensities of these peaks, one can determine the arrangement of atoms in the crystalline material. Not only can Laue diffraction provide geometrical information about the crystalline structure, it can also be extended to measure the intensity of X-ray diffraction to determine the strain tensor in sample materials [4]. The strain tensor information will then help scientists/engineers to determine possible flaws in materials.

The data from an XRD experiment consists of a sequence of digital image files. The XRD data analysis software actually deals with a large amount of data, typically, digital images in the hundreds or even thousands. Existing analysis software processes these images individually sequentially. This is a very time consuming process, normally requiring days to complete the processing of an entire set of data from one XRD experiment. A number of existing software packages for XRD data analysis have been developed. These include the 3D X-ray Micro-diffraction Analysis Software Package [16] at the Advanced Photon Source (APS), which was developed by scientists at the Oak Ridge National Laboratory, and the X-ray Micro-diffraction Analysis Software (XMAS) [15] at the Advanced Light Source (ALS) in Lawrence Berkeley National Laboratory (LBNL). The common feature of these two packages is that they both are Windows-based software, implemented in the Interactive Data Language (IDL) [10]. Both can process a large amount of XRD data sequentially.

Synchrotron time is valuable and it is often difficult to get a scheduled beam time. Data analysis using existing software means that researchers are typically doing the analysis after their time on the synchrotron. More timely analysis could help researchers make decisions on subsequent experiments and gain significant insight into the materials that they are studying. Performance improvement in the analysis software could make the analysis more useful.

Synchrotron experiments can produce several megabytes of image data per second. One can think of the output from such an experiment as a "stream of images". As part of a project funded by Canada's Advanced Research and Innovation Network (CANARIE), we are involved in a project to study the feasibility of processing such "streams" of images. The goal of the Active Network Interchange for Scientific Experimentation (ANISE) project is to develop a high speed network and processing systems to achieve near real-time data analysis for synchrotron experiments. Being able to process this data in near real-time would enable users to make sound decisions about issues on time-sensitive materials under study.

Our proposed approach is essentially one of *stream processing*. The objective is to be able to stream images, as they are generated, to a processing center across CANARIE's network and return results to a user. In this paper we report on our initial efforts in developing stream processing software capable of analyzing such images. Our development efforts make use of *IBM InfoSphere Streams*, a software environment for developing streaming applications that can receive and process data in multiple streams and in parallel. Our initial efforts have been on the development of software to perform background filtering on XRD images, one of the central computational steps in the analysis of XRD images. We describe our techniques for developing stream processing components and our experiences with *InfoSphere Streams*. We report the performance of our program for processing real-time problems of different sizes.

The paper is organized as follows. In Section 2 we provide a brief introduction and an overview of our stream processing platform - *InfoSphere Streams* and its development environment *SPADE*. In Section 3 we introduce the sequential algorithm for the two dimensional (2D) background removal. In Section 4 we outline our pipeline implementation in some detail and include a number of diagrams illustrating how the background processing is decomposed into elements in order to be processed in a pipelined manner. Experimental results are presented in Section 5 and we provide some concluding comments in Section 6.

## 2. Stream Processing Platform

Stream processing is a data-centric programming model. It applies to applications where the data to the application can be viewed as a stream of inputs and where the application may entail a limited form of parallel processing. Given a set of data which can be structured as a stream of *tuples*, a series of operations (processing elements) are applied to each tuple or some set of tuples in the stream. Processing elements are usually pipelined, much like an assembly line. Pipelining does not reduce the time for individual instruction execution. Instead, it improves instruction and data throughput. The perfect condition is that all the pipe stages, i.e. processing elements, are balanced. In such case, the speedup from pipelining equals the number of pipe stages.

Our implementation platform is *IBM InfoSphere Streams* [7] targeting clusters of multi-cores. Streams is an emerging software platform in support of parallel and high performance stream processing. It consists of an application development framework called *Stream Processing Application Declarative Engine* (SPADE) [9] and a runtime system that can facilitate the running of SPADE applications on a distributed set of hosts.

In SPADE, a stream processing application is modeled as a dataflow graph. It is composed of data streams and operators. A data stream is a running sequence of tuples. A tuple is an individual piece of data in a stream. Typically, the data in a tuple represents the state of something at a specific point in time. For example, a row of pixel values in an XRD image, say produced by an XRD detector while scanning on an area of the material being tested, could be consider a tuple. An operator takes in a stream, manipulates the tuple data and generates the results in a form of an output stream. Multiple operators can work on different tuples of a stream or multiple streams. Figure 1 shows a dataflow graph for a SPADE application example [8]. The icons in the figure represents operators, and the lines represent streams.

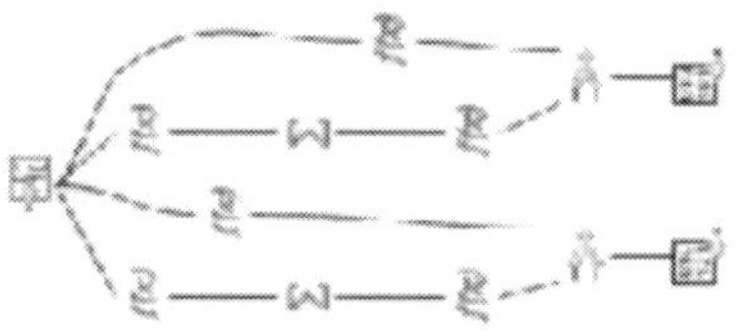

Figure 1: A dataflow graph for a simple SPADE application

When a SPADE application is compiled, the operators and streams are organized into a set of individual execution units called Processing Elements (PEs). The application's structure is described in a Job Description Language (JDL) file, including details about each PE, such as which binary file to load and execute, command-line parameters, scheduling restrictions, stream formats, and internal operator dataflow graph. To run an application, a user needs to submit a job to the Streams runtime system. The Streams runtime system provides a collection of services such as *Resource Manager*, *Scheduler* and *Host Controller* to manage the running of SPADE applications on a distributed set of hosts. Detailed report on Streams runtime system is refereed to [8, 9].

### *2.1. Streams operators*

As a programming interface for end-users to operate on data streams, Streams makes available a toolkit of built-in operators to manipulate data streams. A subset of operators currently supported by Streams include:

**Source** : A Source operator is used for creating a stream of data from an external source.

**Sink** : A Sink operator is used for converting a stream into a flow of tuples that can be used by components that are not part of Streams.

**Functor** : A Functor operator is used for performing tuple-level manipulations such as filtering, projection, mapping, attribute creation and transformation.

**Aggregate** : An Aggregate operator is used for grouping and summarization of incoming tuples. This operator supports a large number of grouping mechanism and summarization functions though sliding or tumbling windows.

**Join** : A Join operation is used for correlating two streams.

**Split** : A Split operator is used for splitting a stream into multiple output streams, based on a split condition.

**Barrier** : A Barrier operator is used as a synchronization point.

SPADE is equipped with a facility to extend the underlying system with user-defined operators (UDOPs), which are operators specialized to both wrap legacy libraries and provide customized processing. In Section 4 we will introduce the UDOPs that we have created in our application, as well as the built-in operators that are used to structure the data streams.

## 3. The Sequential Procedure

We describe a set of procedures for parsing X-ray diffraction (XRD) source data and removing the background noise. XRD is one of the important measurements by X-ray radiation with numerous applications [5]. The high-intensity synchrotron X-ray sources allow diffraction experiments to be performed on very small crystals or on large biological molecules. Background can be produced from a number of sources such as fluorescent radiation emitted by the specimen, diffraction of a continuous spectrum of wavelengths, etc. Estimation of background is a primary requirement for many types of further analysis.

There are different approaches for estimating and removing the background from X-ray diffraction patterns. For example, Kajfosz and Kwiatek [11] utilize a non-polynomial method for approximation of background in X-ray spectra. A polynomial filter [6] is used by Dinnebier and the smoothing procedure [2] was introduced by Brükner for 1D powder diffraction patterns. These methods are usually extended to treat 2D XRD images. For instance the combination of Brükner's 1D procedure and the 2D median filter utilized in XMAS [14, 15] for fitting the background of 2D X-ray microdiffraction patterns. In this work, we consider a practical algorithm which uses a parabolic filter to estimate the background of a 2D XRD image, given in Algorithm 1. It is an extension of the polynomial method for 1D powder diffraction patterns to handle 2D XRD images. Based on the knowledge provided by [16], it has been implemented as a sequential C procedure [12] in the Science Studio project [13].

**Algorithm 1: Remove 2D XRD background using a parabolic filter [12]**

**Input** : file $F$ encoding a $2D$ XRD image $S$ of dimension sizes $r \times c$, a resizing factor $e$, a filter size $w$

**Output** : image $S$ freed of background noise

$S \leftarrow \mathsf{Parse}(F)$;

$N \leftarrow \mathsf{Resize}(S, r/e, c/e)$;

$M \leftarrow \mathsf{GenerateFilter}(w)$;

$R \leftarrow \mathsf{ComputeRatio}(P, M)$ ; // $O(r\,c\,(\frac{w}{e})^2)$

$B \leftarrow \mathsf{EstimateBackground}(R, M)$ ; // $O(r\,c\,(\frac{w}{e})^2)$

$B \leftarrow \mathsf{Resize}(B, r, c)$;

$C \leftarrow \mathsf{Subtract}(S, B)$;

**return** $C$;

In Algorithm 1, the source of the 2D XRD image is from a file $F$. The function Parse reads the data from $F$ and converts it to a $r \times c$ matrix $S$ of energy intensity values. The data type of energy intensities is `double`. Usually the size of $S$ is large, for instance, $2084 \times 2084$. In order to reduce the cost in removing the background, one can first scale down the dimension of $S$ by a factor of $e$. Let the smaller image be $N$ with dimension sizes $s \times t$. Thus we have $s = r/e$ and $t = c/e$. The functions in Lines 3 to 5 will work on $N$ and estimate its background.

In Line 3 of Algorithm 1, a *filter* or *mask* defined by a square matrix $M$ of order $w$, is generated by a parabolic function which models the distribution of the background noise in the neighborhood of a pixel. The mask $M$ is then applied to each pixel in $N$ for calculating the minimal ratios $R$ (a matrix of $s \times t$) by the function ComputeRatio. In this operation, for each pixel $(i, j)$ in $N$, with $0 \leq i < s,\ 0 \leq j < t$, we compute the following minimum ratio

$$\mathrm{r}_{i,j} := \min\left\{\mathrm{n}_{i+h,j+k} / \mathrm{m}_{f+h,g+k} \;\middle|\; \begin{cases} -f \leq h < w - f \\ -g \leq k < w - g \end{cases} \begin{cases} -i \leq h < r/e - i \\ -j \leq k < c/e - j \end{cases}\right\}, \quad (1)$$

where $(f, g)$ are the coordinates of the centre of $M$.

After obtaining the minimum ratios, we apply the mask $M$ to each point in $R$ to estimate the background by EstimateBackground. To do so, for each point $(i, j)$ in $R$ $(0 \leq i < s,\ 0 \leq j < t)$, we compute the following maximum product

$$\mathrm{b}_{i,j} := \max\left\{\mathrm{r}_{i+h,j+k}\, \mathrm{m}_{f+h,g+k} \;\middle|\; \begin{cases} -f \leq h < w - f \\ -g \leq k < w - g \end{cases} \begin{cases} -i \leq h < r/e - i \\ -j \leq k < c/e - j \end{cases}\right\}. \quad (2)$$

The smallest value between $\mathrm{b}_{i,j}$ and $\mathrm{n}_{i,j}$ is the background of the pixel $(i, j)$ in $N$. Once the background values $B$ are computed, we increase its dimension sizes to the original ones. At the end, we subtract (pixel-wise) the background from the source image $S$.

It is trivial to prove that ComputeRatio and EstimateBackground are the most costly operations in Algorithm 1. For each, the number of operations in terms of $+, \times, /$ and *comparison* is bounded by $O(r\,c\,(\frac{w}{e})^2)$.

## 4. Pipelining Implementation

In this section we present our design and implementation of a fine-grained pipelining version of Algorithm 1 in SPADE. According to the stream processing model introduced in Section 2, we structure the data involved in the computations in Algorithm 1 into streams and organize the operations into operators which can work concurrently on elements of data (set of tuples) in the streams. Please refer to Algorithm 2 in Appendix for our SPADE implementation.

The list of sub-figures in Figure 2 illustrates the data streams and operators of Algorithm 2 with an example. In this example, the width of the image is $c = 10$. The resizing factor is $e = 2$ and the order of the parabolic filter is $w = 3$. The quantities $c, e, w$ are defined in Section 3. The steps in the following description correspond to the operations in Algorithm 2.

(1) First, a user-defined source operator Parse is used to parse the image data from the input binary file and turn it into tuples of intensity values of type `double`, as shown in Figure 2 Subfigure (1) . A tuple represents a row of the input 2D image data. The sequence of those tuples forms a *stream* called `Stream_s`. In this example, as mentioned above, the number of pixels in each tuple is 10.

(2) A built-in operator Aggregate is used to make a *tumbling window*, called `Window_A` of size $e = 2$ from `Stream_s`, The Aggregate operator of SPADE aggregates a number of tuples from the incoming stream into one tuple (like a window) of the output stream. Figure 2 shows a state where `Window_A` consists of two input tuples $\{s_{00}, \cdots, s_{09}\}$ and $\{s_{10}, \cdots, s_{19}\}$.

(3) Each `Window_A` tuple is passed to Resize, a user-defined operator (UDOP), so as to scale down each dimension size of the image by a factor of $e = 2$. This generates `Stream_n` where each tuple has 5 data elements. For instance, at the state shown in Figure 2, the Resize

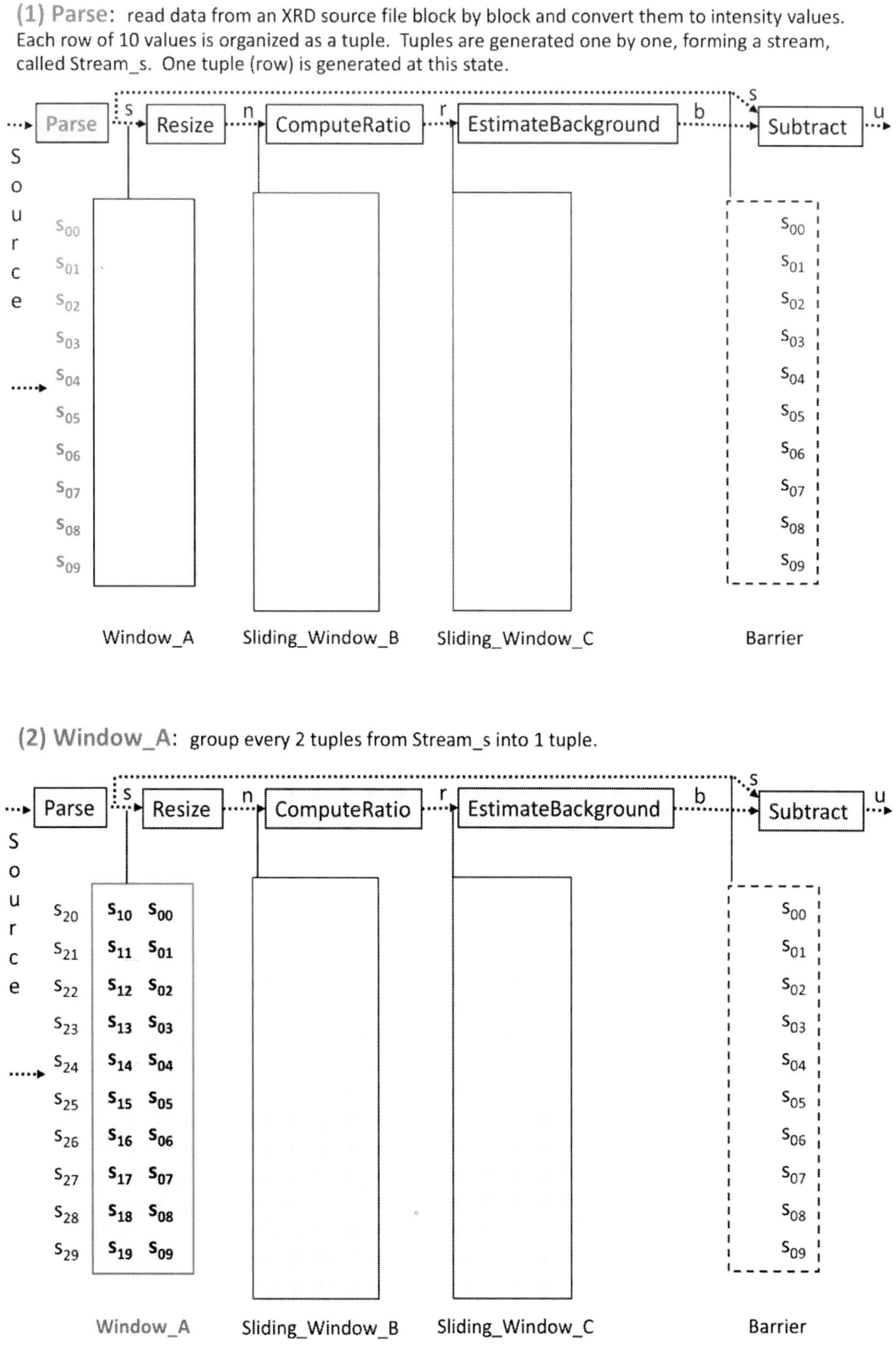


Figure 2: Stream processing model of Algorithm 1 (Subfigures (1) and (2)).

(3) Resize: take in a Window_A and scale down the 2 tuples of size 10 into a tuple of size 5, generating Stream_n. E.g. $n_{04} = (s_{08} + s_{09} + s_{18} + s_{19}) / 4$ .

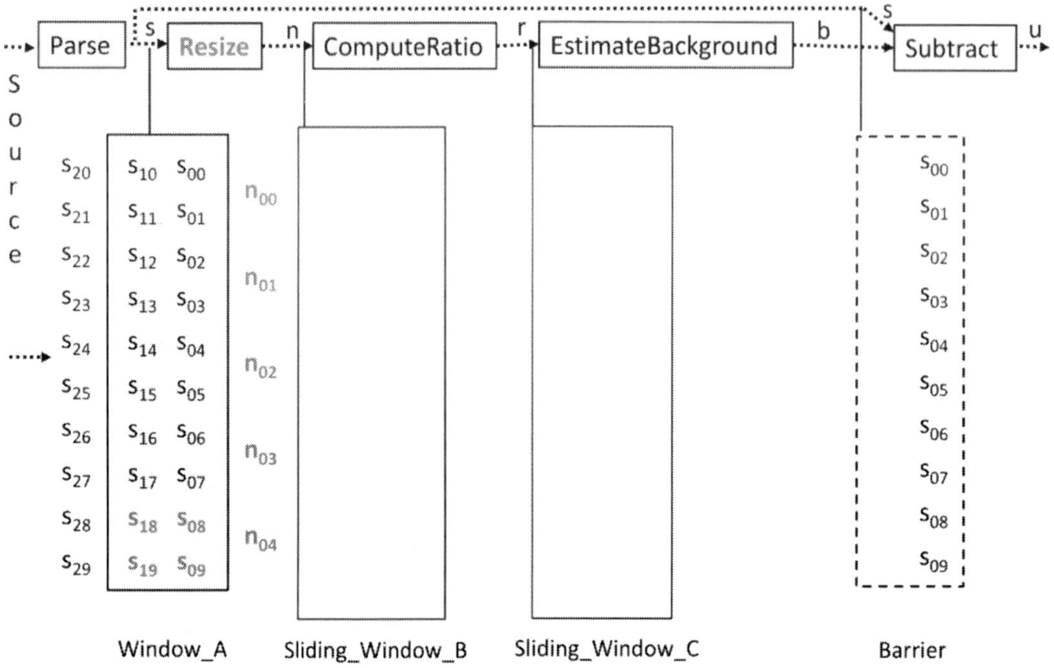


(4) Sliding_Window_B: inject Stream_n and create windows with each having 3 tuples and a sliding factor of 1. E.g. the 1st window contains Tuples 0, 1 and 2; The 2nd window contains Tuples 1, 2 and 3.

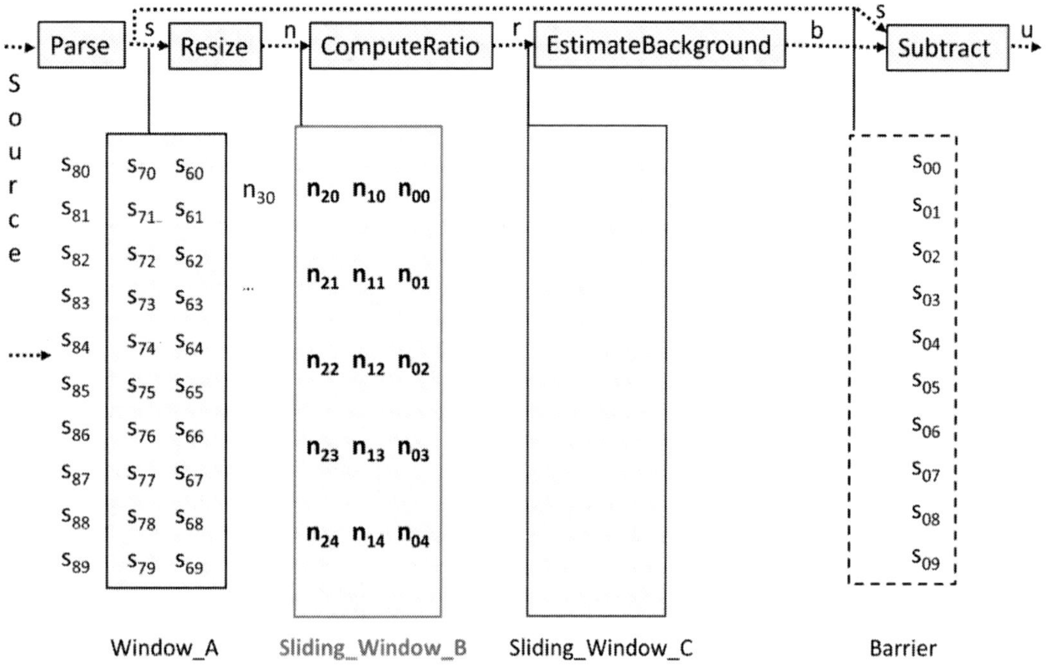


Figure 2: Stream processing model of Algorithm 1 (Subfigures (3) and (4)).

**(5) Compute Ratio:** for each pixel in a Stream_n tuple, compute the minimum ratio among its neighbours w.r.t a mask M of size 3x3 with the center of M, $m_{11}$, over this pixel.
E.g. $r_{04} = \min\{ n_{03}/m_{10}, n_{04}/m_{11}, n_{13}/m_{20}, n_{14}/m_{21} \}$.

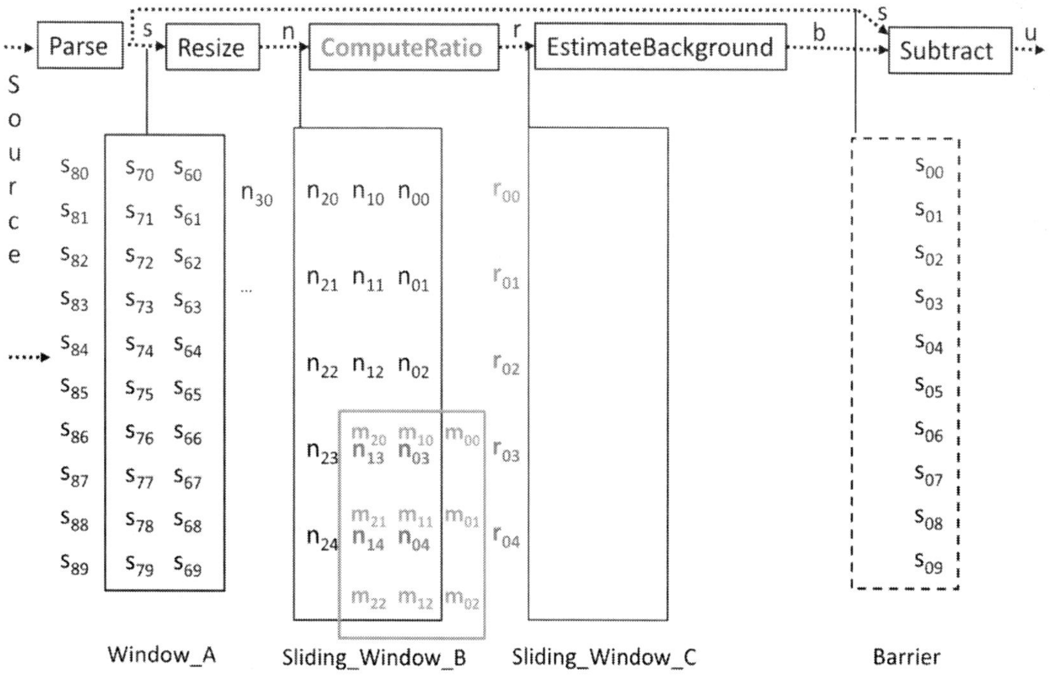


**(6) Sliding Window C:** take in Stream_r and create windows with each having 3 tuples and a sliding factor of 1.

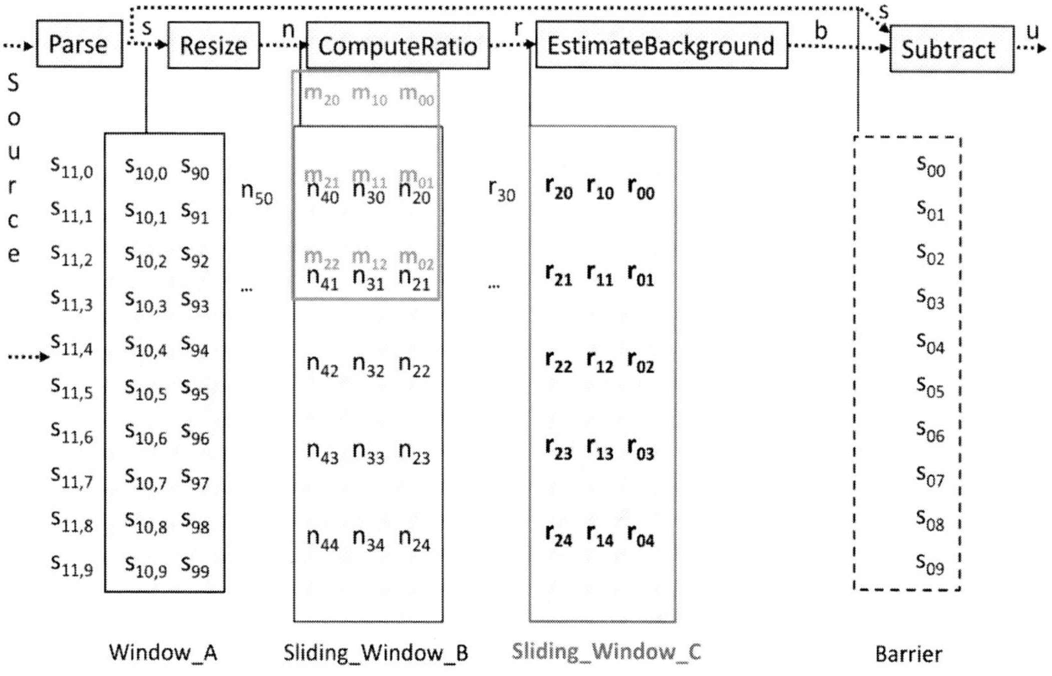


Figure 2: Stream processing model of Algorithm 1 (Subfigures (5) and (6)).

(7) Compute Background: for each pixel in Stream_r, define its background value as the maximum product of the ratio and the mask value among its neighbours with the center of mask, $m_{11}$, over this pixel. E.g. $b_{04} = \max\{ r_{03}m_{10}, r_{04}m_{11}, r_{13}m_{20}, r_{14}m_{21} \}$. One tuple of 5 background values is then scaled up to 2 tuples of size 10, which are submitted to Stream_b.

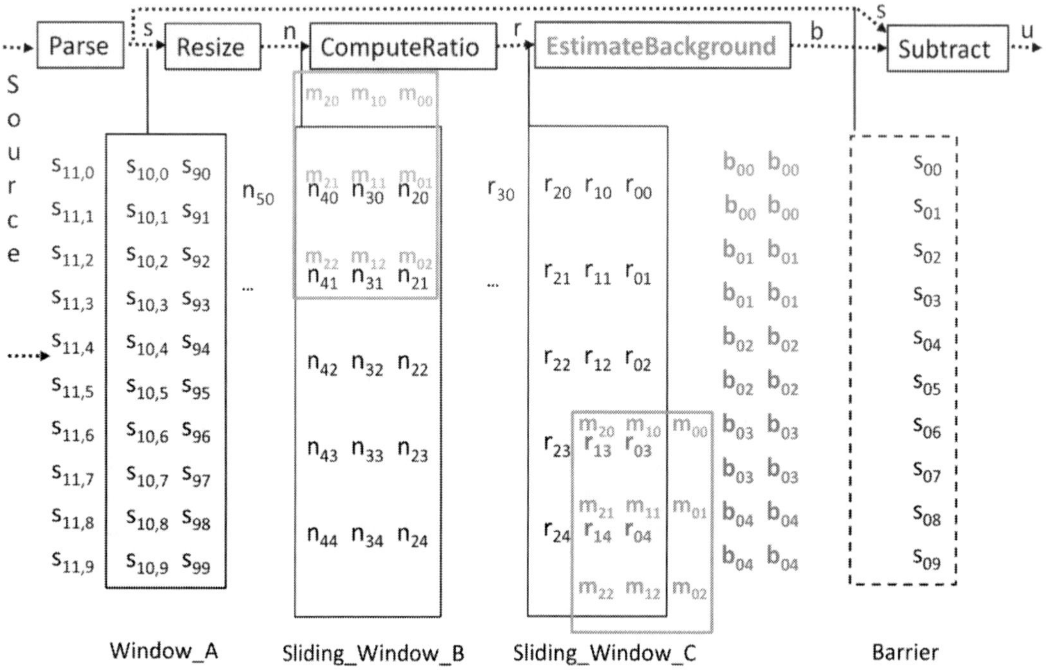


(8) Barrier: synchronize and bundle Stream_s and Stream_b in order to match the original pixel values and their corresponding background values.

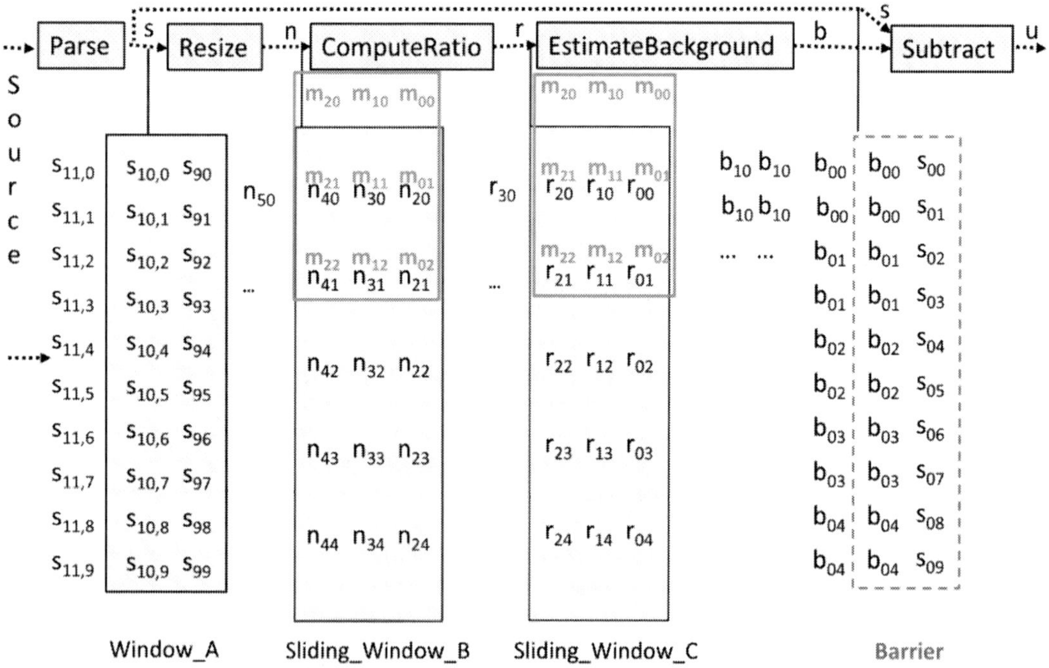


Figure 2: Stream processing model of Algorithm 1 (Subfigures (7) and (8)).

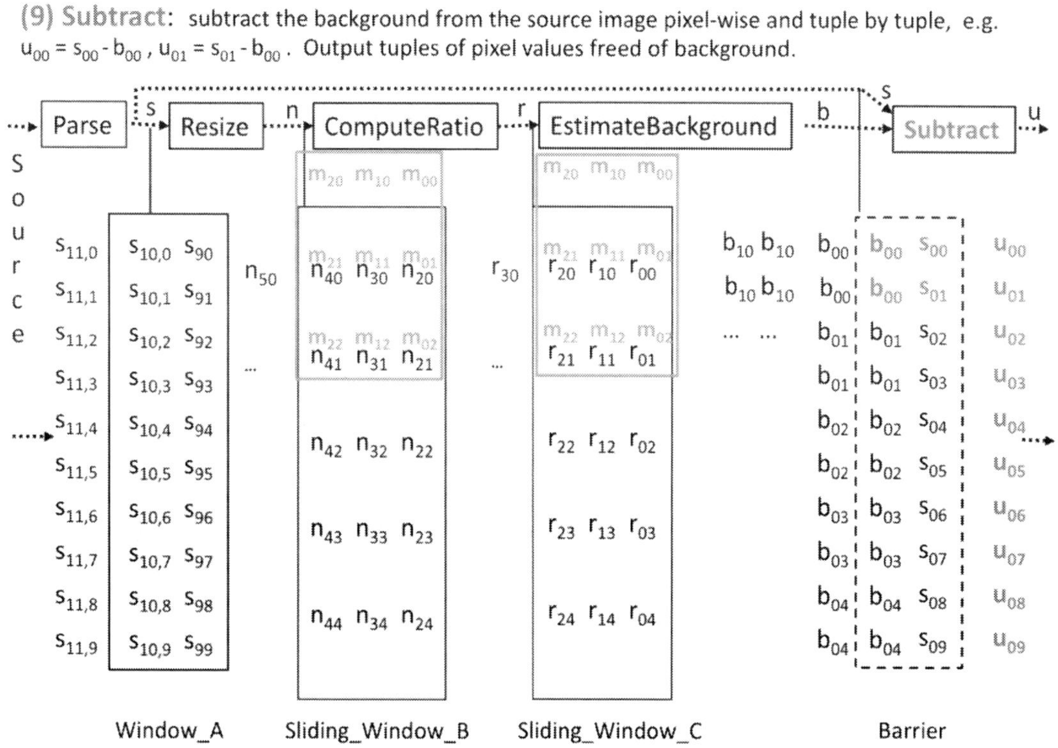


Figure 2: Stream processing model of Algorithm 1 (Subfigure (9)).

operator injects a `Window_A` tuple with data $\{s_{00}, \cdots, s_{09}, s_{10}, \cdots, s_{19}\}$. It then computes $n_{00} = (s_{00} + s_{10} + s_{01} + s_{11})/4, \cdots$, and $n_{04} = (s_{08} + s_{18} + s_{09} + s_{19})/4$. A new tuple of $\{n_{00}, \cdots, n_{04}\}$ will be emitted at the end. As next `Window_A` arrives, the Resize operator will repeat the same operation on it.

(4) Here, another `Aggregate` is used to make a *sliding window* of size $w = 3$ with a *sliding factor* of 1, called `Sliding_Window_B`. It is created by accumulating 3 tuples of the incoming stream. As a new tuple arrives, the oldest tuple get discarded. Every time a new tuple arrives, the aggregation operation is computed and a tuple with the aggregated results is emitted. For instance, in Figure 2 Subfigure (4), the contents in the current `Sliding_Window_B` are $\{n_{00}, \cdots, n_{04};\ n_{10}, \cdots, n_{14};\ n_{20}, \cdots, n_{24}\}$. As a new tuple $\{n_{30}, \cdots, n_{34}\}$ arrives, a new `Sliding_Window_B` will have $\{n_{10}, \cdots, n_{14};\ n_{20}, \cdots, n_{24};\ n_{30}, \cdots, n_{34}\}$.

(5) Every `Sliding_Window_B` is passed to ComputeRatio, a UDOP. The operator ComputeRatio has at hand a mask $M$ representing a parabolic filter, see Section 3. From a `Sliding_Window_B`, the operator ComputeRatio computes a single tuple of size $c/e$, called a `Stream_r` tuple, except for the first and last `Sliding_Window_B`'s which generate two `Stream_r` tuples. Each of these output tuples is computed according to Formula (1).

In the example shown in Figure 2 Subfigure (5), where the order of the mask (i.e. the square matrix encoding it) is $w = 3$, one can see the first `Sliding_Window_B`. The first `Stream_r` tuple $\{\mathrm{r}_{00}, \cdots, \mathrm{r}_{04}\}$ is computed from the incoming tuples $\{\mathrm{n}_{00}, \cdots, \mathrm{n}_{04}\}$ and $\{\mathrm{n}_{10}, \cdots, \mathrm{n}_{14}\}$. Subfigure (5) also shows the view when Formula (1) is applied to compute $\mathrm{r}_{04}$. First, we locate the center of $M$, $\mathrm{m}_{11}$, on top of $\mathrm{n}_{04}$. Then, by Formula (1), $\mathrm{r}_{04}$ is the minimum ratio among its neighbors w.r.t the mask $M$, that is, $\mathrm{r} = \min\{\mathrm{n}_{03}/\mathrm{m}_{10},\ \mathrm{n}_{04}/\mathrm{m}_{11},\ \mathrm{n}_{13}/\mathrm{m}_{20},\ \mathrm{n}_{14}/\mathrm{m}_{21}\}$. Likewise, the second `Stream_r` tuple

$\{r_{10}, \cdots, r_{14}\}$ is computed from the three incoming tuples in the same `Sliding_Window_B`.

Recall that the next `Sliding_Window_B` is obtained by discarding the tuple $\{n_{00}, \cdots, n_{04}\}$ and adding the tuple $\{n_{30}, \cdots, n_{34}\}$. In this case, the three incoming tuples are used to generate $\{r_{20}, \cdots, r_{24}\}$. More generally, and except for the last one, the three incoming tuples of `Sliding_Window_B` are used to produce one `Stream_r` tuple. Finally, the last `Sliding_Window_B` produces two `Stream_r` tuples, similarly to the first `Sliding_Window_B`. More precisely, in this latter case, the three incoming tuples generate one `Stream_r` tuple; then the last two incoming tuples generate the last `Stream_r` tuple.

(6) A `Sliding_Window_C` of size 3 with a sliding factor of 1 is created from `Stream_r` in the same manner as in Step (4).

(7) We develop a UDOP EstimateBackground with the same mask $M$ as in Step (5) to compute the background values of the pixels from each `Sliding_Window_C` according to Formula (2). `Sliding_Window_C`'s with `Stream_r` tuples are used in the same manner as `Sliding_Window_B`'s with `Stream_n` tuples are used in Step (5). After the background values of an incoming tuple are computed, for instance in Figure 2 Subfigure (7), the values of $b_{00}, b_{01}, b_{02}, b_{03}, b_{04}$ for the first `Stream_r` tuple, we scale them up by a factor of 2 resulting in two output tuples of size 10, each containing $\{b_{00}, b_{00}, b_{01}, b_{01}, b_{02}, b_{02}, b_{03}, b_{03}, b_{04}, b_{04}\}$. Thus, the operator EstimateBackground generates a sequence of tuples for the background values, called `Stream_b`.

(8) In Figure 2 Subfigure (8), a `Barrier` is used to synchronize and bundle `Stream_s` and `Stream_b` in order to match the original pixel values with their corresponding background values. This produces a stream where each tuple consists of two fields, one for a row of original pixel values and one for their corresponding background values.

(9) The above synchronized stream is passed to a UDOP `Subtract` to subtract the background from the original values pixel-wise for each tuple, producing a stream of data freed of background noise.

To this point, the stream of XRD data freed of background can be either passed to other operators in the same stream processing system for further analysis such as blob searching and peak fitting or output to an external device such as a file by the built-in Sink operator as in Algorithm 2 in Appendix. Overall, all the operators run concurrently on the elements in streams of data, until the end of the XRD source data. Since our pipelining implementation executes in the same computational order as the sequential algorithm, its correctness follows. It is noticeable that in this design there are two expensive operators ComputeRatio and EstimateBackground which workloads are rather balanced.

## 5. Experimentation

We have InfoSphere Streams Version 1.0 installed and configured on an Intel Core2 Quad CPU Q9550 (2.83 GHz, 8 GB RAM and 6 MB L2 cache) running with Red Hat Enterprise Linux 5.3 operating system. Given $P$ image files, our program is parameterized so that it can set up $P$ pipelines to process them. The $P$ pipelines share the computing resources. Streams runtime system schedules the jobs and manages the resources, which is transparent to a user's application.

The XRD source files in our tests are from a set of synchrotron experimental results in [3]. The size of each 2D XRD source file is 8 MB. It encodes a row-major layout matrix of $2084 \times 2084$ pixel values in binary format. We set the resizing factor $e$ to 2. The larger the resizing factor, the less the accuracy of the estimated background. According to our analysis in Section 3, the larger the filter size $w$, the higher the computation cost. Moreover, what values of $w$ that are suitable depend on the type of X-ray diffraction patterns. As reported in [2] and [6], the Full Width at Half Maximum (FWHM) of a Bragg's peak is in general between 10 and 40. We have used three different sizes of parabolic filters in our experimentation: 25, 45 and 85, which correspond to FWHM of values 10, 20 and 40. This allows us to test the performance of our pipelining implementation for different sizes of problems. Meanwhile it provides information on the effect of using different sizes of filters to help with users' analysis. In Appendix, the sub-figures in Figure 4 show an XRD source image and the output images with the background removed by parabolic filters of size 25, 45 and 85, respectively.

Table 1: Timing and speedup for processing $P$ files using $P$ pipelines on a quad-core machine.

| | $w = 25$ | | | $w = 45$ | | | $w = 85$ | | |
|---|---|---|---|---|---|---|---|---|---|
| P | Seq. (s) | Pipe. (s) | Sp. | Seq. (s) | Pipe. (s) | Sp. | Seq. (s) | Pipe. (s) | Sp. |
| 1 | 4.93 | 2.82 | 1.7 | 16.1 | 8.4 | 1.9 | 55.8 | 26.4 | 2.1 |
| 2 | 9.86 | 3.18 | 3.1 | 32.2 | 9.0 | 3.6 | 111.6 | 27.5 | 4.1 |
| 3 | 14.8 | 4.81 | 3.1 | 48.3 | 10.8 | 4.5 | 167.4 | 35.8 | 4.7 |
| 4 | 19.7 | 5.78 | 3.4 | 64.4 | 14.5 | 4.4 | 223.2 | 45.4 | 4.9 |
| 5 | 24.7 | 6.45 | 3.8 | 80.5 | 18.0 | 4.5 | 279.0 | 57.3 | 4.9 |
| 6 | 29.6 | 8.11 | 3.6 | 96.6 | 21.5 | 4.5 | 334.8 | 66.1 | 5.1 |
| 8 | 39.4 | 10.1 | 3.9 | 128.8 | 28.6 | 4.5 | 446.4 | 93.9 | 4.8 |
| 10 | 49.3 | 13.3 | 3.7 | 161.0 | 36.2 | 4.4 | 558.0 | 112.2 | 5.0 |
| 12 | 59.2 | 15.6 | 3.8 | 193.2 | 42.7 | 4.5 | 669.6 | 135.6 | 4.9 |
| 14 | 69.0 | 18.2 | 3.8 | 259.2 | 49.4 | 5.2 | 781.2 | 158.6 | 4.9 |
| 16 | 78.9 | 20.8 | 3.8 | 257.6 | 56.4 | 4.6 | 892.8 | 181.9 | 4.9 |

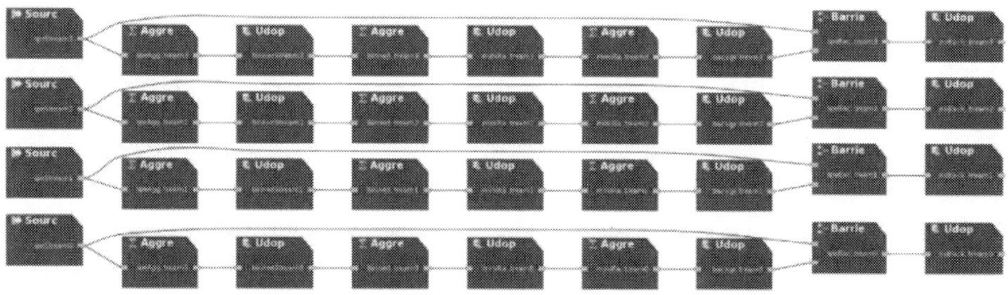

Figure 3: Processing 4 XRD images using 4 pipelines in InfoSphere Streams.

We have obtained timings for processing $P$ files using 3 different sizes of filters by $P$ concurrent pipelines with $P$ from 1 to 16 on the quad-core machine. Figure 3 is the Streams Live Graph captured by Streams Studio for processing 4 XRD source files by 4 pipelines concurrently. The timing results and the speedup factors are summarized in Table 1. `Seq.` denotes the sequential program. `Pipe.` is our pipelining implementation. The speedup (Sp.) factor of the pipelining program is computed w.r.t the timing for processing the same number of files by the sequential C program.

The timing for processing a XRD image by our program increases w.r.t the size of the filter. The sequential program takes about 5 seconds using a filter of size 25, 16 seconds using one of size 45 and about 55 seconds using one of size 85. By examining the quality of the output images users can choose a proper size, as the example shown in Figure 4 in Appendix.

We have developed a tool for measuring the timing of concurrent executions in Streams, as follows. For each pipeline, we create an auxiliary stream and use the Sink operator of SPADE to write the time for two special tuples into two separated files. One is the time when the first tuple from the source file is produced, called *starting time*. The other one is the time when the last tuple of the result stream is emitted, called *ending time*. For processing $P$ files using $P$ pipelines, there are $P$ records of starting time and $P$ records of ending time. We define the difference between the latest ending time and the earliest starting time to be the timing for a concurrent execution.

The speedup for processing one image using one pipeline is from 1.7 for a small-size problem to 2.1 for a large-size problem. The speedup of using one pipeline is obtained by the two costly but balanced operators, ComputeRatio and EstimateBackground. This result agrees well with our analysis in Section 3 and 4. To process 2 files using 2 pipelines, the parallelism is improved by the concurrent execution of two pipelines. We gain a speedup factor of 3.1 for the small-size problem, and 4.1 for the large one. The cases for processing many files (3 to 16) using the same number of pipelines create sufficient parallelism and potentially hide memory latency. They have achieved super-linear speedup on the quad-core machine.

In summary, our pipelining implementation performs well for all sizes of the problems that we have tested, and even better for the large one. It helps to improve the throughput of our XRD data processing by an order of more than the number of cores on a quad-core machine. It demonstrates high potential of making good use of multi-cores for processing large volume of XRD image data in InfoSphere Streams.

## 6. Concluding Remarks

Our initial work on stream processing of XRD images has been encouraging. While full analysis of individual XRD images requires further computation, we feel that the approach will result in a useful stream processing model for these images. Our overall objective was to develop analysis of XRD images where images acquired from a detector could be sent for analysis and results sent to the researcher nearly as quickly as they could be pulled from the detector. Our preliminary work suggests that the streaming approach and the use of the *InfoSphere Streams* platform, has the potential to achieve this goal.

Further extension of this work include achieving performance enhancement at two levels. The first level is at the implementation details of the main algorithms. In particular, developing more fined-grained user defined operators to allow flexibility in paralleling lower steps. The second level is realizing faster file loading mechanism. Image files' reading performances are bounded by the underlying file system throughput. Developing faster file loading techniques, will achieve gain in performance at the pipelining level and the algorithmic level.

In long term, this work provides the seed for the development of more general library of operators for processing images. In particular, operators, such as the background removal algorithm illustrated in this paper, could form part of a more general library of streaming operators for processing images, derived from a varieties of detectors. This will open a framework of collaboration among scientists around streaming technology and reusable assets.

## 7. Acknowledgments

We would like to acknowledge the support for this project from CANARIE and for the networking infrastructure that they provide to enable us to even begin to think about streaming image data. We would also like to thanks all those involved in the Science Studio project.

## References


[1] Active Network Interchange for Scientific Experimentation (ANISE) Project 2010 http://www.anise-project.com/ (University of Western Ontario, London ON, Canada)

[2] Brückner S 2000 Estimation of the Background in Powder Diffraction Patterns through a Robust Smoothing Procedure *Journal of Applied Crystallography* **33** 3.2

[3] Chao J et al. 2009 Study of Residual elastic- and plastic-deformation in uniaxial tensile strained nickel-based Alloy 600 samples by polychromatic X-ray microdiffraction (PXM) and neutron diffraction methods *Materials Science and Engineering: A* Volume 524, Issues 1-2, Pages 20-27

[4] Chuang J and Ice G 1999 Automated indexing for texture and strain measurement with broad-band pass x-ray micro-beams *Journal of Applied Physics* No. 9, Vol. 86

[5] Cullity B D and Stock S R 2001 *Elements of X-Ray Diffraction* ISBN 0201610914, Prentice Hall

[6] Dinnebier R 2003 FWHM Optimized Polynomial Smoothing Filters: A Practical Approach *Powder Diff.*, Vol. 18, No. 3

[7] IBM 2010 *IBM InfoSphere Streams* http://www-01.ibm.com/software/data/infosphere/streams/

[8] IBM 2010 *IBM InfoSphere Streams Version 1.0.10 Installation and Administration Guide*

[9] IBM 2010 *IBM InfoSphere Streams SPADE – Programming Model and Language Reference*

[10] Interactive Data Language (IDL) website: http://www.ittvis.com/ProductServices/IDL.aspx

[11] Kajfosz J and Kwiatek W 1987 Nonpolynomial Approximation of Background in X-ray Spectra *Nuclear Instruments and Methods in physics Research* B22 78-81 (North-Holand, Amsterdam)

[12] Sherry N 2009 C Program in Science Studio Project (University of Western Ontario, London ON, Canada)

[13] The Science Studio Project 2009 http://sciencestudioproject.com/ (University of Western Ontario and the Canadian Light Source)

[14] Tamura N 2010 Fortran Procedure "backfit_bruckner" in XMAS (the Advanced Light Source, Lawrence Berkeley National Laboratory, USA)

[15] Advanced Light Source 2010 *X-ray Microdiffraction Analysis Software (XMAS)* (Lawrence Berkeley National Laboratory, USA) http://xraysweb.lbl.gov/microdif/user_resources.htm

[16] Advanced Photon Source 2009 *3D X-ray Microdiffraction Analysis Software Package* (Argonne National Laboratory, USA) http://www.aps.anl.gov/Sectors/33_34/microdiff/downloads/

[17] Woolfson M M 1997 *An Introduction to X-Ray crystallography* 2nd edition, Cambridge, New York, Cambridge University Press


## 8. Appendix

**Algorithm 2: Stream processing model of Algorithm 1 in SPADE**

**Input** : file $F$ encoding a $2D$ XRD image $S$ of dimension sizes $r \times c$, a resizing factor $e$, a filter size $w$

**Output** : image $S$ freed of background noise

```
1  stream s (stuple : DoubleList)
   := Source () ["file : //F", udfbinformat= "Parse"] { };
2  stream Window_A (Atuple : DoubleList)
   := Aggregate (s ⟨count(e)⟩) [ ] {Col(stuple)};
3  stream n (ntuple : DoubleList)
   := Udop (Window_A) ["Resize"] { };
4  stream Sliding_Window_B (Btuple : DoubleList)
   := Aggregate (n ⟨count(w), count(1)⟩) [ ] {Col(ntuple)};
5  stream r (rtuple : DoubleList)
   := Udop (Sliding_Window_B) ["ComputeRatio"] { };
6  stream Sliding_Window_C (Ctuple : DoubleList)
   := Aggregate (r ⟨count(w), count(1)⟩) [ ] {Col(rtuple)};
7  stream b (btuple : DoubleList)
   := Udop (Sliding_Window_C) ["EstimateBackground"] { };
8  stream sb (stuple : DoubleList, btuple : DoubleList)
   := Barrier (s; b) [ ] {$1.stuple, $2.btuple};
9  stream u (utuple : DoubleList)
   := Udop (sb) ["Subtract"] { };
10 Nil := Sink (u) ["file : //F.out"] { };
```

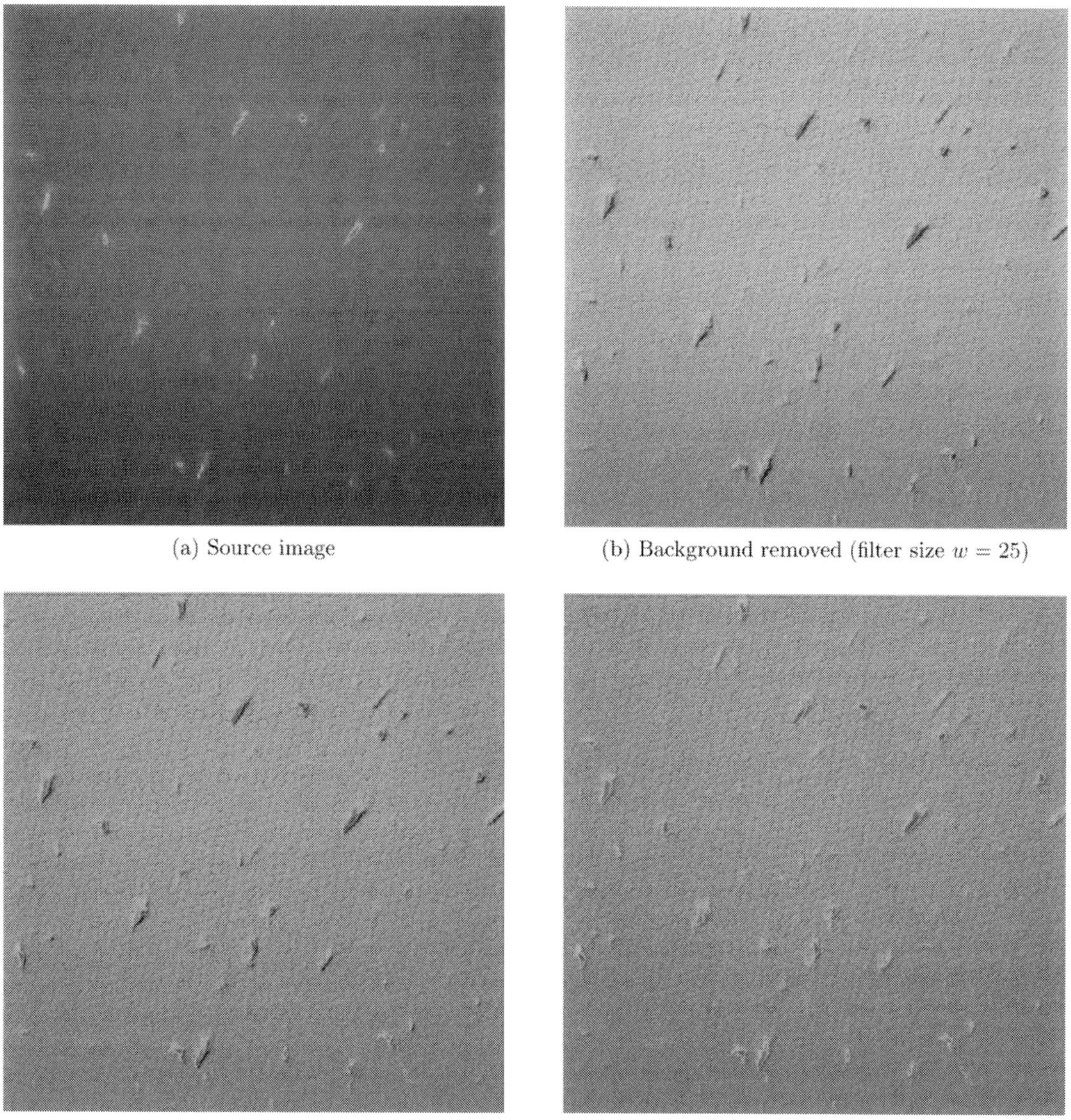

(a) Source image

(b) Background removed (filter size $w = 25$)

(c) Background removed (filter size $w = 45$)

(d) Background removed (filter size $w = 85$)

Figure 4: An XRD image example.

# Parallel Branch and Bound Algorithm - A comparison between serial, OpenMP and MPI implementations

**Lucio Barreto and Michael Bauer**

Department of Computer Science

Middlesex College - The University of Western Ontario – London, ON – Canada

{lbarret6, bauer}@csd.uwo.ca

**Abstract**. This paper presents a comparison of an extended version of the regular Branch and Bound algorithm previously implemented in serial with a new parallel implementation, using both MPI (distributed memory parallel model) and OpenMP (shared memory parallel model). The branch-and-bound algorithm is an enumerative optimization technique, where finding a solution to a mixed integer programming (MIP) problem is based on the construction of a tree where nodes represent candidate problems and branches represent the new restrictions to be considered. Through this tree all integer solutions of the feasible region of the problem are listed explicitly or implicitly ensuring that all the optimal solutions will be found. A common approach to solve such problems is to convert sub-problems of the mixed integer problem to linear programming problems, thereby eliminating some of the integer constraints, and then trying to solve that problem using an existing linear program approach. The paper describes the general branch and bound algorithm used and provides details on the implementation and the results of the comparison.

## 1. Introduction

Integer Programming problems (IP) are special cases of optimization problems where the variables can only assume integer values. Mixed Integer Programming problems (MIP) are special cases where only some of the variables are restricted to integer values. Optimization problems with integer variables can also be linear or nonlinear, depending on the terms of their objective function and their constraints. However, in general, the terms IP and MIP are almost always associated with problems that have linear features. The optimization algorithms considered in this work focus on mixed integer programming, where some variables assume integer values and while other variables take on continuous values.

In considering only problems with linear constraints and linear objective functions, there are important differences between IP and MIP in relation to Linear Programming (LP):

- LP – there are necessary and sufficient conditions of optimality proved theoretically that can be used to efficiently test whether a given feasible solution is an optimal solution. These conditions are used to develop algebraic methods such as the simplex method.
- IP/MIP – there are no known optimality conditions to test whether a given feasible solution is optimal. It is necessary to perform the implicit or explicit comparison of all feasible solutions of the problem.

Approaches to solve IP and MIP problems are numerous, usually specialized in accordance with a particular application. We can divide these approaches into two broad families, with distinct characteristics:

- Heuristic Optimization – may get good results for problems in which classical optimization methods might fail (such as many integer variables and very complex constraints), but there is no guarantee of optimality. Examples of such approaches are genetic algorithms [5][8] and tabu search [3][4].
- Classical Optimization – for convex problems it is guaranteed that the solution is optimal. In this family there are the methods of enumeration, such as the zero-one enumeration algorithm [1] and the branch-and-bound algorithm [2], used in this work.

We focus specifically on using a branch-and-bound approach for finding an optimal solution to MIP problems, including being able to find all optimal solutions when there are multiple ones. The branch-and-bound algorithm is an enumerative optimization technique, where finding a solution to an MIP problem is based on the construction of a tree where nodes represent candidate problems (sub-problems) and branches represent the new restrictions to be considered. Through this tree, all integer solutions of the feasible region of the problem are listed explicitly or implicitly ensuring that all the optimal solutions will be found.

A common approach to solve such problems is to convert the sub-problems that arise in solving integer linear programming problems to linear programming problems, that is, with no integer constraints, and to solve those problems using an existing linear programming solver. The resulting solution can be used to set bounds on the possible solutions or to help select variables to create restrictions that can be used to create sub-problems. The paper presents a comparison of a branch and bound algorithm implemented with three different approaches: serial, shared memory model (OpenMP) and distributed memory model (MPI). The implemented branch and bound algorithm uses the GNU/GLPK [9] as the Linear Programming (LP) solver to find optimal solutions to mixed integer linear problems.

The reminder of the paper is organized as follows. Section 2 describes some important theoretical aspects of branch-and-bound. Section 3 describes the branch-and-bound algorithm itself. Section 4 describes the computational implementation. Section 5 shows a comparison of the approaches. Section 6 presents conclusions about this work and finally Section 7 identifies future work.

## 2. Theoretical Aspects

The branch-and-bound algorithm is an enumerative technique, in which a solution is found based on the construction of a tree in which nodes represent the problem candidates and branches represent the new restrictions to be considered. Through this tree, all integer solutions of the problem feasible region are listed explicitly or implicitly ensuring that all the optimal solutions will be found.

The overall structure of the branch-and-bound algorithm has three key elements, separation, relaxation and pruning [2]. Separation uses the tactic of "divide and conquer" in order to solve the problem (P). In order to find the solution of P, it is decomposed into two or more descendant sub-problems, generating a list of candidate problems (CP). At a subsequent step in the algorithm, a candidate is selected from the list of candidate problems and the algorithm tries to solve that problem. If a solution cannot be found to that problem, that problem is again decomposed and its descendants are added to the list of candidate problems. If the selected problem can be solved, then a new solution is obtained. The objective function value of this new solution is then compared with the value of the incumbent solution, which is the best feasible solution known so far. If the new solution is better than the incumbent solution, it becomes the new incumbent. Then, the algorithm returns to the list and selects the next candidate. This procedure is repeated until the list is empty, and the solution of the problem is taken as the final incumbent solution.

The usual way to carry out separation of an integer programming problem is through contradictory constraints in a single integer variable (separation variable or branching variable). Thus, from the original problem (called node zero), two new descendant sub-problems are created, which are easier to solve than the original one, since a constraint was added to the separation variable. Each generated node has an associated candidate sub-problem and each branch indicates the addition of a constraint related to the variable used in the separation. Therefore, as the algorithm moves down in the tree, the viable region of the generated descendants becomes more restricted.

The most common approach to relaxation is the elimination of the integral constraints (integer variables) where the integer or mixed integer problem is converted to a general LP. Relaxation assumes that the original integer variables may take fractional values and then the resulting LP problem is solved. The obtained optimal solution usually has several variables with non-integer values. Among these variables, one must be selected for separation.

Once the separation and inclusion of new descendants in candidates list is completed, the algorithm must select from among the stored candidates the next to be evaluated and, if necessary, separated successively until the linear problem solution becomes integer, or unfeasible or worse than the incumbent, meaning that the candidate sub-problem can be removed from the list (pruned), not producing any more descendants. This procedure repeats until the list of candidates' sub-problems becomes empty.

## 3. Branch and bound Algorithm

Consider the MIP (P) whose general form is given by:

$$(P)\quad \begin{array}{lll} \text{minimize} & z = \mathbf{c}^{\mathrm{T}}\mathbf{x} & (1) \\ \text{subject to} & \mathbf{A}\mathbf{x} \leq \mathbf{b} & (2) \\ & x_i \text{ integer } \forall i \in I & \\ & x_j \text{ real } \forall j \in J & \end{array}$$

Where **c** is the vector of costs, **x** is the vector of variables integers ($x_i$) and real ($x_j$), **A** is the constraint coefficient matrix, **b** is the vector of right side of the inequalities, $I$ is the set of integer variables and $J$ is the set of continuous variables. The branch-and-bound algorithm to solve ($P$) has the following steps:

1. **Start-up**: set the number of active nodes $n = 0$, set the first incumbent and initialize the list of candidate sub-problems with the original problem ($P$).
2. **Convergence test**: if the candidate list is empty then this means that the process is over and the current incumbent solution is the optimal solution of the problem. Otherwise, continue.
3. **Candidate selection**: among the candidate sub-problems not yet pruned, choose the one that will be the next to be evaluated and remove it from the list. Solve the LP problem related to the relaxed selected problem $\left(CP_R^k\right)$ and store the optimal solution as a lower bound for all its descendants, $z_{\inf}^k = z^*\left(CP_R^k\right)$.
4. **Pruning tests**: the sub-problem $\left(CP^k\right)$ may be pruned if it meets one of the following conditions:
   a) if $(CP_R^k)$ has no feasible solution;
   b) if $z_{\inf}^k > z^*$, where $z^*$ is the actual incumbent value;
   c) if the optimal solution of $(CP_R^k)$ is integer and feasible in $\left(CP^k\right)$. In this case, if the optimal value is lower than the incumbent value, make $z^* = z_{\inf}^k$ and apply the previous test (b) for all candidates' sub-problems not yet pruned.

   If the candidate sub-problem $\left(CP^k\right)$ was pruned, return to the Step 2.

5. **Separation**: from the sub-problem $(CP^k)$, select a variable for separation from those that are integer and still have continuous value. For the chosen variable $x_i$, whose current value is $x_i^*$, generate two new descendant sub-problems and add them to the candidate list. The new sub-problems are generated by adding to $(CP^k)$ the following restrictions:

$$(PC^{n+1}): x_i \leq \left[x_i^*\right] \tag{3}$$

$$(PC^{n+2}): x_i \geq \left[x_i^*\right]+1 \tag{4}$$

where $\left[x_i^*\right]$ is the largest integer not greater than $x_i^*$. Set $n = n+2$ and return to Step 3.

### 3.1. Performance Enhancement

The algorithm efficiency is directly related to the method of selecting the next candidate sub-problem that will be evaluated (Step 3) and the separation variable (Step 5). Moreover, the existence of a good initial incumbent increases the effectiveness of the pruning test (b), thereby reducing the number of candidate sub-problems that need to be evaluated.

Although there is no systematic technique to determine which one of the candidate sub-problems leads more quickly to the solution, some heuristic rules can be used, such as search by creation order, by depth, by the relaxed solution value, by the solution estimated value, among others.

A very common technique used to select the candidate sub-problem is the LIFO rule (Last In, First Out), which produces a depth search. This strategy allows that the descendant sub-problem be solved from the previous problem (because they differ in only by a single variable) and minimizes the memory requirements to store the candidates' information. The way which these selections are made directly influences the number of nodes that need to be evaluated, determining, thus, the computational effort.

On the other hand, there are methods that use estimates of the objective function value to select from among all candidates the most promising sub-problem. This rule minimizes the total number of problems to be evaluated, but at the same time can drastically increase the memory requirements, since the successive LP problems may not have the same similarity that exists in the depth search.

### 3.2. Variable Selection

When the solution of the relaxed candidate $(PC_R^j)$ has several integer variables with continuous value, one of them must be selected to be separated. An inadequate variable choice implies evaluation of several descendants' sub-problems that could be eliminated by pruning their predecessor. There is no systematic technique to identify the optimal separation variable, but there are empirical rules that indicate which variables are most attractive. If simplicity is the goal, the separation variable can be selected from a predetermined sequence using the objective function coefficients, highest to lowest cost, for example, or being more specific depending on the problem's features. A more effective alternative is based on the search for the variable that has the highest value in terms of the estimated increase of the objective function, which can be obtained from one of the following techniques:

[MAX;MAX]: always choose the variable which causes the greatest degradation in the objective function in order to quickly obtain a descendant sub-problem that might be pruned. For the candidate sub-problem node k, the selection method for the separation variable j follows the expression below:

$$\max\left\{\max(P_i^- f_i^j, P_i^+(1-f_i^j))\right\}$$

[MAX;MIN]: choose the variable which the minimum variation causes the maximum impact. This ensures that both descendants contribute to pruning the sub-problem. For the candidate sub-problem node k, the selection method for the separation variable j follows the expression below:

$$\max\{\min(P_i^- f_i^j, P_i^+(1-f_i^j))\}$$

[MIN;MAX]: choose the variable which the maximum variation causes the minimum impact. Similarly, both descendants contribute to pruning this sub-problem. For the candidate sub-problem node k, the selection method for the separation variable j follows the expression below:

$$\min\{\max(P_i^- f_i^j, P_i^+(1-f_i^j))\}$$

[MIN;MIN]: always choose the variable which causes the lowest degradation in the objective function in order to quickly obtain a viable solution. This is the most conservative method. For the candidate sub-problem node k, the selection method for the separation variable j follows the expression below:

$$\min\{\min(P_i^- f_i^j, P_i^+(1-f_i^j))\}$$

Another issue involved in variable selection is the possibility of separating a variable which was assigned an integer value. It often occurs in a branch-and-bound execution that a specific variable gets an integer value prematurely, i.e., without any limits to impose that integer value. In this case, according to Step 5 of the branch-and-bound algorithm, this variable would not be selected for separation (because it is integer), and if all others variables were also integers, the solution would be considered viable, not generating descendants. This would generate a fail in order to guarantee the reaching of all alternative solutions to the problem. By separating the variables with integer values which still have a lower bound minor than its upper bound, the algorithm guarantees that all integer variables involved in the problem will be evaluated in an implicit or explicit way, exhausting all possible separation opportunities and allowing that all alternative solutions to the problem may be found.

### 3.3. Multiple Optimal Solutions

Some problems may have multiple optimal integer solutions, such as optimization models that describe the power networks expansion problem for both transmission [12] and distribution [6][7]. In this case, obtaining all the solutions is very important. Although the objective function value is the same for all solutions, one solution may be greater than the other when taking into account other factors that were not explicitly included in the optimization model. If these other factors have some relevance, one may also want to retain sub-optimal solutions. This can be easily accomplished when solutions found are stored in an ordered list, as used in this work.

To ensure that all solutions are found, it is necessary to make changes in the pruning (Step 4) and separation (Step 5) processes. In Step 4 (c), one solution is considered integer when all the integer variables cannot assume other values, i.e., when they all have inequality constraints with upper and lower bounds both integers and identical. Thus, the Step 4 (c) of the conventional algorithm should be replaced by:

c) if the optimal solution of $(CP_R^k)$ is feasible in $(CP^k)$ and all integer variables in the optimal solution of $(CP_R^k)$ have no degree of freedom – i.e. $\mathbf{x}_k^{\text{lb}} = \mathbf{x}_k^{\text{ub}}$, $\forall k \in I$, where $\mathbf{x}_k^{\text{lb}}$ and $\mathbf{x}_k^{\text{ub}}$ are the lower and the upper bound vectors of integer variables of candidate problem $(CP_R^k)$, respectively. In this case, if the optimal value is lower than the incumbent value, make

$z^* = z_{\inf}^k$ and apply the previous test (b) for all candidates' sub-problems not pruned yet. Otherwise, if the optimal value is equal to the incumbent value, store the new alternative solution.

Notice that the components of vectors $\mathbf{x}_k^{\mathrm{lb}}$ and $\mathbf{x}_k^{\mathrm{ub}}$ are always integer values generated by the separation process, according to equations (3) and (4). In Step 5, in addition to using the variables that have continuous values, it is also necessary to consider separating the integer variables that have some degree of freedom, i.e., with $x_i^{\mathrm{lb}} < x_i^{\mathrm{ub}}$.

## 4. Computational Implementation

The branch-and-bound algorithm was implemented in C++ and was adapted to three different versions: one serial and two parallel. The two parallel implementations respectively used OpenMP [11] and MPI [10]. The parallel versions are based on the serial approach, just making use of the necessary parallel calls. However, all approaches use the same programming logic.

In the system, several parameters may be configured, such as:

- maximum size of the branch-and-bound tree;
- different tolerances for integer and equality;
- selection criteria of the separation variable;
- criteria to select a node;
- minimum depth to open the tree[1];
- occupation percentage of the available space in the tree to exchange sort order of the active nodes in order to preserve available memory;
- possibility of separation in integer variables.

The central data structure used in the branch-and-bound algorithm stores all the information that characterizes each candidate sub-problem (active node). The information associated with each node of the solution tree is divided into two groups:

- active nodes – information related to the candidate sub-problems;
- tree nodes – information of the predecessors nodes that were separated (using a contradictory constraint) generating the active nodes.

In this work, each node stores only the information relating to the last separation; the rest of the information is associated with the tree nodes. This procedure avoids data redundancy.

The candidate sub-problems, when generated, are inserted into the lists of active nodes; several lists are maintained depending on a particular sort criterion. The available ordering possibilities on active nodes are as follows:

- node number (LIFO, Last In First Out);
- node depth;
- node relaxed solution (LP value);
- node estimated solution (using pseudo-costs);
- node estimated solution weighted with depth (deepest nodes are more important).

[1] Before adopting the chosen strategy for selecting the candidate node, the program will expand the tree breadth-first, until all active nodes have the same depth as the minimum chosen in the configuration.

Since all lists are circularly-linked, the algorithm can select the node that has any of the extreme values (minimum or maximum, in each of the criteria) directly. Moreover, as all lists are simultaneously updated, it is possible to switch between any criteria during the solution process without the need for reordering, which have high computational cost.

Whenever a candidate node is pruned (satisfying some of the criteria presented in Step 4 of the algorithm in Section 3) it is removed from the list of active nodes, together with all its predecessors that do not possess active descendants, recursively, from the node that was pruned towards the source node of the tree.

As an example, consider the problem P:

$$(P)\begin{cases} \min & v = 5n_{12} + 2n_{13} + 2n_{23} + \beta \\ s.a. & 350n_{12} + 400n_{13} + \beta \geq 400 \\ & 350n_{12} + 210n_{23} + \beta \geq 200 \\ & \beta \geq 0 \\ & n_{ij} \geq 0 \quad \forall ij \\ & n_{12}, n_{13} \text{ e } n_{23} \text{ integers} \end{cases}$$

The branch-and-bound algorithm for this problem can be represented by Figure 1, where each node is represented for a circle with a number that indicates its generation's order (index $i$ of the algorithm). In the side of each node there is the relaxed solution and if it that solution is "integer" or worse than the incumbent, it is related to a pruned node, represented by a colourful circle.

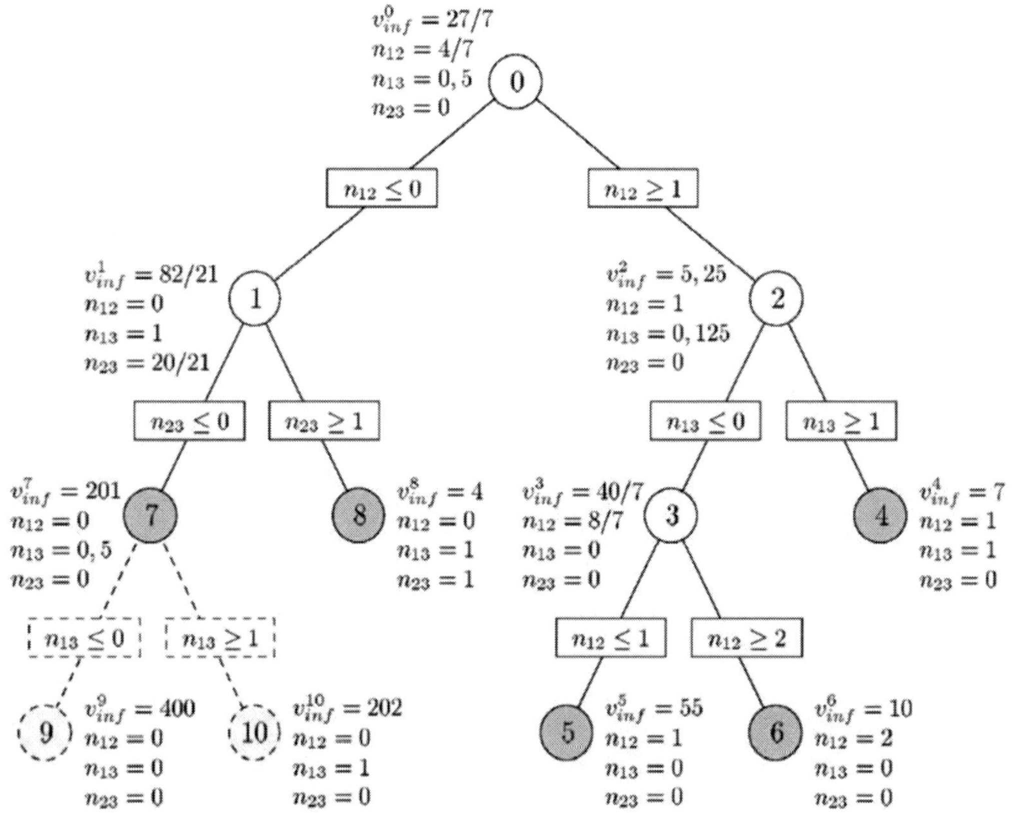


**Figure 1**. Branch-and-bound intermediate tree.

The constraints that are added for each separation are shown inside the rectangles in the respective branch. Thus, for any node, the associated problem can be determined going back on the tree towards the root node (node 0) and adding the correspondent constraints for that path. For example, the candidate problem represented by node 7 is the original problem *(P)* with the addition of the following constraints:

$$n_{12} \leq 0$$
$$n_{23} \leq 0$$

The tree starts with Node 0 through the linear problem resolution *($P_R$)*, which is obtained from the relaxation of the integer constraints of the expression 1.

$$(P_R) \begin{cases} \min \ v = 5n_{12} + 2n_{13} + 2n_{23} + \beta \\ s.a. \ \ 350n_{12} + 400n_{13} + \beta \geq 400 \\ \quad\quad 350n_{12} + 210n_{23} + \beta \geq 200 \\ \quad\quad \beta \geq 0 \\ \quad\quad n_{ij} \geq 0 \quad \forall ij \end{cases}$$

The linear problem solution presents two variables with continuous values: $n_{12} = 4/7$ and $n_{13} = 0.5$. Thus, it does not satisfy the pruning criterion and the process continues with the separation of the problem. The first non-integer variable ( $n_{12}$ ) is chosen, as the separation variable. In this case, the two descendants' nodes are generated by the constraints $n_{12} \leq 0$ (Node 1) and $n_{12} \geq 1$ (Node 2). In this part of the process there are two candidates that may be chosen to be evaluated, represented by Nodes 1 and 2. Supposing that the selecting criterion is to evaluate the last generated node. So, Node 2 will be the chosen one. The equivalent relaxed problem solution – relaxed problem with the constraint $n_{12} \geq 1$ – still has a variable with continuous value ( $n_{13} = 0.125$ ). Then, the problem of the Node 2 should be separated and, in this case, only one variable can be selected, ( $n_{13}$ ).

After the generation of the Node 2 descendants (Nodes 3 and 4), Node 4 is selected and evaluated. Thus, the first integer solution for the original problem is obtained and this solution becomes the incumbent ( $v^* = v^4 = v^4_{\text{inf}} = 7$ ). As the incumbent value was changed, the remained candidates (Nodes 1 and 3) are evaluated, but it is not possible to make any pruning. Again, the last generated candidate is selected (Node 3) and its variable $n_{12}$ is separated. Both descendants of the Node 3 (Nodes 5 and 6) have integer solutions that are discarded since they are worse than the existing incumbent ( $v^5 = v^5_{\text{inf}} = 55$ and $v^6 = v^6_{\text{inf}} = 10$ ).

The algorithm now goes back to its candidates' list and selects the only available node (Node 1) which is separated in its variable $n_{23}$ generating Nodes 7 and 8. Taking the last one to be evaluated, we obtained an integer solution better than the incumbent, which is updated to $v^* = v^8 = v^8_{\text{inf}} = 4$. Next, Node 7 is evaluated and discarded (with its descendants) since its lower bound is greater than the incumbent ( $v^7_{\text{inf}} = 201 > v^*$ ). Nodes 9 and 10 are represented only to complete the enumeration.

In the way that the process evolved, it was necessary to solve 9 linear problems to guarantee the enumeration of all possible alternatives. The optimal solution for the original problem was obtained after 8 LPs:

$$v^* = 4$$
$$n_{12}^* = 0$$
$$n_{13}^* = 1$$
$$n_{23}^* = 1$$

An important issue regarding performance in enumeration algorithms is related to how good the initial incumbent solution is. A good initial solution significantly increases the efficiency of the pruning test (4b) and so, reduces the number of candidates that need to be evaluated. For the example of Figure 1, an initial incumbent $v^* = 5$, would be enough to immediately prune Node 2 (and all its descendants), reducing then the total number of evaluations.

Considering the data structures used and the criteria for node ordering and selection of the separation variable, the serial branch-and-bound algorithm was implemented according to the flowcharts of Figures 2 and 3. A summary of the parallel implementation can be seen in Figure 4.

## 4.1. Parallel Implementation

### *4.1.1. MPI*

In the MPI implementation, the master node initializes the program, creates a base tree with a list of active nodes and sends those active nodes to the slaves whenever they need a problem to work on. The master also controls the process, including list of solutions, statistics, etc.

While the master is creating a base tree, the slaves also create a copy of the same base tree. Each slave receives an index for a node from the master, then initializes its own tree and starts the B&B. All the calculation is done by the slaves. Once a solution is found, it is sent to the master who updates all slaves.

The master node waits for any message from the slaves and it does no other calculations during the B&B process. Whenever a slave finishes its job, it sends a message to the master asking for a new node and process continues.

As the slaves work in different nodes with their own tree, there is no possibility of conflict in accessing the same part of the memory.

### *4.1.2. OPENMP*

In the OpenMP implementation, there is no concept of master and slaves. The cpu0 node is the one responsible for initializing the program and creating a common tree with a list of active nodes. In this implementation, the nodes have a flag indicating whether or not they are being used. After this creation, all nodes start the B&B requesting the first available node from the list of active nodes. So, each slave works in different part of the same tree. Of course, as the memory is shared, once something happens, the variables are automatically updated.

With both implementations there are possibilities for further improvements and those will be discussed in the Sections 6 and 7.

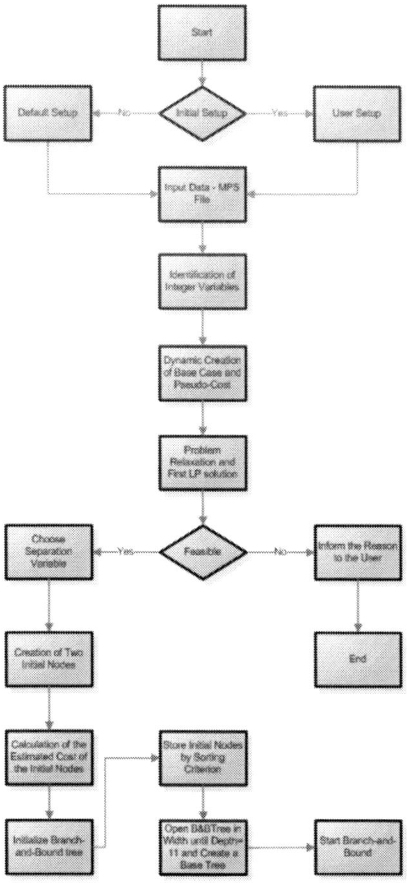


Figure 2. Serial branch-and-bound algorithm – Part 1

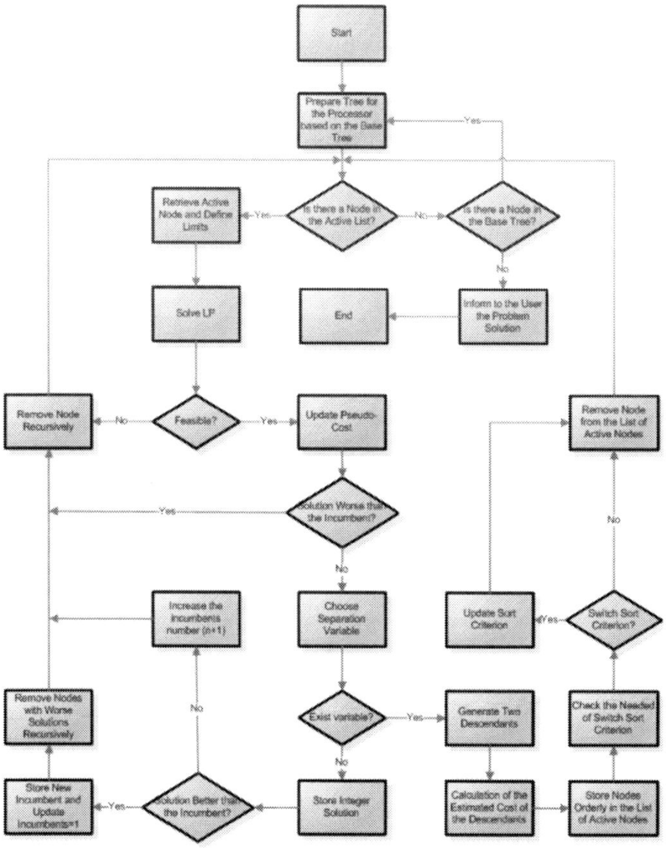


Figure 3. Serial branch-and-bound algorithm – Part 2.

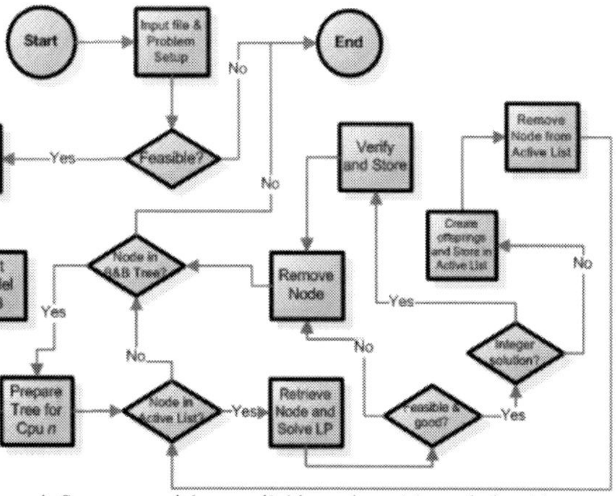


Figure 4. Summary of the parallel branch-and-bound algorithm

In the next section we present results comparing the three implementations for solving the same MIP.

## 5. Results

The initial comparison of the algorithms to date made use of an optimization problem previously presented [6][7]. The problem is a simplified three-phase electrical distribution network consisting of 18 nodes (2 substations and 16 nodes with loads) and 24 branches operating under 13800V. The topology of this network is shown in Figure 5 in which rectangles denote the substations and circles are the nodes where loads are concentrated. Branches drawn as continuous lines denote the initial network (those with a single line are part of the fixed network and those with double lines are candidates for replacement), and branches drawn as dashed lines are candidates for addition (and are not part of the initial network). The search space has approximately $2^{118}$ combinations. The results are illustrated in Figure 6. The problem is fully detailed in the above reference.

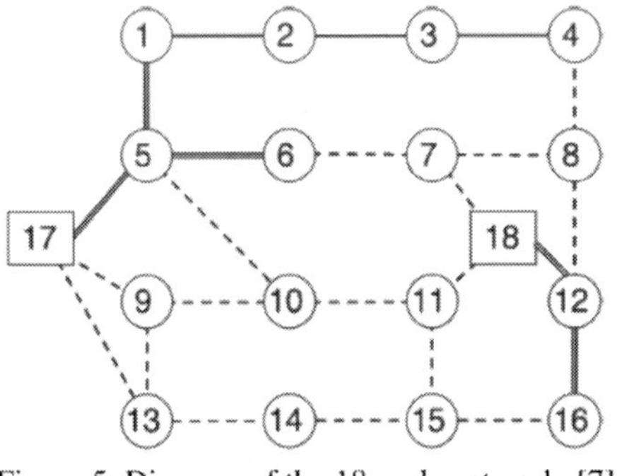


Figure 5: Diagram of the 18-node network. [7]

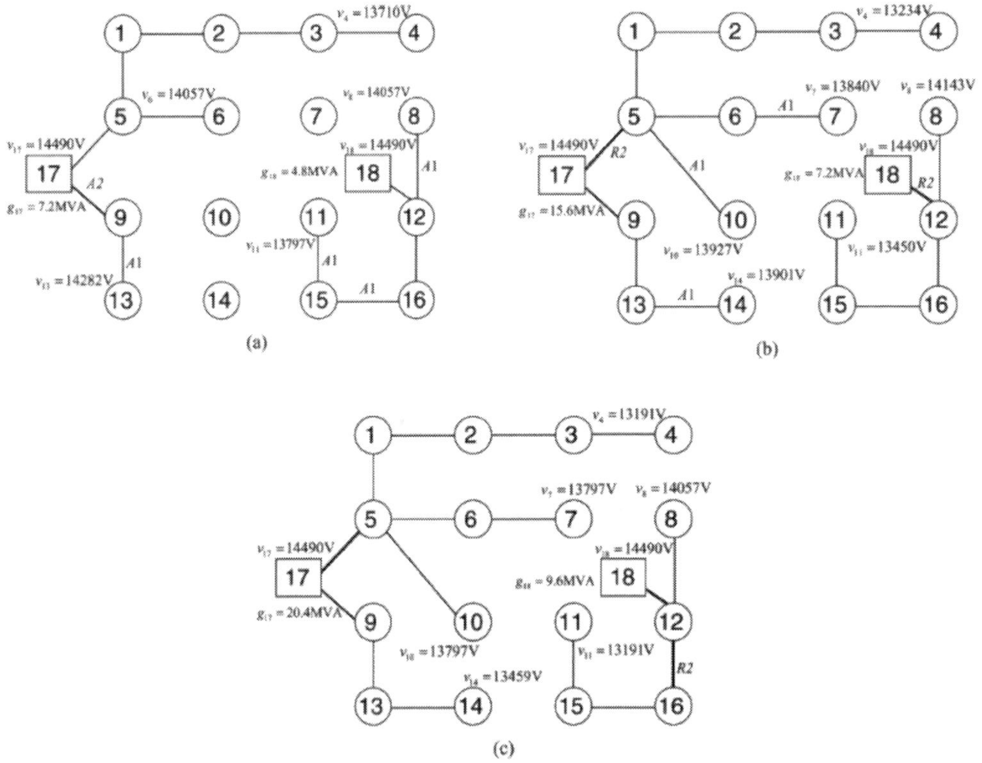


Figure 6: 18-node network - Solution with multistage expansion. [7]

The results for the tests were carried out on two different systems:

- HP Linux cluster running XC 3.1 with 267 nodes, 4 cores each, connected via Myrinet 2g (narwhal.sharcnet.ca) for the MPI and Serial Implementation.
- SUSE Enterprise 10 Linux Cluster running with 128 nodes, 1 core per socket, NUMA (silky.sharcnet.ca) for the OpenMP.

Even though these are different systems with different architectures, the goal of the experiment was to compare the wall time of the parallel implementations with the serial one to support the idea that classic optimization methods can still be used in modern distributed systems for many different areas.

Table 2: Results for the Electrical Distribution Network – Serial Approach

| #LPs | Wall Time (s) |
|---|---|
| 70240 | 7324 |

Table 3: Results for the Electrical Distribution Network – Parallel Approaches

| #Proc | #LPs | | Wall Time (s) | | Speedup | |
|---|---|---|---|---|---|---|
| | MPI | OpenMP | MPI | OpenMP | MPI | OpenMP |
| 2 | 91230 | 65321 | 6155 | 6654 | 1.2x | 1.1x |
| 5 | 86745 | 47652 | 4623 | 6236 | 1.6x | 1.2x |
| 10 | 84451 | 39567 | 2757 | 5712 | 2.7x | 1.3x |
| 50 | 72344 | 26541 | 976 | 3854 | 7.5x | 1.9x |
| 100 | 65645 | 21454 | 806 | 3562 | 9.1x | 2.1x |

Table 2 summarizes the execution of the solution to the problem for the serial implementation. The results show the required number of linear programming problems (sub-problems) (#LPs) and the total time required in seconds. The time was measured by the algorithm and corresponds to the total wall clock time.

Table 3 summarizes the execution for both parallel implementations: MPI and OpenMP. The results show the number of processors used (#Proc), the required number of linear programming problems (sub-problems) (#LPs) for each approach related to #Proc, as well as the total time spent across all processors and the calculated speedup. For the tests we used the [MAX;MAX] separation variable selection strategy.

The difference between the number of LPs in the parallel executions is not related to a different strategy, but is due to several other factors. In the MPI implementation, as the memory is distributed and the processors have to exchange messages, each node is creating its own tree and because of that it could take more time to find a good incumbent solution. Many unnecessary nodes are evaluated before finding a good one, since the algorithm has to ensure the enumeration of all possible solutions without pruning anything that could be promising. On the other hand, in the OpenMP approach, fewer nodes were evaluated because the algorithm always found good solutions earlier than the MPI (for this particular problem). Also, the OpenMP implementation is more similar to the serial one. Unfortunately, other issues with the OpenMP implementation were found and are reported in the conclusions.

## 6. Conclusions

Based on the initial experimental results some conclusions can be drawn:

- The serial implementation works reasonably well.

- The OpenMP implementation always generated fewer sub-problems than the serial, but the speedup was not as good as the MPI implementation. Unfortunately, the chosen LP solver (GLPK) [9] was not "thread safe", which made it impossible to run a shared memory model approach without the use of some explicit mechanisms of control in the implementation. This drastically decreased the performance. Also, the use of a flag in order to control the list of active nodes to avoid race condition has a cost that decreases performance, as well. We plan to address those problems in future work.
- The MPI implementation seemed to provide a more robust way to split the branch and bound search. Though it generated more sub-problems for some cases, it took the least time in all executions and always ran perfectly. It is important to mention here that before the parallel computation starts, the algorithm prepares a base tree for the MPI implementation and the nodes are sent to the slaves whenever they need. This base tree is initially static, which means that there are a specific number of nodes available to be evaluated. When this number is reached, no more nodes are available to be sent and the allocated processors become idle, waiting the others to finish the job. This is a problem that will be addressed in future work.

As indicated, these results are based on our particular problem and, while complex, additional experiments need to be run to better compare the algorithms and approaches to parallel implementation as well as to help improve the implementation. Our initial impression is that the MPI approach works reasonable well and with much less headache than the OpenMP one.

## 7. Future Work

Our next steps include finding a better way to evaluate nodes in parallel and finding a better way to keep all processors working full time. We also want to test the algorithm using others LP solvers that are thread-safe and faster. Finally, we intend to compare the implementations to several other problems including some from other domains, testing the different separation variable choice that we had implemented as well.

## References

[1] Balas, E. (1965). "An additive algorithm for solving linear programs with zero-one variables", Operations Research, Vol. 13, No. 4, pp. 517–546.

[2] Geoffrion, A.M. and R.E. Marsten (1972). "Integer Programming Algorithms: A Framework and State-of-the-Art Survey", Management Science, Vol. 18, No. 9, pp. 465–491.

[3] Glover, F. (1989). "Tabu Search – Part I", ORSA Journal on Computing, Vol. 1, No. 3, pp. 190–206.

[4] Glover, F. (1990). "Tabu Search – Part II", ORSA Journal on Computing, Vol. 2, No. 1, pp. 4–32.

[5] Goldberg, D.E. (1989). "Genetic Algorithms in Search, Optimization and Machine Learning", Addison-Wesley Professional, 432 p.

[6] Haffner, S., L.F. Pereira, L.A. Pereira and L. Barreto (2008a). "Multistage model for distribution expansion planning with distributed generation - Part I: problem formulation". IEEE Transactions on Power Delivery, Vol. 23, No. 2, pp. 915–923.

[7] Haffner, S., L.F. Pereira, L.A. Pereira and L. Barreto (2008b). "Multistage model for distribution expansion planning with distributed generation - Part II: numerical results". IEEE Transactions on Power Delivery, Vol. 23, No. 2, pp. 924–929.

[8] Holland, J.H. (1992). "Adaptation in Natural and Artificial Systems: An Introductory Analysis with Applications to Biology, Control, and Artificial Intelligence", The MIT Press, 228 p.

[9] Makhorin, A., (2001). "GLPK Linear Programming Kit Manual" GLPK documentation,

Moscow Aviation Institute, Moscow, Russia, February 2001, available in http://www.gnu.org/software/glpk/glpk.html.
[10] MPI-forum - http://www.mpi-forum.org/
[11] OpenMP - http://www.openmp.org
[12] Romero, R., Monticelli, A., Garcia, A. e Haffner, S. (2002). Test systems and mathematical models for transmission network expansion planning, IEE Proc.-Gener. Transm Distrib., Vol. 1491, No. 1, pp. 27-36.

Journal of Physics: Conference Series **256** (2010) 012019 doi:10.1088/1742-6596/256/1/012019

# An Adaptive Priority Tuning System for Optimized Local CPU Scheduling using BOINC Clients

**Adel B. Mnaouer**

College of Computer Engineering and Information Technology
Dar Al-Uloom University
Email: adelm@dau.edu.sa

**Colin Ragoonath**

Dept. of Information and Communication Technology
University of Trinidad and Tobago
Email: colinragoonath@gmail.com

**Abstract**. Volunteer Computing (VC) is a Distributed Computing model which utilizes idle CPU cycles from computing resources donated by volunteers who are connected through the Internet to form a very large-scale, loosely coupled High Performance Computing environment. Distributed Volunteer Computing environments such as the BOINC framework is concerned mainly with the efficient scheduling of the available resources to the applications which require them. The BOINC framework thus contains a number of scheduling policies/algorithms both on the server-side and on the client which work together to maximize the available resources and to provide a degree of QoS in an environment which is highly volatile. This paper focuses on the BOINC client and introduces an adaptive priority tuning client side middleware application which improves the execution times of Work Units (WUs) while maintaining an acceptable Maximum Response Time (MRT) for the end user. We have conducted extensive experimentation of the proposed system and the results show clear speedup of BOINC applications using our optimized middleware as opposed to running using the original BOINC client.

## 1. Introduction

Internet-based volunteer computing (VC) environments are distributed client-server computing frameworks which seek to provide a means where participants (volunteers) wishing to contribute the idle computing resources of their computers can be facilitated. BOINC (Berkeley Open Infrastructure for Network Computing) [1], [6] is one such distributed computing environment and is used by many research projects in molecular biology, medicine, physics, chemistry, mathematics, astronomy [4] and many other fields of study requiring enormous amounts of computing resources to solve large problems using parallel processing techniques.

BOINC is the classic client-server distributed environment, with each of these components containing a set of scheduling policies/algorithms which enable the system to function in a cohesive manner despite the many drawbacks of operating in such a large, distributed, open environment. In these distributed heterogeneous parallel computing systems, computing resources vary with different levels of availability and reliability, also projects tend to vary in terms of job length, sensitivity to errors in the results and in result verification requirements (BOINC uses replication of results for comparison before final acceptance) [8].

Most of the previous research into optimizing the BOINC system is focused on job scheduling within the BOINC framework and neglects external factors which may contribute positively to the speedup of jobs within the system. BOINC runs on top of an underlying Operating System and there are advantages which could be gained by improving the interaction between BOINC and the underlying system. Currently BOINC jobs are run at the lowest priority from the perspective of the Operating System (OS), and even if BOINC applications are run in the highest priority within the BOINC framework, at the OS level those jobs are at the lowest priority.

It is understood that the creators of the BOINC system never intended for BOINC processes to actively compete for computing resources with the local processes, but to use the idle resources available to it. But this concept leaves the BOINC processes at a disadvantage especially when the user's preferences are set to lower values for resource usage. The current BOINC client is static (See 'figure 1') and does not adapt to the constantly changing nature of the system.

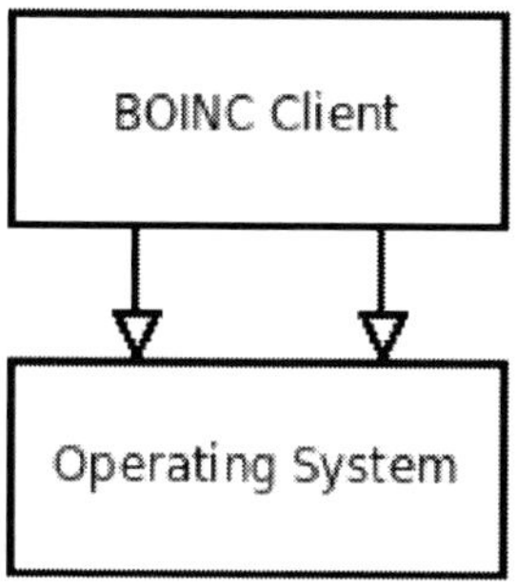


**Figure 1:** Static BOINC Client

If a user sets usage for the BOINC system to 50% CPU time the BOINC system would constantly request 50% CPU time even if the CPU is underutilized and BOINC is in danger of missing the deadline set by the project. This is not a very efficient way of utilizing the donated computing resources and our Adaptive Priority Tuning system seeks to address this issue resulting in a more dynamic client (See 'figure 2') which improves the Quality of Service (QoS) to the projects that use this platform.

This paper is organized as follows: Section 2 describes work which relates to the work done in this paper. Section 3 presents a short overview of how the BOINC client works. The Local Optimization Middleware is described in Section 4. In Section 5 we describe the experiment design used to test the Optimization Middleware. Section 6 is where our results are presented and discussed. In Section 7 we conclude this paper.

## 2. Related Works

BOINC server-side scheduling have been studied by a number of researchers in [9] the authors use a method based on a distributed, evolutionary algorithm capable of searching a large space of scheduling policies looking for policies that outperform manually-designed scheduling policies. In [5] the authors increase throughput for BOINC projects by intelligently coordinating schedules for volunteers. They achieve this, by developing a constraint optimization (COP) mapping that pursues high throughput while trying to adhere to the volunteer's preferences.

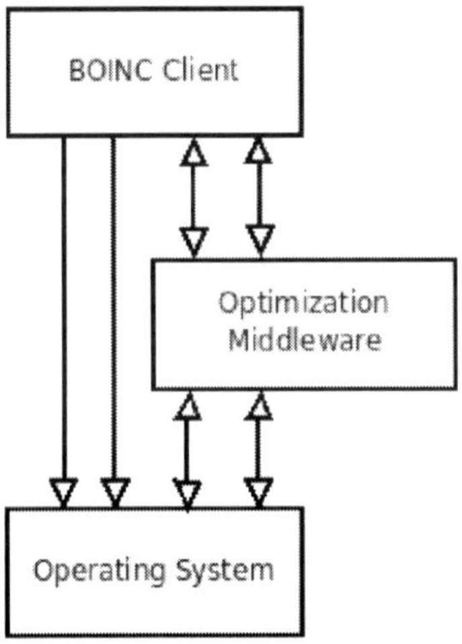


**Figure 2:** Optimized BOINC Client

Research into the operation of the BOINC client is focused mainly on its work fetch policy and its CPU scheduling policy, in [7] the authors evaluate a combination of CPU Scheduling policies, Work Fetch policies, Work Send policies and Job Completion Estimation policies to determine the performance of the local client with a new set of metrics i.e. waste due to deadline misses, idleness, share violation and monotony. In [3] the authors describes the issues involved in local scheduling for volunteer computing, and they present policies that have proven to work well in the real world. They also pointed out that client and server scheduling are not well integrated but plan to remedy this by having the client send information about queued and in-progress work, including completion time estimates. The server will then use this information to do a deadline-scheduling based simulation to decide what jobs, if any, can safely be sent.

## 3. How the BOINC client works

The BOINC architecture consists of the BOINC client and the BOINC server components. The BOINC client is responsible for running the application for the particular project/s to which the host is attached. These applications are linked with a runtime system whose functions include process control, checkpoint control and graphics [2]. The client performs CPU scheduling for BOINC jobs (performed on top of the Operating System's scheduler, these jobs are run at the lowest (Linux) nice value (-19)). This point is very important to this research since our algorithm seeks to dynamically adjust this value to the ever changing environment which exists within a host system.

The BOINC client is also responsible for all network communication between the host system and the project server. The client communicates with the project server via HTTP. The requests are sent as an XML document and includes a description of the host hardware and availability, a list of completed jobs, and a request for a certain amount (expressed in CPU time) of additional work. The reply message includes a list of new jobs (each described by an XML element which lists the application, input and out files, including a set of data servers from which each file can be downloaded) [3].

Volunteers install the BOINC client software onto their system and attach to the projects of their interest using the client. Once installed on the system the client transparently contacts the project server to request work to process (Work Units WU). The BOINC client is responsible for a number of important tasks these include

- Network communication with scheduling and data servers.
- Executes and monitors applications and enforces user preferences.
- Implements a Local Scheduling Policy to achieve the following goals:

  - To maximize resource usage (i.e. keep all processors busy).
  - To satisfy result deadlines.
  - To respect the participant's resource share allocation among projects.
  - To maintain a minimal "variety" among projects.
- Handles request to provide graphics and provide heartbeat functionality.
- Implements a scheduling policy based on a dynamic "resource debt" to each project [4].

Some hosts have intermittent physical network connections (for example, portable computers or those with modem connections). Such computers may connect only every few days. During a period of network connection, BOINC attempts to download enough work to keep the computer busy until the next connection [3].

*The BOINC client local scheduling*
The client is responsible for implementing two related scheduling policies:
**Work Fetch Policy:** this policy is responsible for deciding when to ask a project for more work, which of the projects to ask and how much work to ask for [3].
**CPU Scheduling Policy:** this policy deals with deciding which of the currently runnable jobs should be run. And for the jobs which are preempted, which of these should be kept in memory [3].

As mentioned earlier both of these policies are enforced within the BOINC client but whatever decision is made e.g. if it is decided that a job X should be run at the highest priority by the client that decision does not translate into job X actually being run at a higher priority by the OS. Our Adaptive Priority Tuning mechanism seeks to address this issue in a reasonable manner while maintaining a Maximum Response Time (MRT) which is acceptable to the end user.

## 4. The Optimization Middleware Algorithm

In this section we describe the Optimization Algorithm with the aid of pseudo code, for simplicity we have decomposed the algorithm into four sections each of which is responsible for particular tasks being completed. Our algorithm is a simple algorithm which seeks to maintain an acceptable MRT for the user while providing quality processor time to the BOINC applications.

BOINC was designed with a number of entry points into the core-client and the algorithm takes advantage of this fact to make the client more dynamic and to adjust to the changes in its environment.

On startup the algorithm determines this is the first time it is running (since the core-client can run without our middleware present) if it's the first run a copy of the global_prefs_override.xml file is made. This copy would be used as a reference (since it contains the user's original preferences) for the middleware in its decision making process since the original would be constantly changing as processing proceeds and the original settings of the user, would be lost. Both the original (now dynamic and is required by BOINC) and the copy (static and is required by the middleware) would then be read into memory upon further iterations of the algorithm (see 'Table 1' below).

During the beginning of the optimize() function the middleware determines if any BOINC applications are currently running. If no BOINC applications are running the middleware exits, however if there are BOINC applications currently running the middleware determines some key user preference settings and system state such as whether the BOINC client should continue processing if the system is currently running on batteries. If the system is running on batteries then the middleware updates the global_prefs_override.xml file and then forces the BOINC client to reread the file by calling its --read_global_prefs_override() function so that the new settings can take effect, which is to set the percent CPU usage to one quarter of what the user settings contain. The middleware also renice the nice values under which the running BOINC applications are being executed to the lowest setting (in Linux 19), these settings are meant to extend the battery life of the computer while still getting some work done. This process is completed in the **doUpdate()** function 'Table1' above, the middleware then sleeps for 4 seconds which is important to meet the MRT criteria.

**Table 1:** Algorithm 1 – Optimized Middleware onStartup() function

```
on startup()
if This is first run then
   Make a copy of the global prefs override.xml file
   Read the original and the copy of global_prefs_override.xml files and store in variables
   SET the firstrun indicator variable to 1 {indicates firstrun has occurred}
else
   Read original and copy of global_prefs_override.xml files and store in variables
end if
```

If the computer is not running on batteries then the **optimize**() function gathers system information such as the current CPU idle time, the user preference settings for CPU usage and whether the BOINC

**Table 2:** Algorithm 2 – Optimized Middleware optimize() function

```
if BOINC not running then
  Quit Program
end if
else BOINC is running
while true do
   if computer is on battery AND user preference is set to run on battery AND to    run
    while user is active then
      UPDATE values in a new global_prefs_override.xml file
      SET the nice value variable to the lowest nice level(19)
      DO an update() {see below for details}
      sleep for 4 seconds {4 seconds is chosen to meet the MRT criteria}
  end if
  if computer is not on battery AND user preference is set to run while user is  active
  then
     get CPUidletime(), user preferences and the status {indicates settings of the  last
      iteration of the algorithm}
  then
     SELECT appropriate optimization logic – {Algorithm 3}
     SET values in a new global_prefs_override.xml file
     SET the nice value variable based on – {Algorithm 3}
     SET the new status variable value {this indicates what was previously done}
     DO an update() {see below for details}
     sleep for 4 seconds {4 seconds is chosen to meet the MRT criteria}
  end if
end while
```

client should run while the user is active, and the status (which indicates what was done before) see 'Table 3' for details. Once these pieces of information have been gathered the **optimize()** algorithm then selects an appropriate optimization logic, see 'Table 3' for further details.

**Table 3:** Algorithm 3 - Optimized Middleware example of Optimization Logic

**if** user preference for CPU usage is ≥ *X* **AND** idle time of CPU is ≤ *Y* **AND** status == Z
*{Y chosen after several observations and experiments to*
*determine an optimized value to release resources to other*
*programs}*
**then**
  **SET** the cpu usage limit to random value within a specified range for these values
  **SET** new nice value
  **UPDATE** status variable
**end if**

In the optimization logic algorithm 'Table 3':

- **X** is the threshold value for CPU usage set by the end user and can have a value between 1% and 100%.
- **Y** is the threshold value for the current idle time of the CPU and its value ranges as follows:
  - Less than 17% CPU idle time
  - Between greater than 18% and less than and equal to 40% CPU idle time
  - Greater **than** 40% CPU idle time
- **Z** indicates what was done before by the system, this is important since this value is used to maintain the MRT for the end user. It's can be either:
  - 0: indicates the middleware used is least aggressive algorithm for the previous iteration taking into account the amount of free resources available.
  - 1: indicates the middleware used a medium aggressive algorithm for the previous iteration again taking into account the amount of free resources available.
  - 2: indicates the middleware used the most aggressive algorithm for the previous iteration taking into account the available resources.

The optimization logic algorithm compares the user preference for CPU usage to a predetermined value (we decided to set these values after several experimental observations and selected the performance we determined to be optimum for our experiments), it also looks at the current idle time of the CPU and what was done during the last iteration (status). Based on these values **X, Y, and Z** the middleware sets the CPU usage limit variable to a random value which is generated by the system but falls within a specified range for the user preferred CPU usage and current idle time of the CPU for these values. It then writes this new value for the CPU usage to the global_prefs_override.xml file for the BOINC client to be updated with this new value during the call to the doUpdate() function, it also determines a new value for the nice local variable and the status local variable according to the algorithm logic. The optimize() algorithm then calls the doUpdate() function.

The doUpdate() function is an update process which simply makes a call to the core-client's --read_global_prefs_overide() function, this forces the core-client to reread the global_prefs_override.xml which now contains the updated information (see doUpdate() Function 'Table 4'). It then sets the nice value to the new value for each running BOINC application as determined by the optimization logic algorithm.

'Figure 3' shows the State diagram of the Optimization Middleware describing the decisions the middleware makes to ensure optimum performance of the BOINC client taking into consideration the user's preference and the MRT constraint.

**Table 4:** Algorithm 4 - Optimized Middleware doUpdate() function

```
query BOINC client to get the following parameters:
    estimated CPU time remaining, current CPU time, fraction done then
            use these parameters to determine the following:
                percent CPU time to completion {see equation 2}
                days to deadline (deadline date – today's date)
          for each BOINC application
              if (days to deadline < 2 and fraction done < 0.8 and percent to  complete <
                  estimated CPU time remaining)
              set the nice value to the highest (-19)
          else
              call BOINC client's read global_prefs_override() function
              to reread the global_prefs_override.xml file
          then
              for each BOINC application not running in high priority do
                SET the nice value to the new value as per the algorithm's choice
        end for
end doUpdate
```

## 5. Our Experimental Design

We conducted a series of simulation experiments to test the optimization middleware and compared the results with simulation results using the plain BOINC client. In order to test the middleware we needed to process the same combination or Work Units repeatedly recording the Wall Time spent processing each WU for each test.

**Table 5:** Hardware Specification

| Model | Specification |
| --- | --- |
| 2 Dell Precision Workstations | O.S. Ununtu 9.10<br>CPU: Intel Xeon Dual Core 3.00 GHz.<br>RAM: 2048 G.B.<br>Hard Disk: 160 G.B. |
| 2 IBM 6223MCI Workstations | O.S. Ubuntu 9.10<br>CPU: Intel Xeon HT 3.89 GHz.<br>Hyper Thread Technology Enabled<br>RAM: 2048 M.B.<br>Hard Disk: 160 G.B. |

Scripts were developed to intermittently consume system resources so as to imitate the effects of a user on the system. Also we decided to test the middleware using hardware with different numbers of CPUs in order to study the effects of the multi-core architectures which are available today.

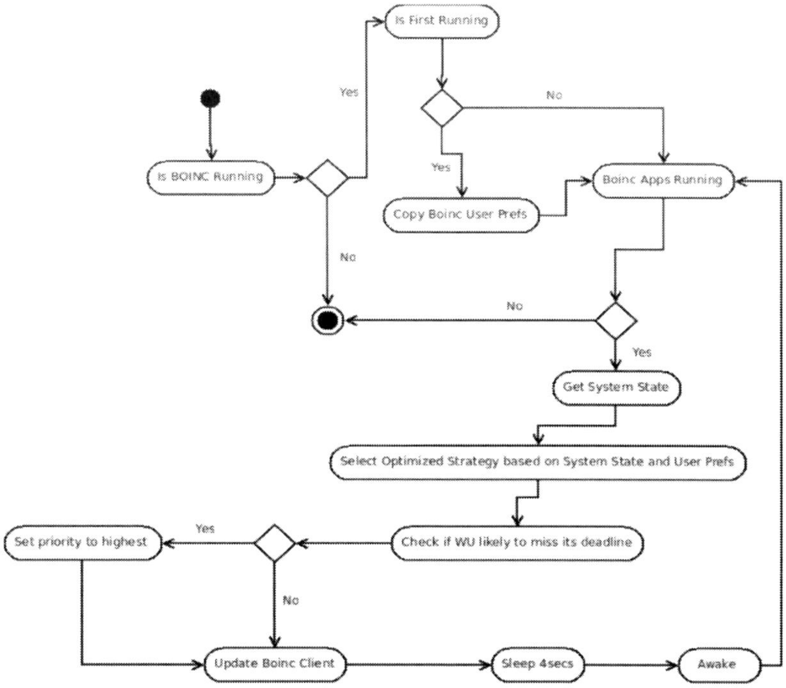


**Figure 3:** BOINC Optimized Middleware diagram

Tests were designed to be run for the user's preference settings for CPU times at intervals of [40%, 60%, 80% and 90%].

**Table 6:** Experiment Configuration

| Station | Architecture | WU Description |
|---|---|---|
| WKS 1 | IBM Single Core with HT | Einstein@home |
| WKS 2 | IBM Single Core with HT | Einstein@home<br>Edges@home |
| WKS 3 | Dual Core Dell Precision | Einstein@home<br>Milkyway@home |
| WKS 4 | Dual Core Dell Precision | Einstein@home<br>Milkyway@home<br>Rosetta@home<br>Charmm@home |

A minimum of two repetitions were done for each interval level for both the optimization middleware and without the optimization middleware (plain BOINC client). The results were then normalized by finding the average for each run at the particular interval for the specific architecture (i.e. dual core or single core with Hyper Threading HT). The 'Table 5' details the model of workstations used for this experiment along with the hardware specifications and Operating System.

In 'Table 6' a brief description of the system architecture along with the project name for the WU selected for processing is given.

NOTE: The repeated project names for the different workstations (WKS) does not indicate different WUs from the projects, but means the same WUs are used on the different machines.

To determine the speedup gained at each setting the following equation is used:

**ETP** = execution time of BOINC plain client
**ETO** = execution time of BOINC client with optimization middleware

$$speedup = ETP/ETO \quad (1)$$

Experiments were also conducted to test the performance of the optimization middleware when a project was likely to miss its deadline. The setup for these simulations was the same as for the previous simulations except the workstation clock was adjusted so that the BOINC client would calculate an application would miss its deadline. The selected applications for this simulation were Einstein@home (the original one used for all experiments) and a new WU for the Charmm@home application (this WU would be run in normal mode within the BOINC client). The Einstein@home application was the application which the BOINC client would identify as likely to miss its deadline during the experiment, this would cause the BOINC client to run this application in "Highest Priority". It was decided to run this experiment using the Dual Core system (WKS 3) replacing the Milkyway@home application with the Charmm@home WU.

This section describes how the optimization middleware identify WUs which are likely to miss their deadlines. The optimization middleware queries the BOINC client to gather useful information which is then used to determine which WU the BOINC client would most likely run at the Highest Priority.

The following parameters are used in the algorithm:

- estimated CPU time remaining
- current CPU time
- fraction done (see 'Table 4' for usage)

Using the *estimated CPU time remaining* and the *current CPU time*, the middleware calculates the percentage of time needed to complete the WU (*percenttocomplete*) using the following formula:

$$percenttocomplete = 0.2 * (\,rtime + ctime\,) \quad (2)$$

Where : **rtime** represents the *estimated CPU time remaining* and **ctime** denotes the *current CPU time*

## 6. Experimental Results and Analysis

The results of the experiments are now presented and analyzed to demonstrate how well the optimization middleware performed when compared to the plain BOINC client. The 'figure 4' illustrate the speedup for the Einstein@home WU, the graph clearly show that at the lower CPU usage settings the largest speedup is realized, also the graph shows the optimization middleware caused the Wall Processing Time to be almost constant at all the CPU usage settings. This occurred because the middleware made the BOINC client utilize all available resources even though the user settings were at a lower value.

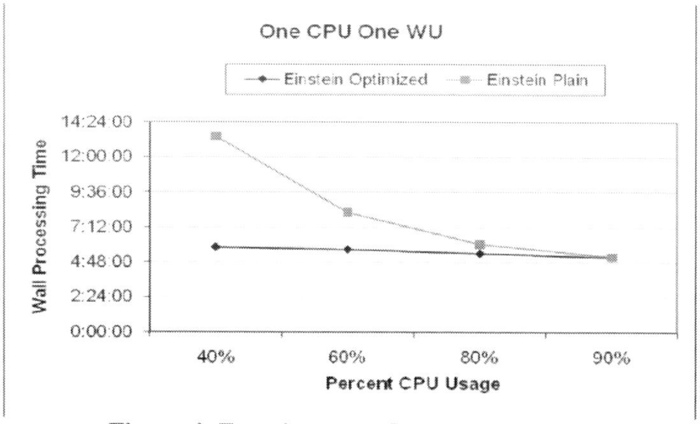


**Figure 4:** Experiment 1 - One WU One CPU

In 'Table 7' the speedup gained at each tested CPU usage level is listed. A speedup value > 1 indicates a realized speedup whereas a value ≤ 1 indicates no realized speedup.

**Table 7:** Speedup for One WU One CPU

| % CPU Usage | Speedup |
|---|---|
| 40 | 2.30 |
| 60 | 1.45 |
| 80 | 1.12 |
| 90 | 1.01 |

According to 'Table 7' as the CPU Usage setting is increased the effects of using the middleware decreases and at 90% there is almost no difference when using the middleware or the plain BOINC client.

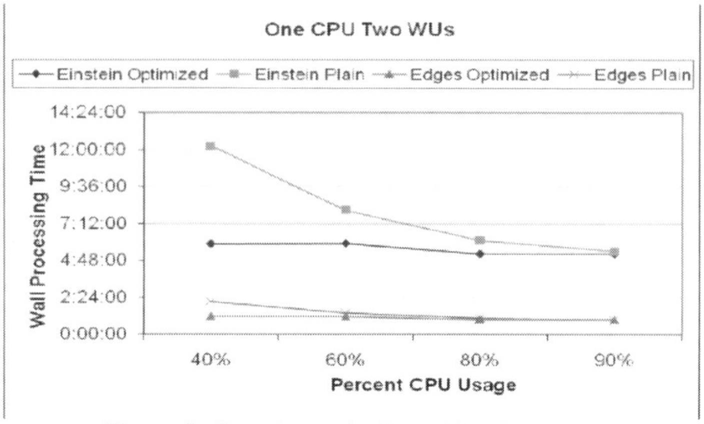


**Figure 5:** Experiment 2 - Two WUs One CPU

The 'figure 5' displays the results for two WUs, Einstein@home (the same WU used in 'figure 4') and Edges@home. In this case the Einstein@home WU requires the larger amount of CPU time while the Edges@home WU requires less CPU processing time which were calculated by their respective project server. Here the optimization middleware caused the Wall Processing Times for both applications to be maintained for all the CPU Usage settings while the plain BOINC Client performs poorly at the lower CPU Usage settings and increases as the CPU Usage settings are increased.

The results in 'Table 8' shows similar speedup values when two WUs are processed on a single CPU system, however the length of the WU in CPU time requirements have an effect on the ability of

**Table 8:** Speedup for Two WUs One CPU

| Percent CPU Usage | Speedup | |
|---|---|---|
| | Einstein | Edges |
| 40 | 2.09 | 1.83 |
| 60 | 1.37 | 1.22 |
| 80 | 1.17 | 1.09 |
| 90 | 1.04 | 0.97 |

the middleware to improve Wall Processing Time when high CPU Usage settings are used. In this case the results show at 80% CPU Usage for the shorter Edges application the speedup realized is insignificant (1.09) and there is no significant speedup beyond that setting.

The 'figure 6' shows the comparison of the **total times** for processing both WUs (Einstein and Edges) using the optimization middleware and the plain BOINC client. Again the curve of the graph indicates the same characteristics exist where speedup values at the lower CPU Usage settings are greater, than for higher CPU Usage settings. The 'Table 9' contains the actual speedup values for the total times comparison in 'figure 6'. According to 'Table 9', as the CPU Usage setting is increased the effects of

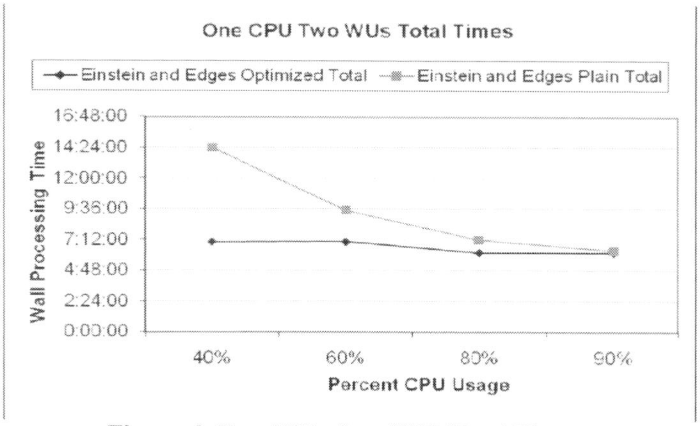


**Figure 6:** Two WUs One CPU (Total Times)

using the middleware decreases and at 90% there is almost no difference when using the middleware or the plain BOINC client. This result is similar to the results which was obtained in the earlier experiments see 'Table 7' and 'Table 8' for confirmation of this trend.

**Table 9:** Speedup for Two WUs One CPU (Total Times)

| % CPU Usage | Speedup |
|---|---|
| 40 | 2.30 |
| 60 | 1.45 |
| 80 | 1.12 |
| 90 | 1.01 |

The graph in 'figure 3' displays the comparison of the Wall Processing times for two WUs (Einstein and Milkyway) but these results are for the system with two CPUs. The graph continues to show the same trend when using the optimization middleware and the plain BOINC client.

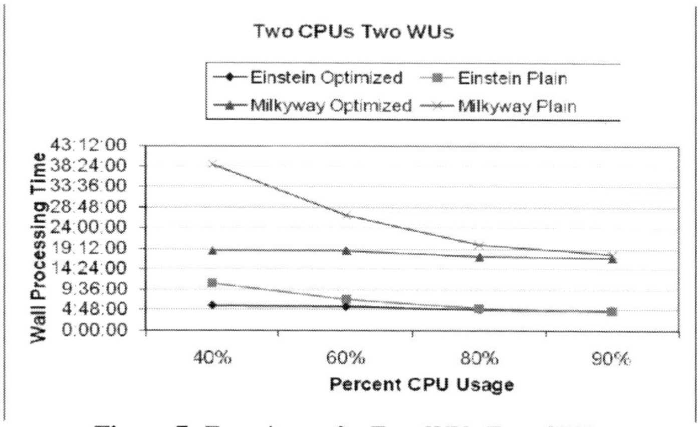


**Figure 7:** Experiment 3 - Two WUs Two CPUs

In this experiment the curve of the graph again indicates the same characteristics which were displayed in the previous simulation results. The speedup values at the lower CPU Usage settings are greater than for higher CPU Usage settings by the user. In 'Table 10' the actual speedup values for the two WUs again show that the speedup gained decreases as the user settings for percent CPU usage is increased.

**Table 10:** Speedup for Two WUs Two CPUs

| Percent CPU Usage | Speedup | |
|---|---|---|
| | Einstein | Milkyway |
| 40 | 1.89 | 2.08 |
| 60 | 1.31 | 1.44 |
| 80 | 1.08 | 1.16 |
| 90 | 0.97 | 1.05 |

An interesting observation in both 'Table 8' and 'Table 10' is that in both cases the shorter of the two jobs in cputime requirements (Edges@home 'Table 8' and Einstein@home 'Table 10') showed no significant speedup at the 80% settings. This may suggest that if a system has WUs of fairly differing lengths, the WUs with the shorter duration would benefit less at the higher CPU usage settings when compared to the larger WUs when using the optimization middleware.

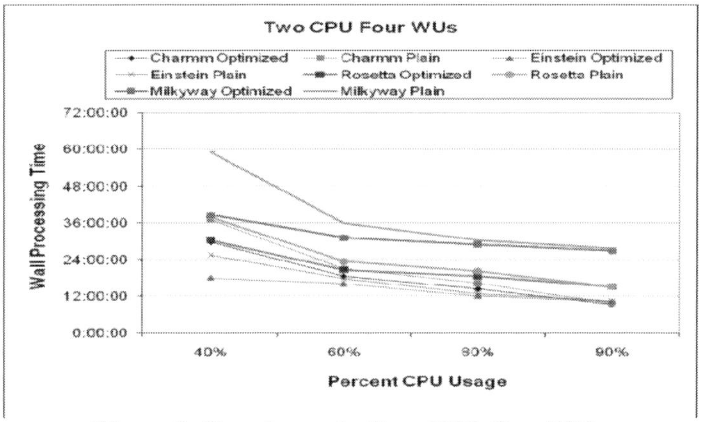


**Figure 8:** Experiment 4 - Two CPUs Four WUs

In this experiment 'figure 8' the curve of the graph again indicates the same characteristics which were displayed in all previous simulation results, this consistency was expected. Again the speedup values at the lower CPU Usage settings are greater than when the CPU Usage settings are increased. In 'Table 11' the actual speedup values for the four WUs using two CPUs confirm the consistency of the middleware to speedup the processing of the WUs at the lower user settings for percent CPU usage.

**Table 11:** Speedup for Four WUs Two CPUs

| Percent CPU Usage | Speedup | | | |
|---|---|---|---|---|
| | Einstein | Milkyway | Rosetta | Charmm |

| | | | | |
|---|---|---|---|---|
| 40 | 1.41 | 1.53 | 1.25 | 1.25 |
| 60 | 1.11 | 1.15 | 1.13 | 1.13 |
| 80 | 1.09 | 1.05 | 1.06 | 1.12 |
| 90 | 1.03 | 1.03 | 1.00 | 1.06 |

All the previous analysis was based on the optimization middleware running while the BOINC client was processing applications that were well within their deadline period. This meant that the WUs were processed at normal priority within the BOINC framework and were supposed to run at the lowest nice value (priority) by the OS, if the optimization middleware was not used. The next simulation result was produced while the BOINC client was running a WU in "Highest Priority" (a job likely to miss its deadline) and another job in normal priority within the framework. The result of this simulation would be analyzed next.

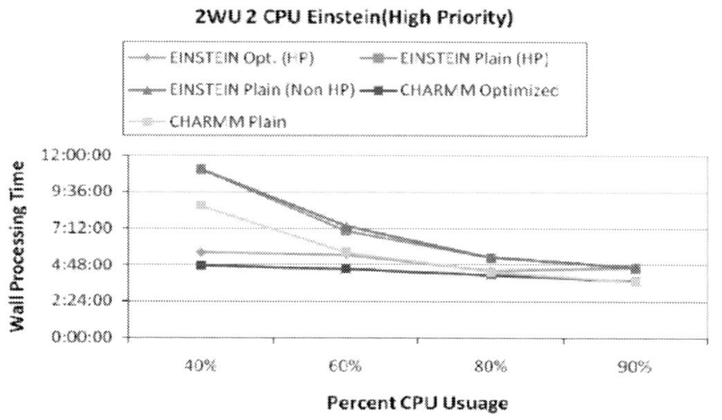


**Figure 9:** Experiment 5 - Two WUs Two CPUs (High Priority)

Before analyzing 'figure 9' it must be noted that the **EINSTEIN Plain (Non HP)** data is taken from 'figure 7' which was the experiment use to determine how the middleware performed using Two WUs on a Two CPU system. This was done in order to compare the performance of the processing of this WU (EINSTEIN) when it was run at normal priority and highest priority within the BOINC framework without the middleware running. Looking at the graph both of these curves are almost identical except for a slight variation at the 60% value which can be attributed the fact that these two curves were produced at two different times during different experiments. But the way both curves closely follow the same path indicates that the plain BOINC client has no influence on increasing the speed at which a WU is processed even if it runs that WU at the highest priority within the client. This result was expected since as stated earlier the BOINC client can only request resources at the specified user settings and even though the BOINC client may have WUs that may miss deadlines and is running those WUs at the highest priority the OS is unaware of this high priority of the BOINC WU and the plain client cannot influence the OS to improve the resource allocation to the BOINC client or to adjust the WU's (likely to miss deadline) priority at the OS level to improve the WU's chances of completing within its deadline period.

The optimization middleware addresses these issues and produces results similar to the previous simulations. Looking at the EINSTEIN Opt. (HP) and the EINSTEIN Plain (HP) (remember this WU was run at the highest priority within the BOINC client) the curve of the graphs indicates similar speedups as in the previous experiments when using the middleware. The CHARMM WU was run at normal priority within the client so it was expected to have the same curve as in the previous simulations which indicate a speedup was realized for that WU.

**Table 12:** Speedup for Two WUs Two CPUs (High Priority)

| Percent CPU Usage | Speedup | |
|---|---|---|
| | Einstein | Charmm |
| 40 | 1.97 | 1.84 |
| 60 | 1.29 | 1.24 |
| 80 | 1.20 | 1.04 |
| 90 | 1.00 | 0.99 |

The 'Table 11' contains the data for experiment 5 and as expected it shows a steady decrease in the speedup realized at each increased CPU Usage setting. It also confirms that the shorter WU benefits less in terms of speedup when using the optimization middleware as the data once again shows the shorter WU (CHARMM) had no significant speedup from the 80% settings.

The experiments clearly indicate the effectiveness of the optimization middleware to decrease the time taken by volunteer systems to process their allocated WUs. This behaviour would mean that the optimized middleware can improve the Quality of Service provided by the BOINC framework to the research projects that use the BOINC platform for their high performance processing needs.

The optimization middleware was stable throughout the experiments and this is indicated by the results which were produced by the different experiments. The middleware was able to improve the performance of jobs which were not in any danger of missing their deadlines and it was equally successful in improving the processing of jobs which were in danger of missing their deadlines.

## 7. Conclusions and Recommendations

The adaptive priority tuning middleware system for optimizing local CPU scheduling using the BOINC client has shown that it is possible to fully utilize the unused processing resource of a volunteer system even if the user has low resource settings for CPU usage. This middleware uses the "User Preferences" settings and the underlying operating system state to dynamically adjust the BOINC client properties so that the client can adjust to the constantly changing nature of its running environment. This concept provides a new approach to enhancing the local execution of BOINC applications to provide a meaningful speedup of jobs while maintaining an acceptable response time to the end-user especially at lower CPU Usage settings.

Results have shown a larger speedup is realized at the lower CPU Usage settings and this speedup gradually decline as the CPU Usage settings are increased eventually resulting in no realizable speedup at settings > 90%. This suggests that setting the CPU Usage at > 90% in the normal BOINC client would result in optimal performance for applications. It also suggests that using the optimization middleware for lower CPU Usage settings < 90% would result in better performance for applications.

The optimization middleware monitors the operating system and is able to adjust certain parameters within the BOINC client which causes the client to change its resource requests to the Operating System. The middleware also is able to adjust the *nice* value of the running BOINC applications which (in Linux 2.6) causes the Operating System to adjust the weighted values of the BOINC applications resulting in larger timeslices for those applications. These features result in a more dynamic BOINC client which is able to adjust to the changing nature of the environment in which it runs. This would benefit the various projects which use the BOINC framework by reducing the Wall Time taken to process the WUs, they would thus be able to reduce the deadline time for their WUs. Another benefit of this research would be in providing a degree of QoS by the BOINC framework to the projects since the speedup would mean results can be returned within the expected time with a reduction in missed deadlines. Further research needs to be done to determine why shorter WUs does not seem to benefit from using the optimization middleware when the CPU usage setting is above 80%.

It is recommended that the middleware be maintained as a separate piece of software and not be integrated within the BOINC client since there are security issues to consider. Also the middleware was designed to be a separate tool from the actual BOINC client so that the end user has the final decision on whether they want to use it or not. This was on purpose since it was felt that users should have the choice to decide whether they wanted to use the middleware or not, also this decision does not violate the spirit of BOINC, which is to only use the unused CPU time on the system.

**References**

[1] Anderson D 2004 *BOINC: A System for Public Resource Computing. GRID*, 4-10
[2] Anderson D, Christensen C and Allen B 2006 *Designing a Runtime System for Volunteer Computing. SC*, 126
[3] Anderson D and McLeod J 2007 *Local Scheduling for Volunteer Computing. IDPS*, 1-8
[4] Anderson D and Reed K 2009 *Celebrating Diversity in Volunteer Computing. HICSS*, 1-8
[5] Atlas J, Estrada T, Decker K and Taufer M 2009 *Balancing Scientist Needs and Volunteer Preferences in Volunteer Computing using Constraint Optimization. CCGRID*, 331-38
[6] *http://boinc.berkeley.edu/* 2010
[7] Anderson D, Kondo D and McLeod J 2007 *Performance Evaluation of Scheduling Policies for Volunteer Computing. eScience*, 415-22
[8] Estrada T, Anderson D and Taufer M 2009 *Performance Prediction and Analysis of BOINC Projects: An Empirical Study with EmBOINC. J. Grid Comput. 7(4)*, 537-54
[9] Estrada T, Fuentes O and Taufer M 2008 *A Distributed Evolutionary Method to Design Scheduling Policies for Volunteer Computing. Conf. Computing Frontiers*, 313-22

Journal of Physics: Conference Series **256** (2010) 012020 doi:10.1088/1742-6596/256/1/012020

# Adaptive Replica Placement in Hierarchical Data Grids

**Mohammad Shorfuzzaman, Peter Graham and Rasit Eskicioglu**

Department of Computer Science, University of Manitoba, Winnipeg, MB R3T 2N2

E-mail: {szaman,pgraham,rasit}@cs.umanitoba.ca

**Abstract.** Data grids support distributed data-intensive applications that need to access massive (multi-terabyte or larger) datasets stored around the world. Ensuring efficient and fast access to such widely distributed datasets is hindered by the high latencies of wide-area networks. To speed up access, data files can be replicated so users can access nearby copies. Replication also provides high data availability, decreased bandwidth consumption, increased fault tolerance, and improved scalability. Since a grid environment is highly dynamic, resource availability, network latency, and users requests may change frequently. To address these issues a dynamic replica placement strategy that adapts to dynamic behavior in data grids is needed. In this paper, we extend our earlier work on popularity-based replica placement proposing a new adaptive algorithm for use in large-scale hierarchical data grids. Our algorithm dynamically adapts the frequency and degree of replication based on data access arrival rate and available storage capacities. We evaluate our algorithm using OptorSim. Our results show that our algorithm can shorten job execution time greatly and reduce bandwidth consumption compared to its non-adaptive counterpart which outperforms other existing replica placement methods.

## 1. Introduction

Grid Computing [1] enables the selection, sharing, and aggregation of a wide variety of geographically distributed resources for solving large-scale computational problems. Data grids [2] provide services and infrastructure for distributed data-intensive applications that need to access massive datasets stored across distributed storage sites. An example of this is the experiments being run at the European Center for Nuclear Research (CERN) using the Large Hadron Collider (LHC) [3]. The CMS [4] and ATLAS [5] experiments using the LHC will collect massive amounts of data (many petabytes per annum) and involve thousands of researchers around the world. In such environments, maintaining a local copy of data on each accessing site is cost prohibitive. Further, storing such amounts of data centrally is impractical due to remote access latency and reliability concerns. Given the characteristics of the wide-area networks underlying many Grid systems and the need to access large volumes of data, scalability, data availability and access speed are key issues.

One way to speed up access in data grids is to replicate datasets at multiple locations [2]. Data replication not only reduces access costs, but also increases data availability [6, 7]. Creating replicas allows the routing of client requests to different replica sites thereby distributing the workload across the replica servers. The network load is also distributed across multiple network paths thereby decreasing the probability of congestion-related performance degradation.

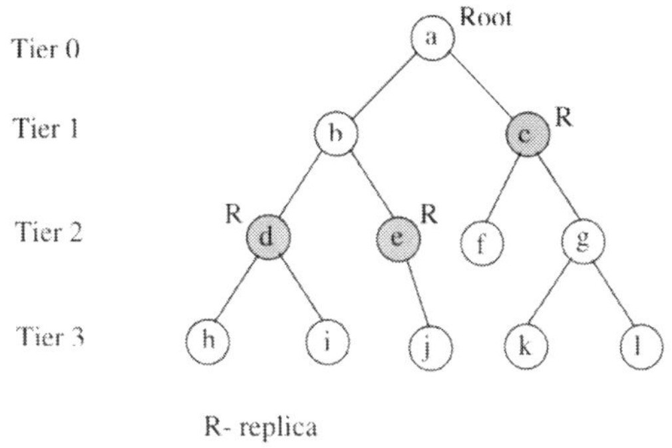


**Figure 1.** An example hierarchical data grid

To maximize the possible gains from file replication, strategic placement of the file replicas is critical [8, 9, 10]. Replication methods can be classified as static or dynamic [11]. With static replication, after a replica is created, it will be stored in the same location until it is deleted. Static replication cannot adapt to changes in file access patterns so the benefits of replication may be limited. On the contrary, dynamic replication automatically creates new replicas for frequently referenced data files or moves the replicas to other sites, as needed, to improve performance.

This paper addresses the problem of replica placement in data grids using an adaptive dynamic placement algorithm that extends our earlier work on Popularity-Based Replica Placement(PBRP) [12] which selectively replicates files based on their popularity. Our adaptive version, APBRP, presented in this paper dynamically adapts the frequency and degree of replication based on request arrival rates from clients and the available storage at replica servers to provide reduced access time and efficient use of bandwidth and storage resources. To evaluate the performance of ABRP, we use the data grid simulator OptorSim [13] and a hierarchical data grid structure reflecting the characteristics of the Compact Muon Spectrometer (CMS) experiments [4] at CERN. Simulations are done varying replica server capacities and data access patterns. We compare APBRP to PBRP and, transitively, to other replica placement algorithms.

The rest of this paper is organized as follows. Section 2 describes the hierarchical data grid model considered. In Section 3, research on replica placement in data grids is reviewed. We present our new algorithm in Section 4. Our simulation methods and performance results are described in Sections 5 and 6, respectively. Finally, Section 7 concludes the paper and suggests some directions for future work.

## 2. Hierarchical Data Grids

As is common in the current literature [14, 3, 15, 16], we focus on *hierarchical* data grids (e.g. [15]). Our example hierarchical data grid is modelled on the Large Hadron Collider (LHC) Computing Grid (LCG) [3] which provides data storage and analysis infrastructure for the thousands of users worldwide using the LHC. The data from the LHC experiments, estimated at roughly 15 Petabytes annually, is organized using a four-tier hierarchy with CERN as the root, or "tier 0" site, which stores all data produced. After initial processing, this data is distributed to the regional tier-1 sites that make data available to many national tier-2 sites. Individual scientists will access data through tier-3 sites (e.g. a local cluster in a University Department). The tier-3 sites is where users request access to their required data.

To minimize data access time and network load, replicas can be placed from the root to regional, national, or institutional centers. Thus, a user at a local site will try to access a replica locally. If a replicas not present, the request will go to the parent node to find a replica there.

Generally, a user request goes up the hierarchy and uses the first replica encountered along the path to the root. For example, Figure 1 shows a four-tiered hierarchical data grid. Assume a user at node $k$ tries to access a data file required by the job running on $k$. The data cannot be found locally, so the request goes to the parent node $g$, where the data is not available either. Finally, the request reaches node $c$, and is served there.

## 3. Related Work

Ranganathan and Foster ([15, 16]) describe and evaluate various replication strategies for hierarchical data grids. These strategies are defined depending on when, where, and how replicas are created and destroyed. They compare six different replication strategies: 1) No Replication: only the root node holds replicas; 2) Best Client: replica is created for the client who accesses the file the most; 3) Cascading: a replica is created somewhere on the path from the root node to the best client; 4) Plain Caching: local copies are stored on initial request; 5) Caching plus Cascading: combines plain caching and cascading; 6) Fast Spread: file copies are stored at each node on the path from the root to the best client. Results show that the cascading strategy reduces response time by 30% over plain caching when data access patterns contain both temporal and geographical locality. When access patterns contain some locality, Fast Spread saves significant bandwidth over other strategies.

Park et al. [17] propose a dynamic replication strategy, called BHR (Bandwidth Hierarchy based Replication), to reduce data access time by avoiding network congestion. BHR accesses replicas located at sites that have the best current bandwidth to a job's execution site. The BHR strategy can thus exploit the 'network-level locality' that occurs naturally in regional (and other) networks.

Two dynamic replication mechanisms are proposed by Tang et al [11] for multi-tier data grids. The Simple Bottom-Up (SBU) algorithm replicates any data file that currently exceeds a pre-defined threshold of access rate as close as possible to the clients without considering historical accesses. The Aggregate Bottom-Up (ABU) algorithm considers the access histories of files used by sibling nodes and aggregates the access information for files so that the frequently accessed files are replicated first. This process is repeated from the leaves to the root. ABU improves access latency significantly at the expense of increased storage use.

Different sites in a data grid may have different service quality requirements making Quality of Service (QoS) another important factor. Wang et al. [18] describe a heuristic QoS-aware algorithm that places replicas to simultaneously improve system performance and satisfy users' quality requirements. Their algorithm, for general graphs, starts by finding the cover set of every server in the network. The algorithm then identifies and deletes super cover sets in the network. Finally, it inserts replicas into the network iteratively until all servers are satisfied. Results indicate that their algorithm finds near-optimal solutions.

In our earlier work [12], we presented the Popularity Based Replica Placement (PBRP) algorithm for hierarchical data grids. Since data access patterns may change over time, any dynamic replication strategy must track file access histories to decide when, what and where to replicate. The "popularity" of a file is determined by its access rate at each location and PBRP is invoked at regular intervals to create replicas as close as possible to those clients that frequently request data files. Replication is done in two phases. The bottom-up aggregation phase aggregates access history records for each file to upper tiers (as in ABU) summing the access counts for records whose nodes are siblings and which refer to the same file. In the second, top-down, phase the aggregated information is used to place replicas. The placement process traverses down the hierarchy as long as the aggregated access count is greater than or equal to a pre-defined popularity threshold (selected empirically based on historical access information). A replica is placed at a node if the threshold prevents further traversal through one or more of its children. An instance of bottom-up aggregation and top-down placement is shown in Figure 2.

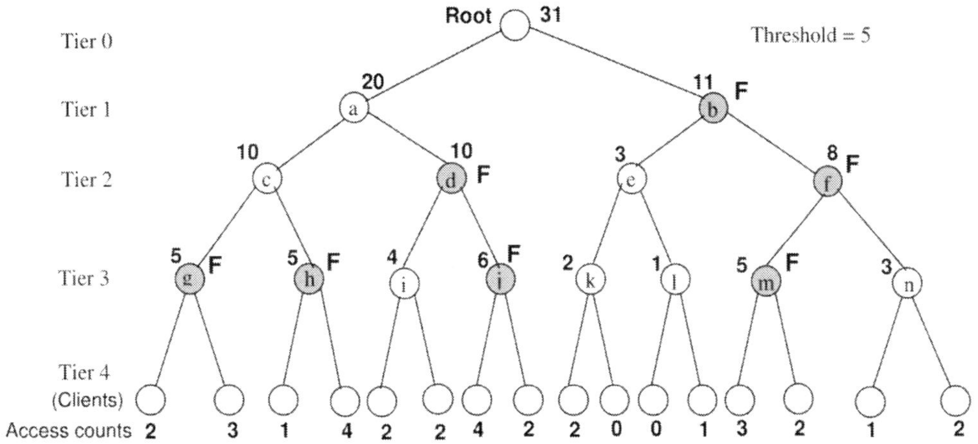


**Figure 2.** Bottom-up aggregation of access counts and top-down placement of replicas

Our simulation results show that PBRP can shorten job execution time significantly and reduce bandwidth consumption compared to other dynamic replication methods including ABU, Fast Spread and Cascading placement. However, the effectiveness of PBRP depends on careful selection of the popularity threshold. The adaptive algorithm proposed in this paper is designed to address this issue by determining the threshold value dynamically based on the data access arrival rates and available storage at replica servers.

## 4. Adaptive Replica Placement

The goal of our Adaptive Popularity Based Replica Placement (APBRP) algorithm is to reduce data access time by dynamically creating replicas for "popular" data files. Placement of the replicas is guided by dynamic adjustment of the threshold value used in PBRP that determines how deep and hence how often files are replicated in different parts of the hierarchy. APBRP sets the threshold dynamically based on client request rate and replica server capacities.

Like PBRP, replication in APBRP is done in two phases. The first phase aggregates access records for each file from lower to upper tiers. In the second phase, replicas are placed from the root to the leaves of the tree using the aggregated information and threshold value. As in PBRP, our algorithm is invoked at regular intervals. The access pattern logs are cleared at the start of each interval to capture evolving pattern changes. A short interval can be chosen for high arrival rates incurring greater overhead but adapting more rapidly to changing access patterns. Determination of the threshold is done independently of placement.

We use access request rate and available server capacity to determine the threshold value. A high access request rate corresponds to frequent accesses by clients which, in PBRP, results in more "popular" files whose access counts exceed the threshold. This in turn results in more replicas which causes additional replication overhead (in terms of both bandwidth and storage use). Increasing the threshold lessens the number of replicas created. On the other hand, when the request rate drops, the number of popular files is less so fewer replicas are created even though the system might be capable (in terms of bandwidth and storage) of supporting more replicas to improve access latency. The available storage capacities of replica servers are also used in adapting the threshold value. The required threshold value can be decreased to create more replicas if replica servers have abundant storage space. The threshold value can also be increased to lessen the number of replicas as the replica servers become full.

Figure 3 illustrates replica placement with dynamic threshold adjustment. First, the variation in request access rate from the users is calculated to determine the change in threshold value

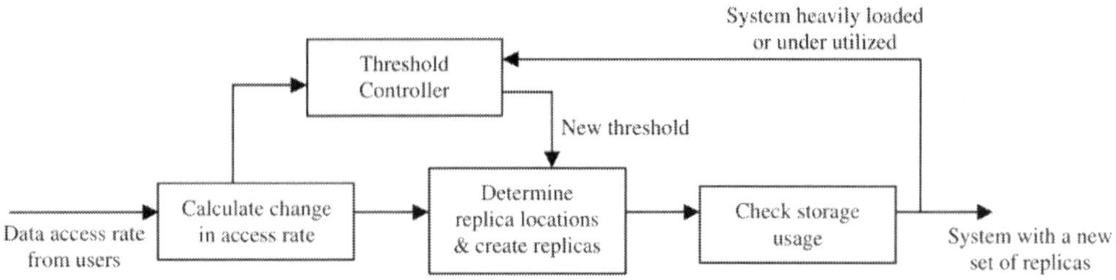


**Figure 3.** Replication with threshold adjustment

(discussed later). Then, the replica placement algorithm uses the new threshold value to determine the replica placement. A storage checker assesses the storage used by all replica servers to decide whether any further adjustment in threshold value is necessary. Feedback is sent to the threshold controller in case of heavily loaded or under-utilized replica servers.

The key challenge in this approach is to determine how much change in the threshold value the system can tolerate based on changes in request arrival rate and available storage capacities of replica servers. Overly aggressive threshold increases will cause the number of replicas to drop which will increase access time and decrease storage utilization. Aggressive threshold decreases will increase the number of replicas and the storage cost. Hence, the threshold value has to be adjusted so that the trade-off between space use and access latency is balanced.

### *4.1. Determining the Initial Threshold*

The initial threshold value is set based on the average aggregated access counts at the replica servers in the bottom tier. The value of the average aggregated access count is calculated by dividing the total number of aggregated access counts for a file at the replica servers in the second to lowest tier of the hierarchy by the number of replica servers at that tier. This is done during the bottom-up aggregation phase. The initial value is then adjusted dynamically based on available storage and user request arrival rates.

### *4.2. Dynamic Adjustment of the Threshold Value*

Changes to the threshold based on variation in the request arrival rate are not applied immediately to avoid reacting to transient changes in the arrival rate. Changes are only made to reflect an ongoing trend over an interval of time. This interval is calculated as a fraction of the length used for sampling the access histories of clients. The threshold value is increased or decreased by the difference between the current and previous average aggregated access counts of replica servers in the lower tier of the hierarchy. For example, assuming that $R_i$ is the new average aggregated access rate in the $i^{th}$ interval and $R_{i-1}$ is the average aggregated access rate in the $(i-1)^{st}$ interval, the threshold value for the $i^{th}$ sampling interval, *threshold*$_i$, is calculated as follows:

$$threshold_i = threshold_{i-1} + \triangle \tag{1}$$

where $\triangle$ is defined as:

$$\triangle = \begin{cases} 0, & \text{for the first interval of access histories} \\ R_i - R_{i-1}, & \text{otherwise} \end{cases} \tag{2}$$

Once the threshold value is updated, the available storage capacities of the replica servers are checked. Further adjustment of the threshold value is done based on the available capacities

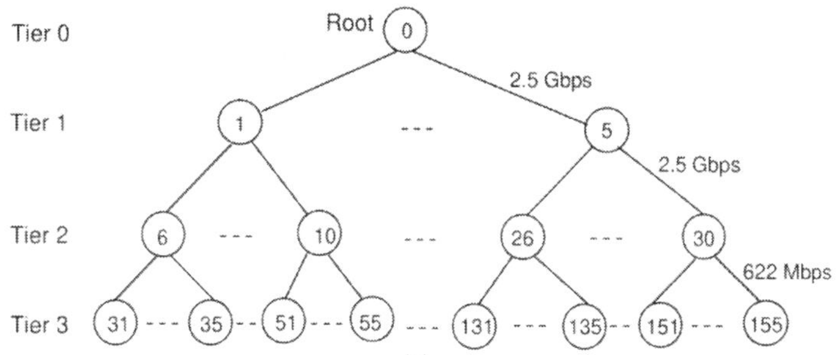


(a)

| Config | Tier 1 (GB per node) | Tier 2 (GB per node) | Tier 3 (GB per node) | Relative storage capacity (%) |
|---|---|---|---|---|
| 1 | 1000 | 300 | 50 | 75 |
| 2 | 500 | 200 | 50 | 55 |
| 3 | 250 | 100 | 50 | 40 |
| 4 | 125 | 50 | 20 | 17.5 |
| 5 | 50 | 25 | 20 | 13.75 |

(b)

**Figure 4.** (a) The simulated data grid topology, (b) Different storage configurations

at the replica servers. The threshold controller (Figure 3) updates the threshold based on the ratio between the desired percentage of storage use and the actual percentage of storage used by replica servers. The new threshold value for the $i^{th}$ interval is thus calculated as:

$$threshold_i(updated) = threshold_i \times ratio \tag{3}$$

In this paper, we define three storage utilization states for when the system is lightly, moderately, and highly loaded corresponding to the percentage of storage capacity used by replica servers (less than 50%, 50% to 75%, and greater than 75%, respectively). If the system is heavily loaded or under utlilized then the calculated ratio is used in updating the threshold.

We also use a popularity-based form of the Least Recently Used (LRU) replacement policy with an added constraint to ensure that replicas created in the current replication interval will not be replaced. This additional constraint is needed to avoid the deletion of newly created replicas by the dynamic replication algorithms. If the available space on a given server is less than the size of the new replica, some "removable" replicas need to be deleted to make room for the new replica. Removable replicas are those created before the current replication interval that are not in use by any client at present.

## 5. Simulation Setup

To evaluate APBRP we used a locally extended version of OptorSim [19] created to support the necessary information gathering needed for the ABRP algorithm.

### *5.1. Simulated Data Grid Topology*

The simulated data grid structure and network bandwidths between sites are based on estimates for the CMS experiment. The topology of the simulated data grid is shown in Figure 4. All data requests are generated by the leaf nodes.The links between nodes show the available network bandwidth. In our simulations, we considered five different storage configurations based on the relative storage capacity of the replica servers, defined as a ratio between the total size of the replica servers ($S$) and the total size of all data files in the system ($F$). There are 2500 data files in the system and each one is 10 GB, so $F$ is approximately 25 TB. The relative storage capacity of the replica servers in the data grid are varied from 75% down to 13%. The detailed storage configurations for all five cases are shown in Figure 4. For example, in configuration 2, each node in tier-1, tier-2, and tier-3 has 500 GB, 200 GB, and 50 GB storage available for replicas, respectively. The absolute capacities are of little interest but the relative capacities affect placement decisions. The total storage size of the replica servers is 13.75 TB and the relative storage capacity is 55% ($\frac{13.75}{25} \times 100$). To assess the impact of file size distribution on the performance of the replication algorithms, we also considered file sizes of 2 GB and 20 GB. We adjusted the number of files so the total size of all data files in the system and the relative storage capacity of the servers were unchanged.

### 5.2. *Data Access Patterns*

Each simulated job requests a set of filename(s) to access. The order in which the files are requested is determined by a selected access pattern. We experimented with five access patterns: sequential (files are accessed in the order stated in the job configuration file), random (files are accessed using a flat random distribution), unitary random (file requests are one element away from previous file request but the direction is random), Gaussian random walk (files are accessed using a Gaussian distribution), and the heavily tailed Zipf distribution: $P_i = \frac{K}{i^s}$, where $P_i$ is the frequency of the $i$th ranked item, $K$ is the popularity of the most frequently accessed data item and $s$ determines the shape of the distribution.

We ran all simulations with 150 jobs of various types. The jobs were submitted with a fixed probability but some jobs were more common than others. A resource broker acted as a meta-scheduler that controlled job scheduling to different client nodes with computing resources based on the estimated data access cost for the current and all queued jobs.

## 6. Simulation Results

The metrics we chose to evaluate the performance of ABPRP were job run time, average bandwidth use, and storage use. Run time is the total time to execute all jobs and is scaled for display. For each data transmission, the bandwidth cost is the data size multiplied by the summation of costs along the data transmission path. The value of the average bandwidth cost is the total bandwidth cost divided by the number of client requests. Storage use is the percentage of the available storage used by the replicas. To determine the effectiveness of our *adaptive* algorithm we considered three scenarios: when the data access rate consistently increases, consistently decreases, and when it fluctuates. For each case, we compared APBRP with its non-adaptive counterpart PBRP for the various replica server configurations.

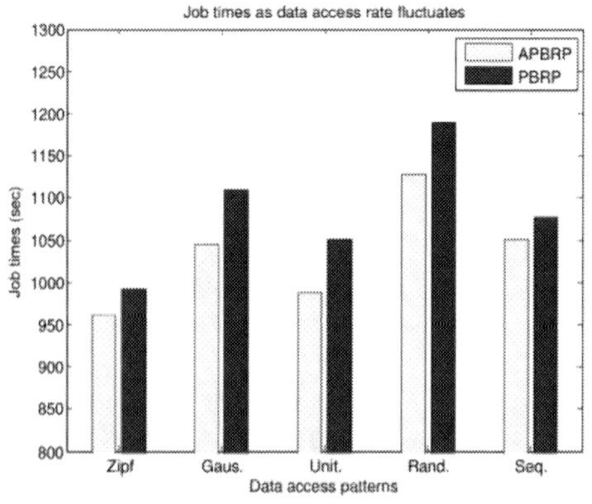


**Figure 5.** Runtime: Config. 1, Fluctuating Rate

Journal of Physics: Conference Series **256** (2010) 012020 doi:10.1088/1742-6596/256/1/012020

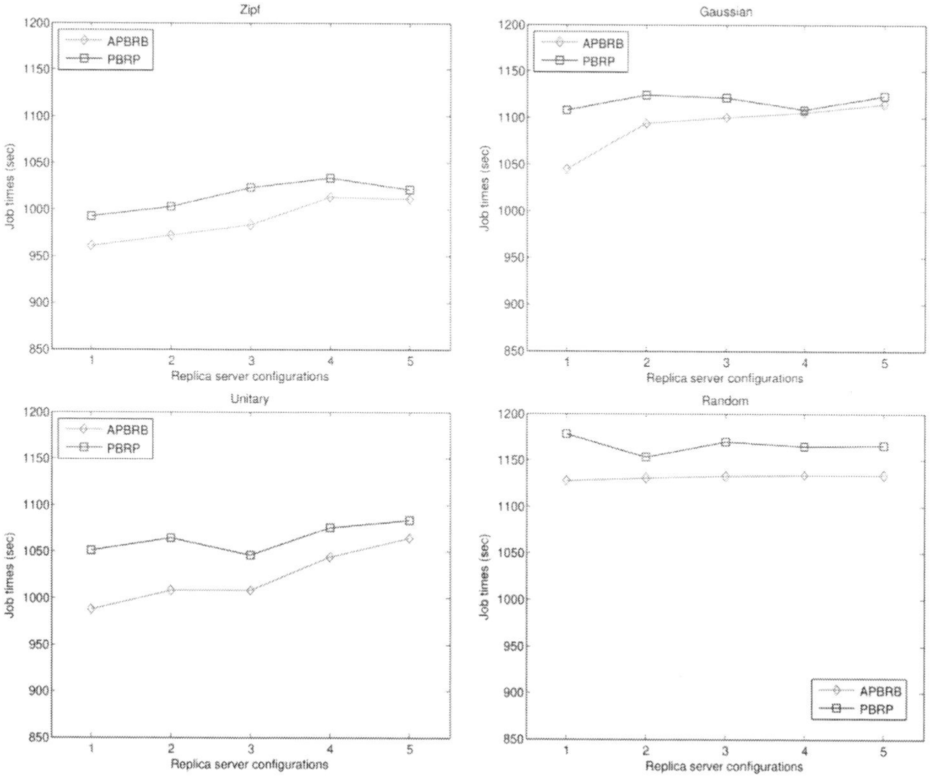


**Figure 6.** Runtime by Pattern & Config, Fluctuating Rate

### *6.1. Job Execution Time*

Figure 5 shows the job execution times of APBRP and PBRP using configuration one (uniform capacities) when the data access arrival rate from the clients is fluctuating. APBRP displays shorter execution times for all data access patterns because APBRP adjusts the threshold value based on the varying access arrival rate which leads to the creation of more replicas which decreases the run time. For example, with a Zipf distribution, the number of replicas created by APBRP and PBRP were 37 and 27, respectively. The benefit of APBRP is most significant when the accesses exhibit some randomness. ABRP clearly adapts well to fluctuating accesses.

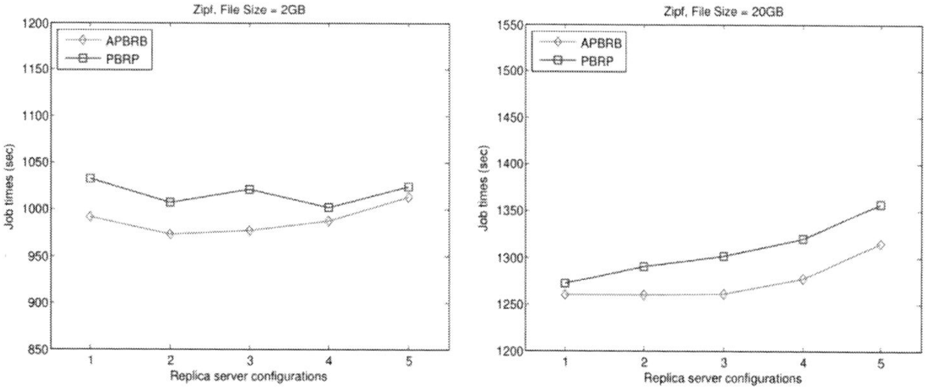


**Figure 7.** Runtime by File Size, Fluctuating Rate

Figure 6 compares the run times for all access patterns and storage configurations as the

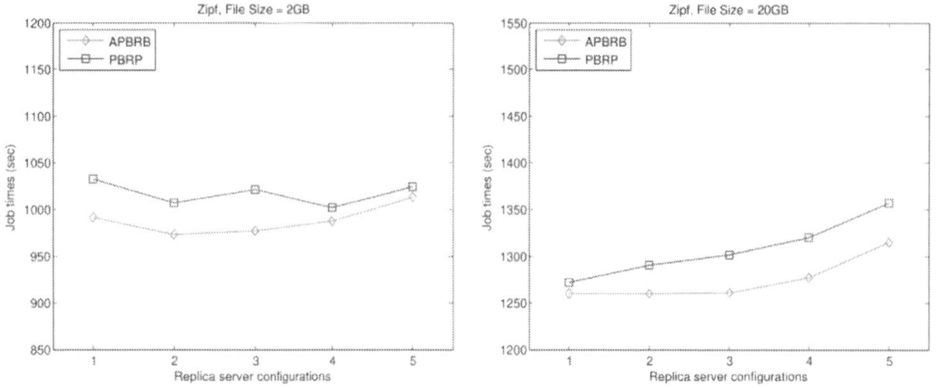


**Figure 8.** Runtime by File Size, Fluctuating Rate

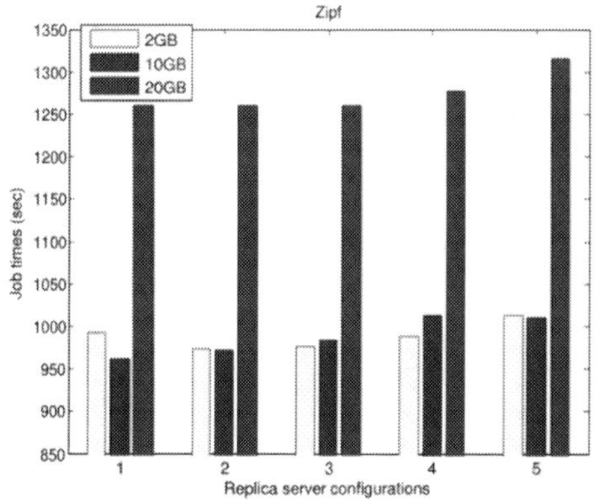


**Figure 9.** Runtime: APBRP by File Size, Fluctuating Rate

access rate fluctuates. With decreasing storage size, the execution times of both techniques are increased but by different amounts. APBRP shows shorter run times for all storage configurations. However, when the available capacity is smaller, the benefit of APBRP over PBRP decreases since there is less available space to allocate additional replicas.

Figure 8 shows the job times of APBRP and PBRP for all storage configurations using the Zipf access pattern with file sizes of 2 GB and 20 GB. In both cases, APBRP displays shorter execution times for all configurations. Unlike the other tested scenarios, the benefit of APBRP over PBRP increases with decreasing storage size when a file size of 20 GB is used. This is due to the increased number of replicas and relatively low replica replacement frequency in APBRP.

Figure 9 compares the run times of APBRP using different file sizes for the Zipf distribution. The execution time with a file size of 20 GB is much higher than the other two cases due to the increased data transfer latency and frequent replacement of new replicas. The performance of the replication algorithm in terms of job execution time is not much impacted when smaller file sizes (2 GB and 10 GB) are used. The job times are quite similar for both cases except for some storage configurations where the use of a 2 GB file size results in shorter execution times.

Figure 10 shows the run times of APBRP and PBRP, again for configuration one, as the client request arrival rate *decreases*. APBRP shows consistently better performance for all data access patterns. This is due to the fact that when the arrival rate decreases consistently, APBRP also decreases the threshold which results in more replicas compared to PBRP. This in turn decreases the execution time. The Zipf distribution shows the lowest run time using APBRP.

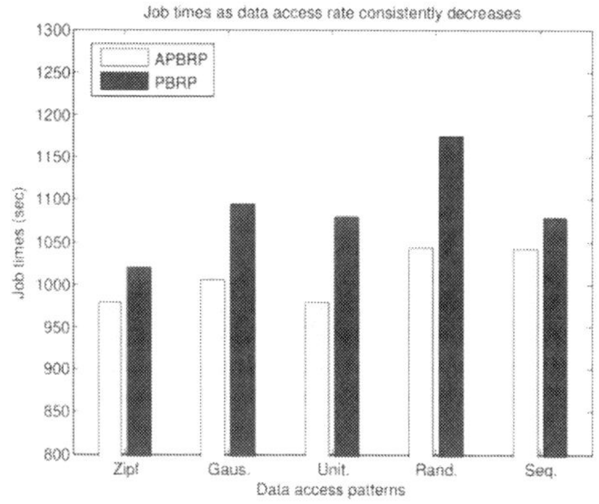


**Figure 10.** Runtime: Config. 1, Decreasing Rate

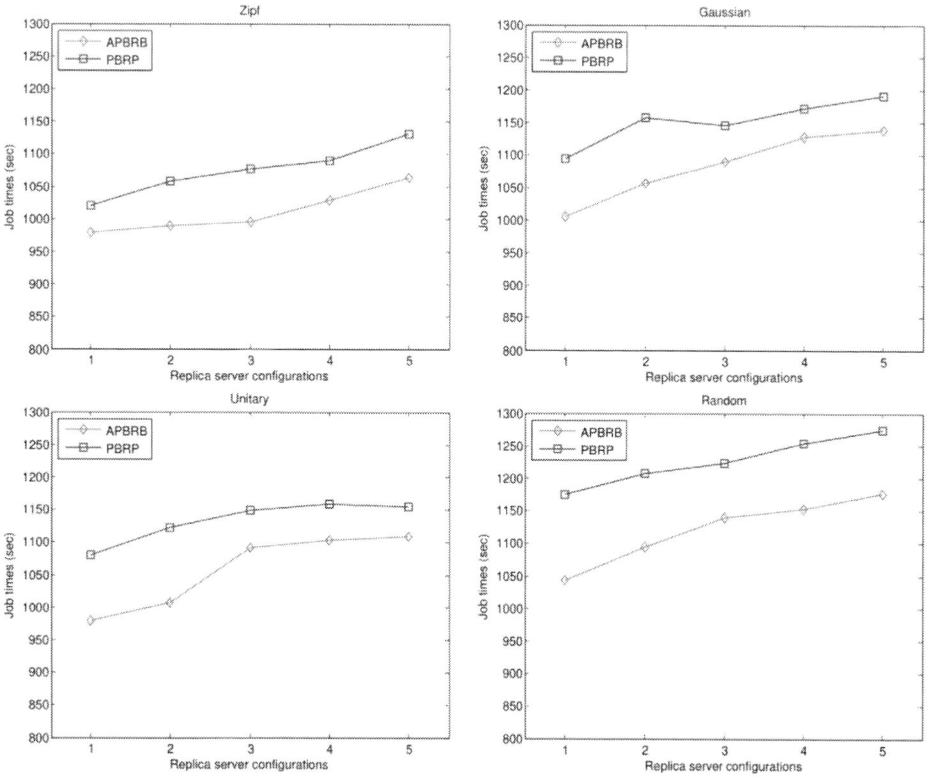


**Figure 11.** Runtime by Pattern & Config, Decreasing Rate

Figure 11 compares the job execution times of the two placement algorithms using different storage resource configurations as the access rate decreases consistently. With decreasing capacity, the execution times for both algorithms increase but by different degrees. APBRP shows better job execution times for all storage configurations. However, when the available storage capacity is smaller, the benefit of APBRP over PBRP decreases as before. This happens because of frequent replica replacement in APBRP when the replica servers have limited storage.

Figure 12 compares the run times of APBRP and PBRP, again for configuration one, as the request arrival rate increases. When the request arrival rate increases, APBRP increases the threshold to limit the number of replicas to a reasonable level based on the storage available (as

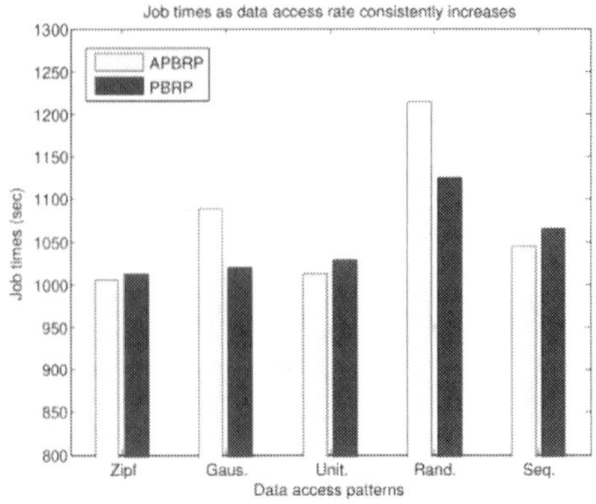


**Figure 12.** Runtime: Config. 1, Increasing Rate

described in section 4.2) while PBRP experiences a drastic increase in the number of replicas (that eventually fills the replica servers).

APBRP performs as well as PBRP or somewhat better for the Zipf, Unitary, and Sequential access patterns but with the Random and Gaussian distributions PBRP performs better due to the significant drop in the number of replicas when using ABRP. For example, PBRP creates approximately 66 and 56 replicas for these two access patterns. These numbers drop to 29 and 25 with APBRP. Overall, APBRP shows less benefit over PBRP, for the tested scenarios, when the client request rate is consistently increasing.

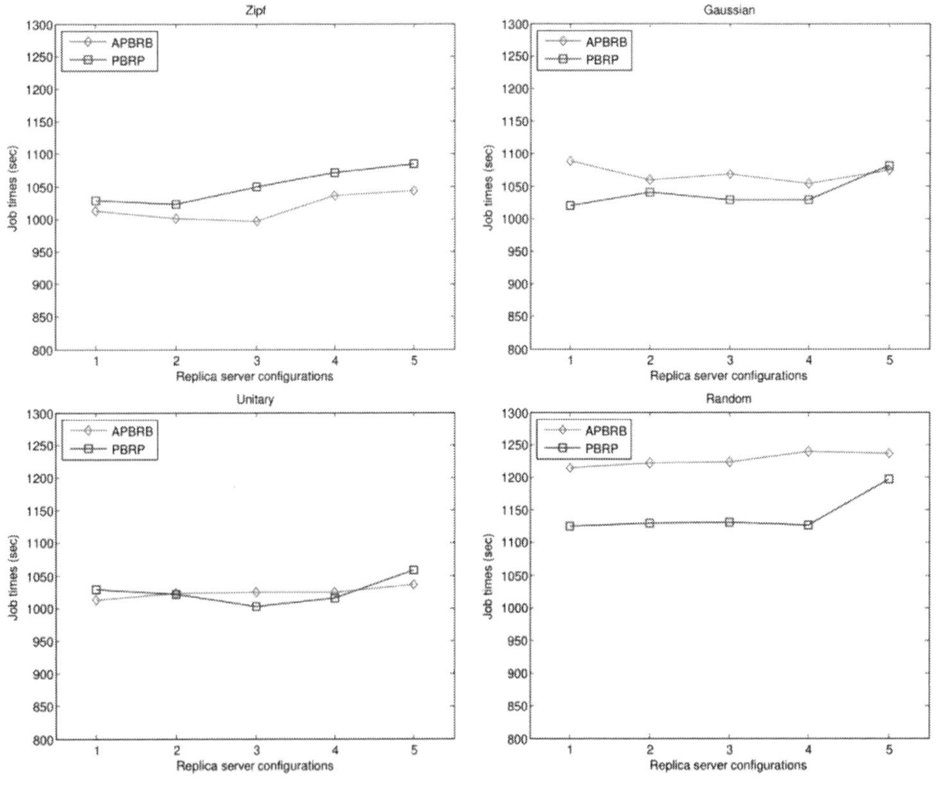


**Figure 13.** Runtime by Pattern & Config, Increasing Rate

Figure 13 compares the job execution times of the two placement algorithms using different

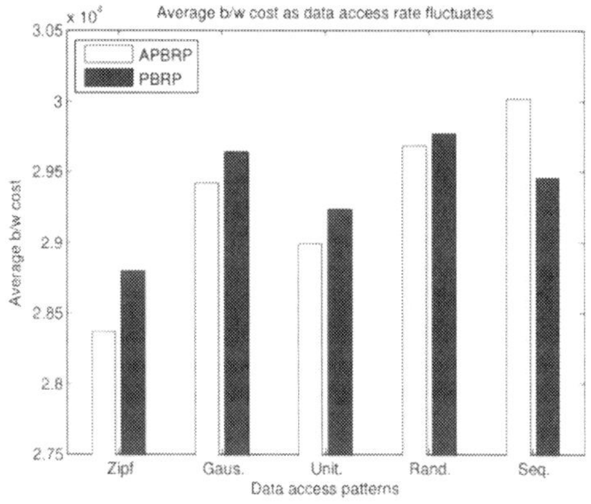


**Figure 14.** Avg. B/W: Config. 1, Fluctuating Rate

storage resource configurations with consistent increase in access rate. With decreasing capacity, the job execution times using both algorithms increase but by different amounts. Adaptive-PBRP maintains better job execution times for all five storage configurations. For the Zipf and Random patterns, the performance difference between the methods reduces as the storage capacities of the servers decrease. This happens because the difference between the number of replicas created by PBRP and Adaptive-PBRP decreases significantly.

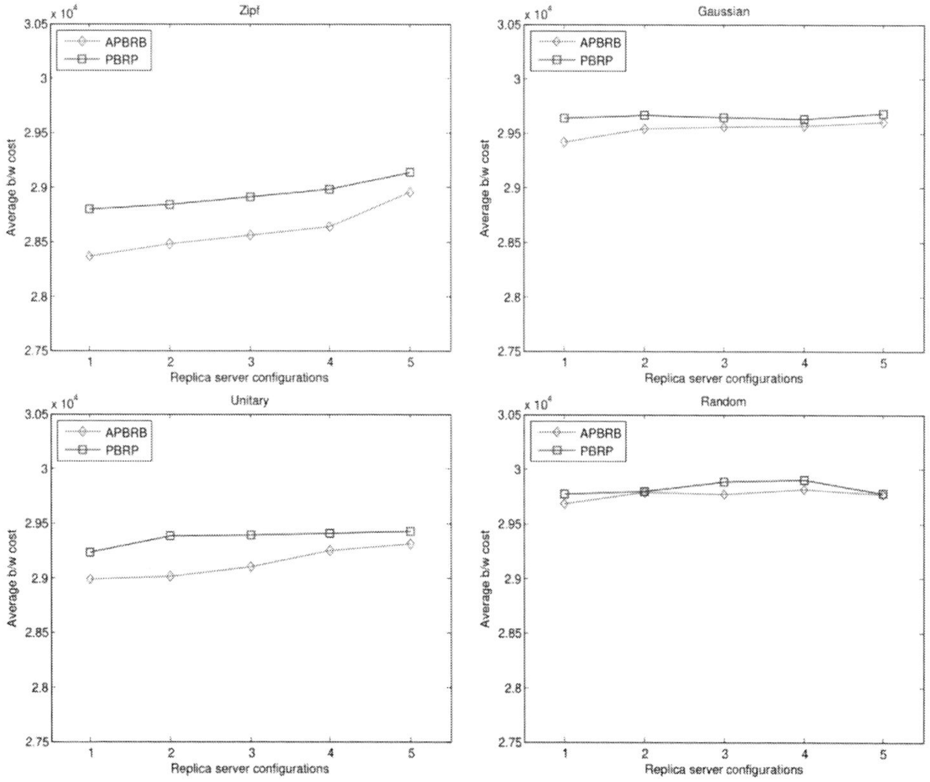


**Figure 15.** B/W by Pattern & Config, Fluctuating Rate

*6.2. Average bandwidth cost*
Figure 14 shows the average bandwidth (B/W) costs of APBRP and PBRP for configuration one as the request access arrival rate fluctuates. APBRP requires less bandwidth for all file access patterns except sequential. ABRP's increased number of replicas close to clients compared to PBRP generally results in decreased bandwidth consumption for accessing files.

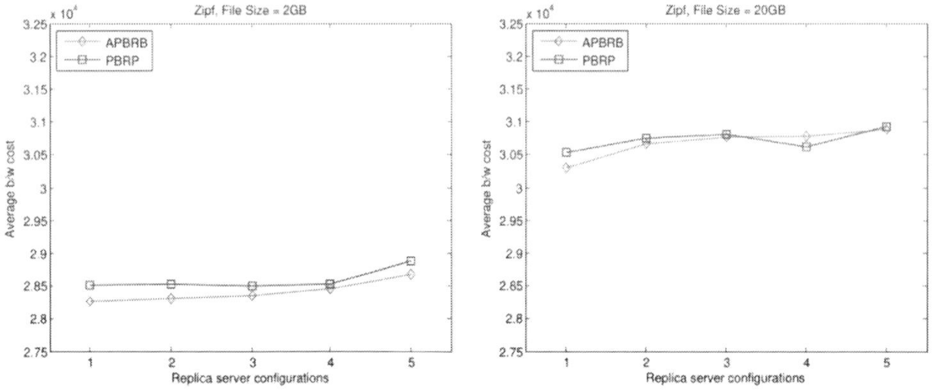


**Figure 16.** Avg. B/W by File Size, Fluctuating Rate

The average bandwidth costs for all storage configurations and access patterns given regular fluctuation in request rates are shown in Figure 15. With decreasing storage size, the bandwidth costs of both techniques are increased but by different amounts. APBRP shows lower bandwidth costs for almost all storage configurations and access patterns but the benefit of APBRP over PBRP decreases as the storage capacities of replica servers are reduced.

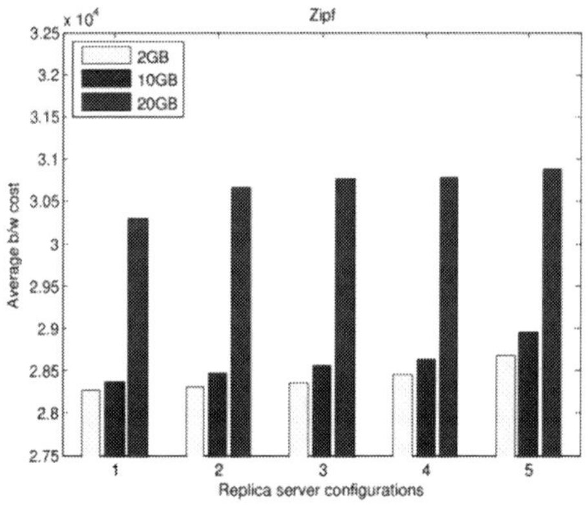


**Figure 17.** Avg. B/W: APBRP by File Size, Fluctuating Rate

Figure 16 shows the average bandwidth costs of APBRP and PBRP for all storage configurations using the Zipf access pattern with file sizes of 2 GB and 20 GB. For both cases, APBRP has less bandwidth cost for all but one configuration. The benefit of APBRP over PBRP decreases somewhat with decreasing available storage. Since there is less available space for storing more replicas, the frequency of replica replacement increases which in turn leads to additional bandwidth consumption.

Figure 17 compares the average bandwidth costs of APBRP using different file sizes for the Zipf distribution. The bandwidth cost for 20 GB files is much higher than the other two cases due to the increased data transfer cost and the frequent replacement of new replicas.

The performance of the replication algorithm in terms of average bandwidth consumed is not significantly impacted for smaller file sizes (2 GB and 10 GB). The bandwidth costs are similar in both cases. However, 2 GB files result in less bandwidth consumption than 10 GB files.

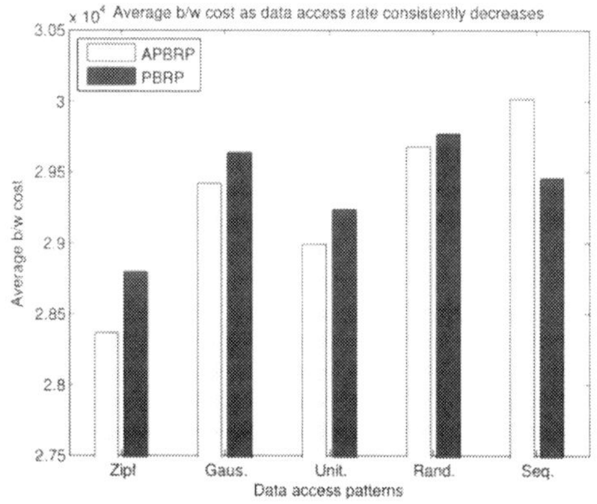


**Figure 18.** Avg. B/W: Config. 1, Decreasing Rate

Figure 18 shows the bandwidth costs for storage configuration one when the request arrival rate is decreasing. APBRP shows better performance for only some access patterns. The benefit of APBRP over PBRP is not terribly significant and the liability for sequential access is great.

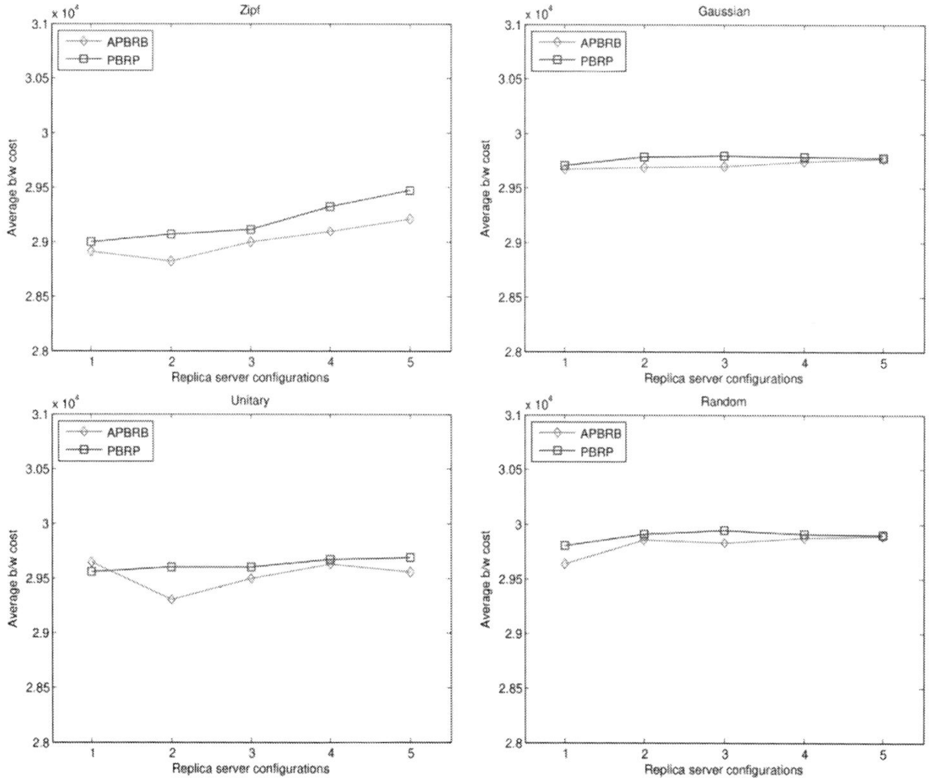


**Figure 19.** Avg. B/W by Pattern & Config, Decreasing Rate

Figure 19 shows the average bandwidth costs of both strategies for the three configurations with consistent decrease in access rate. Adaptive-PBRP again requires less bandwidth compared to PBRP. As the storage size decreases, the bandwidth costs of both the strategies increase.

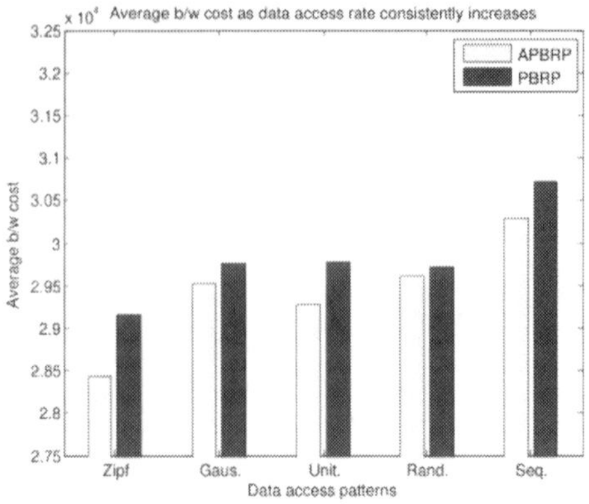


**Figure 20.** Config 1., Avg. B/W: Increasing Rate

Figure 20 shows the bandwidth costs for configuration one as the request arrival rate from the clients increases. As the request arrival rate increases, PBRP experiences a drastic increase in the number of replicas. This in turn results in consumption of extra bandwidth for unnecessary data movement. APBRP adjusts the threshold value to reduce the number of replicas and thus saves significant link bandwidth. For all data access patterns, APBRP requires less bandwidth than PBRP.

### *6.3. Storage Use*

Figure 21 shows the storage use of APBRP and PBRP using server configuration one as the request arrival rate fluctuates. APBRP requires more storage than PBRP for all data access patterns. This is due to the creation of an increased number of replicas using APBRP.

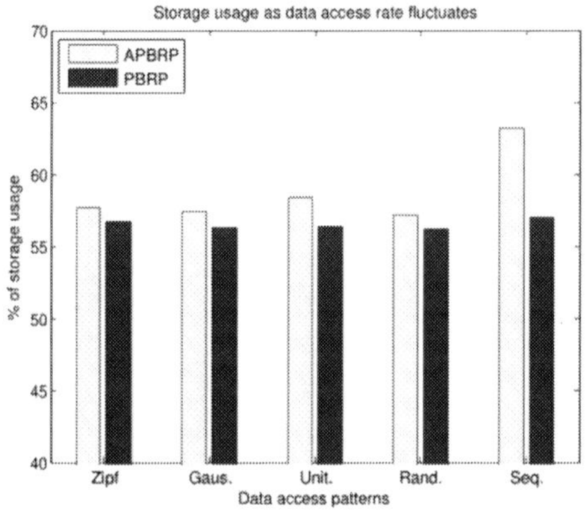


**Figure 21.** Storage Use: Config. 1, Fluctuating Rate

Figure 22 compares the storage use of the two placement algorithms for all storage configurations and access patterns as the request rate fluctuates. With decreasing storage availability, the percentages of storage used for both techniques increases but by different amounts. When the available storage capacity is limited, the difference in storage use for APBRP and PBRP tends to decrease. This happens because of frequent replica replacement in APBRP since the replica servers have limited storage available.

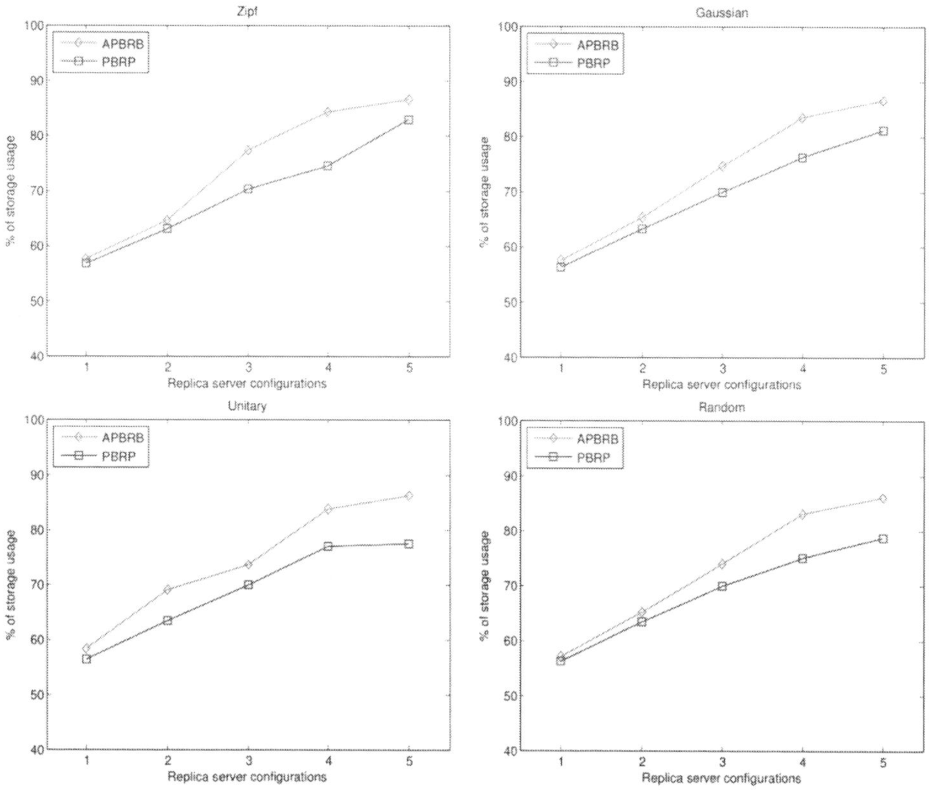


**Figure 22.** Storage Use by Pattern & Config, Fluctuating Rate

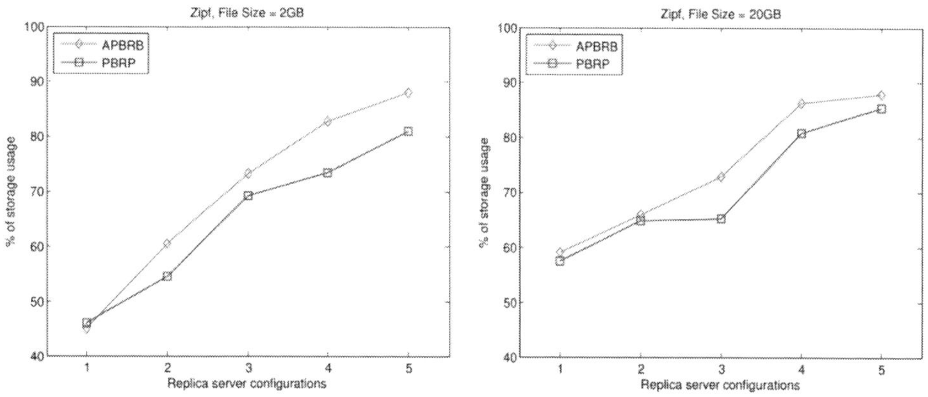


**Figure 23.** Storage Use by File Size, Fluctuating Rate

Figure 23 shows the storage use of APBRP and PBRP for all storage configurations and the Zipf file access pattern with file sizes of 2 GB and 20 GB. In both cases, APBRP has greater storage cost due to its creation of additional replicas when compared to PBRP.

Figure 24 compares the storage costs of APBRP using different file sizes for the Zipf distribution. The performance of APBRP in terms of storage cost is not much impacted by the data file size. Storage use is very similar for all three cases and storage configurations.

Figure 25 (left) shows the storage use of APBRP and PBRP for configuration one as the request arrival rate decreases. When this happens, APBRP decreases the threshold value which results in an increased number of replicas compared to PBRP. This increases the storage use.

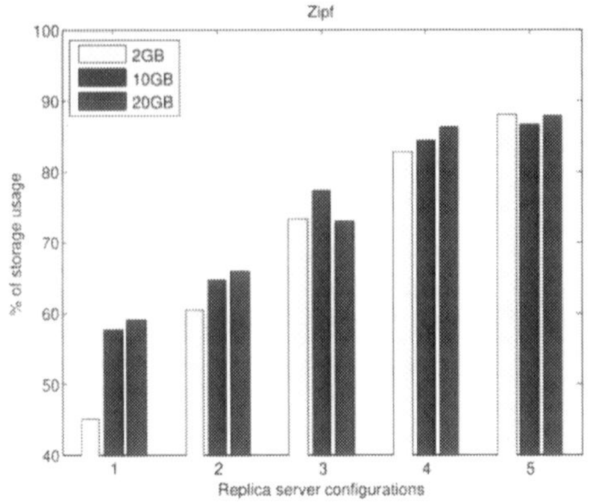


**Figure 24.** Storage Use: APBRP by File Size, Fluctuating Rate

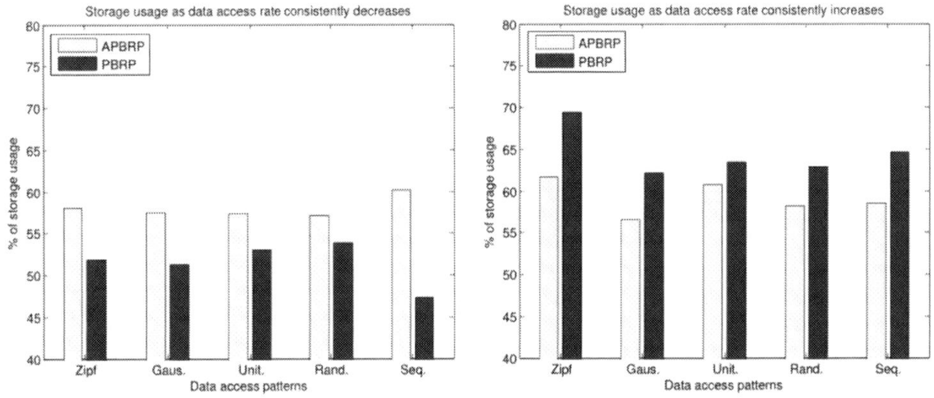


**Figure 25.** Storage Use: Config. 1, Decreasing (left) and Increasing Rate (right).

Figure 25 (right) shows the percentage of storage used by APBRP and PBRP for configuration one as the request arrival rate increases. In PBRP, the number of replicas increases drastically with the increase in request arrival rate whereas APBRP limits the number of replicas to a reasonable level by increasing the threshold value resulting in lower storage costs then PBRP.

## 7. Conclusions and Future Work

In this paper, we have introduced an adaptive popularity-based replica placement algorithm (APBRP). Simulation results show that our algorithm can significantly shorten job execution time and reduce bandwidth consumption at the cost of moderate increases in storage use compared to PBRP. For the tested access patterns and storage configurations, APBRP outperforms PBRP in terms of both job execution time and average bandwidth use. The increased consumption of (relatively cheap) storage space by APBRP makes it a good candidate when access time and bandwidth use are of primary concern. Our earlier results [12] show that PBRP can shorten job execution time significantly and reduce bandwidth consumption compared to other dynamic methods including ABU, Fast Spread and Cascading placement. Thus, transitively, APBRP also performs better than these other non-adaptive dynamic replication methods. The novelty of APBRP is its ability to balance space use and access latency by means of adaptive popularity-based file replication. The improved efficiency of access to data and flexibility to changing access patterns offered by APBRP will help make large-scale data intensive applications practical, potentially running over public wide-area networks.

For future work, we are exploring the incorporation of user QoS requirements into the placement process. We also plan to explore how to exploit network knowledge in making placement decisions. Finally, we will investigate dynamic replica maintenance (e.g. replica consistency) and work to extend our approach to non-hierarchical network models.

## Acknowledgments

This research was supported in part by the Natural Sciences and Engineering Research Council of Canada and by a University of Manitoba Graduate Fellowship.

## References

[1] Kesselman C and Foster I 1998 *The Grid: Blueprint for a New Computing Infrastructure* (Morgan Kaufmann Publishers)
[2] Venugopal S, Buyya R and Ramamohanarao K 2006 *ACM Comp. Surveys* **1**(38) 1–53
[3] Worldwide LHC Computing Grid http://lcg.web.cern.ch/lcg/ Distributed Production Environment for Physics Data Processing
[4] Holtman K 2001 CMS Data Grid System Overview and Requirements CMS Experiment Note 2001/037, CERN
[5] The ATLAS experiment http://atlas.ch/ Particle Physics Experiment at CERN
[6] Ranganathan K, Iamnitchi A and Foster I 2002 *Proceedings of the 2nd IEEE/ACM International Symposium on Cluster Computing and the Grid (CCGRID'02)* pp 376–381
[7] Lamehamedi H, Szymanski B, shentu Z and Deelman E 2002 *Proceedings of the Fifth International Conference on Algorithms and Architectures for Parallel Processing* pp 378–383
[8] Wolfson O and Milo A 1991 *ACM Transactions on Database Systems* **16** 181–205
[9] Kalpakis K, Dasgupta K and Wolfson O 2001 *IEEE Transactions on Parallel and Distributed Systems* **12** 628–637
[10] Tu M, Li P, Ma Q, Yen I and Bastani F 2005 *Proceedings of the 19th IEEE International Parallel and Distributed Processing Symposium (IPDPS'05)* vol 1 p 14
[11] Tang M, Lee B, Yeo C and Tang X 2005 *Future Generation Computing System* **21** 775–790
[12] Shorfuzzaman M, Graham P and Eskicioglu R 2008 *Proc. 9th Intl. Conf. on Parallel & Dist. Computing, Applications and Technologies* pp 524–531
[13] Bell W, Cameron D, Capozza L, Millar A, Stockinger K and Zini F 2002 *Proc. 3rd Intl. IEEE Workshop on Grid Computing (Grid'2002)*
[14] Bell W H, Cameron D G, Carvajal-Schiaffino R, Millar A P, Stockinger K and Zini F 2003 *Proceedings of the 3rd IEEE/ACM International Symposium on Cluster Computing and the Grid*
[15] Ranganathan K and Foster I T 2001 *Proc. Intl. Wksp. on Grid Computing* pp 75–86
[16] Ranganathan K and Foster I 2001 *Proc. of the Intl. Conf. Computing in High Energy and Nuclear Physics (2001).*
[17] Park S, Kim J, Ko Y and Yoon W 2003 *Proc. 2nd Intl. Workshop on Grid and Cooperative Computing*
[18] Wang H, Liu P and Wu J 2006 *Journal of Grid Computing* 96–103
[19] Bell W, Cameron D, Capozza L, Millar P, Stockinger K and Zini F 2003 *Intl. Journal of High Performance Computing Applications* **17**(4) 403–416

# An Ontological-Fuzzy Approach to Advance Reservation in Multi-Cluster Grids

**D.J. Ferreira[1], M.A.R. Dantas[1], Michael A. Bauer[2]**

[1] Federal University of Santa Catarina (UFSC), Department of Informatics and Statistic (INE), 88040-900, Florianopolis - SC, Brazil

[2] University of Western Ontario, Department of Computer Science, London, Ontario, N6A 5B7, Canada

E-mail: [1] {ded,mario}@inf.ufsc.br, [2] {bauer}@csd.uwo.ca

**Abstract.** Advance reservation is an important mechanism for a successful utilization of available resources in distributed multi-cluster environments. This mechanism allows, for example, a user to provide parameters aiming to satisfy requirements related to applications' execution time and temporal dependence. This predictability can lead the system to reach higher levels of QoS. However, the support for advance reservation has been restricted due to the complexity of large scale configurations and also dynamic changes verified in these systems. In this research work it is proposed an advance reservation method, based on a ontology-fuzzy approach. It allows a user to reserve a wide variety of resources and enable large jobs to be reserved among different nodes. In addition, it dynamically verifies the possibility of reservation with the local RMS, avoiding future allocation conflicts. Experimental results of the proposal, through simulation, indicate that the proposed mechanism reached a successful level of flexibility for large jobs and more appropriated distribution of resources in a distributed multi-cluster configuration.

## 1. Introduction

Multiple compute clusters, if well orchestrated as in a grid environment, can represent significant computational power for the execution of several classes of applications, including parallel jobs. One significant challenge in such grid environments is how to efficiently allocate resources on both local and remote clusters to applications. Within a cluster, a local resource management system (RMS) will naturally provide high priority to local applications. Some RMSs also support advanced reservation (AR), such as PBS Professional [1] and LSF [2].

RMS packages that have a built-in AR capability are broadly called meta-schedulers. Meta-schedulers are characterized by receiving users' requests and then scheduling an application to one (or perhaps more) of the clusters to run. Thus, meta-schedulers are intrusive since they can change both local and remote queues. This AR functionality requires efficient design of co-reservation and co-allocation functions. A number of meta-scheduler projects for grid environments have already been proposed with different characteristics, such as Gridway [3] and GridARS [4].

Advance reservation is an important mechanism for ensuring effective utilization of available resources in distributed multi-cluster environments in order to meet a variety of user requirements. AR allows, for example, a user to specify requirements related to an applications execution time, temporal dependencies among processes, types of hardware and software, etc. These specifications can, in turn, enable the system to attain higher levels of service. However, the support for advance reservation has been limited in multi-cluster environments due to the differences in how resources and capabilities are defined among different systems by different administrators within different organizations. This makes matching the user specifications, which can also be done in a variety of ways, to the actual resources difficult.

In this work we introduce an advance reservation method that can allow a user to reserve a wide variety of resources among different clusters (matching). It also dynamically verifies the possibility of any reservation with the local RMS, thus avoiding future allocation conflicts. The approach makes use of a fuzzy-ontological approach for matching user requirements to descriptions of system configurations. An architecture for doing this and interacting with local RMSs is introduced. Experimental results of the proposed approach through simulation suggest that the proposed mechanism can achieve a reasonable level of flexibility for handling large jobs and a more appropriate distribution of resources in a distributed multi-cluster configuration.

## 2. Related Research Work

Co-allocation can be referred as the distribution of processes of a job among several clusters. This mechanism can increase the utilization level of computer resources, reduce turn-around time and computations using numbers of processes larger than processes on any single cluster. The research presented in [5] introduced a scalable co-allocation strategy called the Maximum Bandwidth Adjacent cluster Set (MBAS) strategy. This strategy focuses on improving the usage of networks in high performance computing clusters.

Advance Reservation means to dedicate a set of resources to a user at a specific future time for a specific duration. The work of [6] presents an extension of the path selection process for advance reservations, introducing the notion of time to dynamic routing. It shows and classifies different temporal link metrics.

Co-Reservation, in the scope of multi-clusters, means the coordination of resources in sequence, as a grid configuration. In [7], a mechanism for advanced reservation and co-allocation of grid resources is presented. It takes into account users' interaction to negotiate with a resource provider for a suitable option between application requirements and resource capabilities. The paper focuses on enhanced utilization of resources, maintaining the Quality of Service (QoS) requested and a specific time for performance.

## 3. Approach

A shared coordinated multi-cluster environment that could be able to solve problems in dynamic fashion, crossing different enterprises, can be referred as a computing grid configuration. An application is composed of a group of dependent, or independent, subtasks. These subtasks can be scheduled and distributed through a parallel multi-cluster grid system therefore meeting various objective functions. This resources sharing can improve resource utilization across virtual organization (VOs) and enhance the response time for the application [8].

In this work, we assume that when an application requires the execution of a job, this job is composed of processes, each one with different subtasks. The focus of this paper is to introduce a dynamic advance reservation method for multi-cluster configurations. The method developed supports jobs comprised of a number of processes that could not be executed in a single cluster. A global selection and co-reservation tool interacts directly with a clusters local reservation system. The design and implementation of our approach takes into account the research in [9] and [10].

On the client side, we assume that there is an *Interactive Interface* that receives a request from a user detailing the type and quantity of resources required to execute a job. A semantic query is created and then sent to a server. The server has static semantic information about the resources of each cluster of the environment. The *Selector* matches the query with the information about the clusters resources, except the information about the number of processors. Then it finds all possible scenarios, based on that semantic stored data, and sends it to the *AR_Manager* component.

In the server, the *AR_Manager* accesses information about the reservations already done. Based on the information from the *Selector* and the availability of each selected cluster, it determines how to make the reservation of resources for the job. However, depending on the quantity of processors required, sometimes no single cluster is sufficient for the request. In this case, distributed reservations among clusters (i.e. multi-clusters) are done for the application tasks. This distribution is controlled by a fuzzy method based on research presented in [10], which takes into account bandwidth workload between clusters.

In order to avoid goal conflicts, the *AR_Manager* dynamically requests the reservations with the RMS of each local cluster. Only when the reservation is locally accepted, is its information stored by the *AR_Manager*. Our experiments were done with the Gridsim simulator [11], and assumed that all local RMS involved support advanced reservations.

### *3.1. Server Architecture*

The server side represents the main contribution of this paper. It is the server that decides on the reservation scheduling for users requests. Figure 1 shows the architecture of the server.

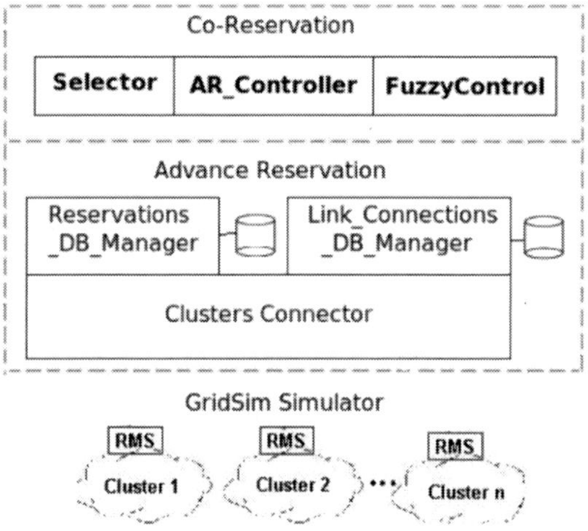


**Figure 1.** Server Architecture

The *Selector* module uses a resource matching approach based on multiple ontologies, an extension of the research in [9]. A reference and query ontology was conceived in order to integrate different ontologies from clusters in different VOs. It treats the diversity and integration of multi-clusters environments statically, keeping semantic information of some cluster parameters, such as operating system version, processor architecture, processor clock speed and amount of memory.

The *FuzzyControl* module is an extension of work in [10], that uses fuzzy logic algorithms to control and to determine the matching of a user's request to that of available resources and communication links in a multi-clusters configuration.

The *AR_Control* is responsible for the co-reservation of users' requests. It takes into account parameters of timing and uses the fuzzy control method to calculate availability of the clusters in the environment. Therefore, it decides which clusters are capable, and preferable, in providing the requested resources.

Finally, the advanced reservation layer has information about global reservations already done and the possible links among clusters. It also has a module for contacting the environment, in order to test local reservations.

## 4. AR_Manager Implementation

### *4.1. Main Concepts*

In this section, we provide details about the implementation of the *AR_Manager* module. In order to understand its design, it is important to clarify some aspects and assumptions considered in the development of the system components.

The *Selector* component allows a user to decide what kinds of resources are required in a cluster to execute a specific job. It offers various features (e.g. operational system, minimum disk space, etc.) that are used as filters to the semantic selection process. It is important to observe that the number of processors is not considered in this phase. When the selection is done, just the clusters that fit the user's requirements remain in the set of clusters available to answer the user's requests. Then, these clusters pass through the final filter, the *AR_Manager* module, which decides the final scenario.

In the advance reservation process, a few more parameters are analyzed in order to get the appropriate configuration. These parameters include the number of processors, the availability of the cluster and, if it is the case that multiple clusters must be used, the workload bandwidth required between clusters.

We assume that the query for one job is a set of resource features required for the execution of that job. Among these features there are the inter-process workload bandwidth required and the number of processors. The latter parameter it is referred as *the size* of the job. In addition, a query for a job also has three types of time parameters: a start time, which is the time a job is ready to run; the *deadline* for the execution; and the *duration* of time the resource should be reserved for the job.

When the *Selector* returns results satisfying the query, it generates a *ReservationJobQuery*, composed of the job query itself, and the set of clusters remaining after the semantic selection phase. In the second phase, a *ReservationJobResult* is created. It has the identifier of the final clusters selected to be reserved, the job size, and an interval. This interval consists of the start time for the reservation and the how long it will spend.

A time-slot based resource management approach is used to keep the available and reserved intervals of each cluster. The idle periods are in duration order so as to return from a search the interval with minimum duration that can answer the job. This behavior reduces fragmentations and leaves major intervals free for bigger jobs.

Another structure maps connections between clusters. A connection is composed of the identifiers from of two connected clusters, the link maximum bandwidth and the workload. These workloads are updated when a reservation is completed. If a reservation requires inter-cluster communication, it also carries a link connection information.

### *4.2. Advance Reservation Activities*

This subsection describes the main activities of the *AR_Manager*, which are illustrated in Figure 2. On the left, the semantic *Selector* gets an answer for a query from the ontology base specifications previously stored. It returns a set of clusters that can probably meet the query individually, but does not take into account information about the quantity of processors required for the job. This set of clusters is arranged in ascending order of cluster size. The job and the set of selected clusters comprise the *ReservationJobQuery*, which is sent to the *AR_Controller*.

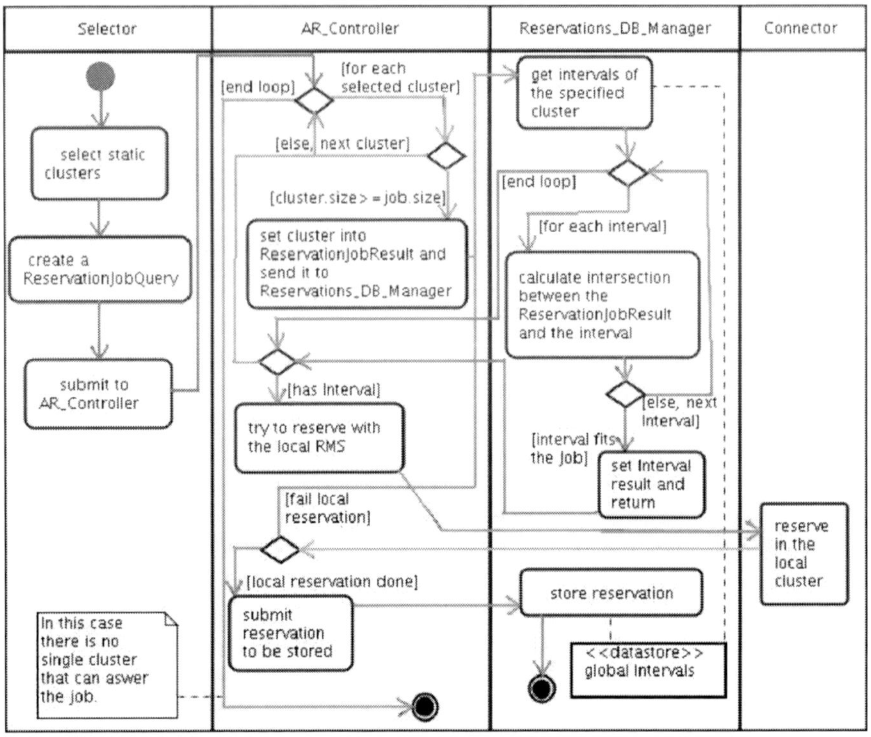


**Figure 2.** Advance Reservation of a single cluster.

The *AR_Controller* provides a feature that allows each selected single cluster to verify if it is able to answer the request alone. First it determines the number of processors available within the cluster. If this number is greater or equal to the required number of processors for the job, the cluster is allocated to the *ReservationJobResult*, and the *Reservations_DB_Manager* component statically analyses the availability of this specific cluster. The *Reservations_DB_Manager* has access to information about global reservations already done and idle intervals of each cluster. An interval fits the job when $max(sI, sJ) - min(dI, dJ) \geq durJ$, where $sI$ is the cluster interval start time, $sJ$ is the job start time, $dI$ is the cluster interval deadline, $dJ$ is the job deadline and $durJ$ is the job duration. The algorithm passes through the cluster intervals in ascending order of duration. Therefore, the interval with minimum duration that fits the job is the one which is selected as result. This behavior leaves major intervals for incoming greater jobs, giving to theses jobs more chance of being allocated in a single cluster.

The interval found is then allocated to the *ReservationJobResult*, and it is sent to the *Connector* component. The *Connector* sends this result to the specific remote local cluster, asking it to try the reservation. If the cluster can reserve the job at any interval, it will do the local reservation and return the interval reserved. The interval result of the *Reservations_DB_Manager* is not always equal to the interval returned by the *Connector*. However, it is important to make the static search before trying it locally, in order to reach higher probability of finding an interval. This avoids unnecessary communication with the local cluster and spending time for nothing.

Finally, the result of the reservation is stored by the *Reservations_DB_Manager*, and the intervals are updated.

Once the clusters are searched in ascending order of amount of processors, the answer is the one with minimum size that has an interval fitting the job. The algorithm lets majors clusters free for incoming jobs of larger sizes. An important feature is that it tries to avoid putting dependent tasks of the job to be executed among different clusters, since that would increase communication time. If at the end of this phase there was no single answer found at all, then the *AR_Controller* tries to distribute the job among the clusters, as is shown next.

### *4.3. Co-Reservation*

In the co-reservation process, job process reservations are distributed among clusters received by the *Selector*. The control of this distribution is done by a fuzzy logic method based on the work presented in [10]. It generates two thresholds:

$\delta$ - Calculated from the job size and the type of communication among clusters within the VO. It controls the job splitting, trying to avoid excess inter-cluster communication, which could increase the time to execute the job;

$\lambda$ - Calculated from the same parameters as $\delta$, plus the job bandwidth workload required for inter-process communication, and the maximum bandwidth between each cluster selected to answer the query. It controls the saturation of the links between clusters. If the link is saturated, it cannot be used by the job.

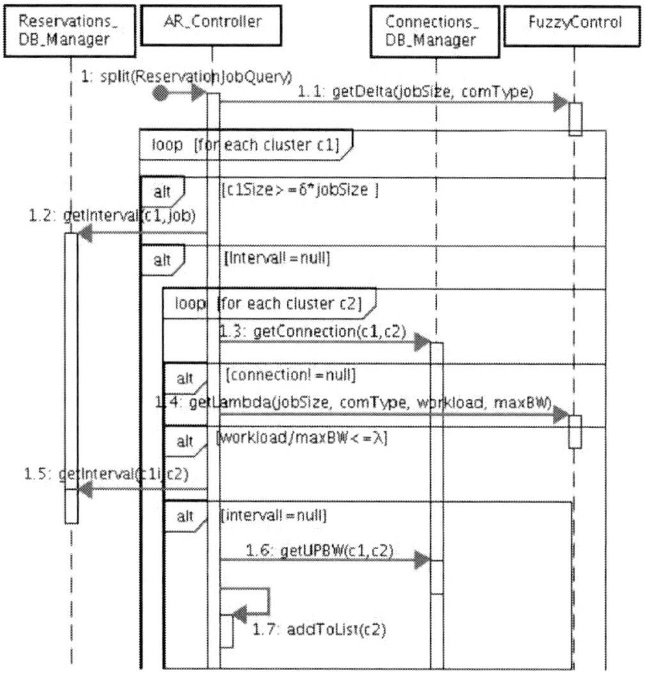


**Figure 3.** Co-allocation for larger jobs

Each cluster of the *ReservationJobQuery* that passes the test of the *FuzzyControl* module is summed to answer the job until reaches the job size, as illustrated in Figure 3. In order to control the job splitting, at least one of the clusters to be reserved must follow the rule $c1Size \geq \delta * jobSize$, where *c1Size* is the cluster. Therefore, it is the first requirement calculated. If the cluster meets this first constraint, the *Reservations_DB_Manager* finds an idle interval *c1i* for the job in the cluster *c1*.

In a second phase, all clusters that have direct connections to cluster *c1* are examined. This is done by the *Connections_DB_Manager*, that has previous information about all connections

**Table 1.** The ontology set utilized for the experiments

| Request's Concepts | VO_B's Concepts | VO_C's Concepts |
|---|---|---|
| min_number_of_cpus | [numero_cpus] | [number_of_cpus] |
| min_virtual_memory_available | [swap_livre_MB] | [free_virtual_memory] |
| min_clock_speed | [velocidade_cpu] | [max_clock_speed] |
| owner | [nome_conta] | [login] |
| number_of_cpus | [numero_cpus] | [number_of_cpus] |
| min_physical_memory_available | [memoria_livre_MB] | [free_physical_memory] |
| min_free_disk_space | [espaco_disco_livre_GB] | [available_space] |
| os_type | [SO, SistemaOperacional] | [os_type] |
| max_load_cpu_15min | | [percentage_load_15min] |
| max_load_cpu_5min | | [percentage_load_5min] |
| max_load_cpu_1min | | [percentage_load_1min] |

among clusters in the environment. These clusters do not have to pass on the share constraint, but only the link saturation constraint. This last constraint defines that the link between *c1* and *c2* must follow the rule $workload/maxBW \geq \lambda$, where *workload* is the inter-process workload required and $maxBW$ is the maximum bandwidth between *c1* and *c2*. As a result, the link is saturated and *c2* cannot be in the final list of clusters.

If there is an interval in *c2* that fits *c1i*, the cluster *c2* is added to a list in UPBW (Unit Process Bandwidth) order [10]. The UPBW is a value that measures the $linkW/c1Size$, where $linkW$ is the current link workload between *c1* and *c2* given by the *Connections_DB_Manager*. When the list size reaches the job size, it creates a reservation for each cluster in the list and then tries to do all the cluster reservations locally, through the *Connector*. If one of the reservations fails, the *AR_Controller* tries a new cluster to be the *c1*.

## 5. Experimental Environment and Results

In order to test the approach, a simulation environment was built with the GridSim simulator [11]. This section describes the environment created under GridSim and describes a case study that illustrates the general behavior of the methods.

### *5.1. Simulation Aspects*

This work is based on a configuration with a central server, remote clients and remote clusters inside a multi-cluster grid environment, as seen in Figure 1. The central server stores cluster information and has the global co-reservation module, which was described above. This module decides in which cluster a job or subtask must run. We assume that all clusters in the multi-cluster system support advanced reservation. In other words, the local schedulers on each cluster are capable of making advanced reservations for local and global jobs.

The GridSim toolkit was used to simulate this multi-cluster configuration. It provides a discrete event framework for simulation of different classes of core Grid entities such as heterogeneous resources, users, jobs, applications, resource brokers, and schedulers. It provides schedulers for single or multiple administrative domains systems configurations such as clusters and Grids.

One important feature of GridSim is that it allows advance reservations. It uses FCFS (first-come-first-served) approach for global scheduling. However, we used the co-reservation module described, which offers the possibility to analyze not only the availability of the local cluster, but also take into account features of the job and resources. Then, the local advanced reservation provided by GridSim is used to check if there is any conflict with local reservations.

In the following experiments, each process is considered as a GridSim gridlet, and a job is composed of various processes.

### *5.2. Environment*

The environment created in the GridSim for experiments is composed of two VOs, and their characteristics are described in Table 2. VO-C has three clusters which are all inter-connected. VO-B has two inter-connected clusters. However, the clusters of VO-C do not have direct connections to the clusters of VO-B.

After creating the multi-cluster configuration, we identified a number of basic concepts to be used in both virtual organizations for requesting resources. The matching of these semantic concepts is done by the tool described in this paper. Table 1 presents how these request concepts are understood in virtual organizations C and B. VO-C represents a Canadian environment, whereas VO-B is a Brazilian configuration.

**Table 2.** Multi-cluster configuration

| VO-C | processors | VO-B | processors |
|---|---|---|---|
| Cluster_01 | 7 | Cluster_1 | 4 |
| Cluster_02 | 4 | Cluster_2 | 4 |
| Cluster_03 | 3 | - | - |

### *5.3. Case Study*

Several experiments are presented and analyzed in order to clarify the sequence and main aspects of a selection and co-reservation in this work. It considers two users, user_0 and user_1, making a request for a reservation. Both users require almost the same clusters characteristics, presented in Figure 4 and 6. The requests differ with respect to the number of processors needed and time characteristics for each user job, as seen in Table 3.

**Table 3.** Required parameters for reservations

| Parameters | user_0 | user_1 |
|---|---|---|
| CPUs | 2 | 6 |
| start time (min) | 10 | 20 |
| duration (min) | 60 | 60 |
| deadline (min) | 80 | 90 |

User_0 requires 2 CPUs. Cluster_03 is the one with minimum size in the environment that can answer this request. Since there were no goal conflicts during the reservation process, it can be answered by a single cluster, Cluster_03, which is locally and globally reserved. The result received by user_0 is shown in Figure 4. On the left part of the figure, there are the parameters required by the user. On the right side, there are a set of clusters that could possible answer the request. These clusters are results from the first phase of the search, when only the resource features are considered. On the bottom of the figure, there is the final result of the reservation, in which the availability of the clusters is analyzed.

On the other hand, user_1 requires 6 CPUs. In the environment, Cluster_01 is a single cluster that has enough CPUs for this request. However, it does not have the minimum disk space required, thus it cannot be an answer to the request. Therefore, it does not appear in the first phase of results for user_0 (Figure 4), or user_1 (Figure 6).

User_1's job has to be distributed among clusters. The algorithm first tests the cluster with the largest size, namely, Cluster_02. But, the other cluster selected that it has connection to is

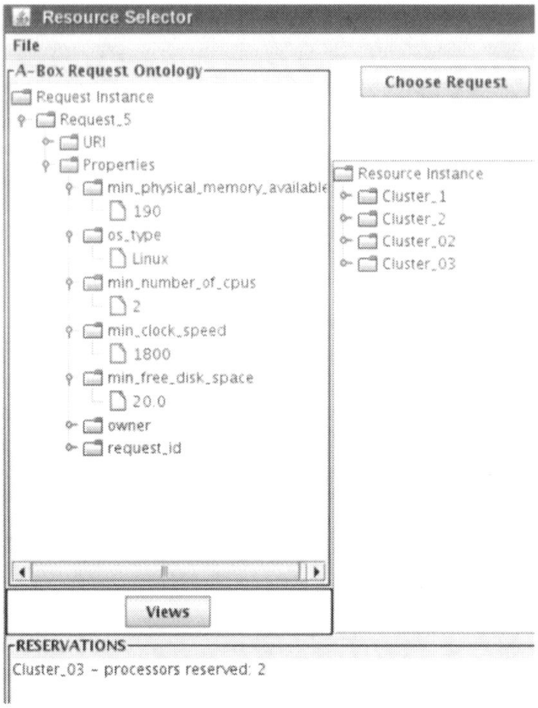


**Figure 4.** Results for user_0

Cluster_03, which has two CPUs occupied by user_0. As a result, the sum of free CPUs available does not match the requirement. Afterwards, the clusters of the second VO are tested, reaching the answer shown in Figure 6.

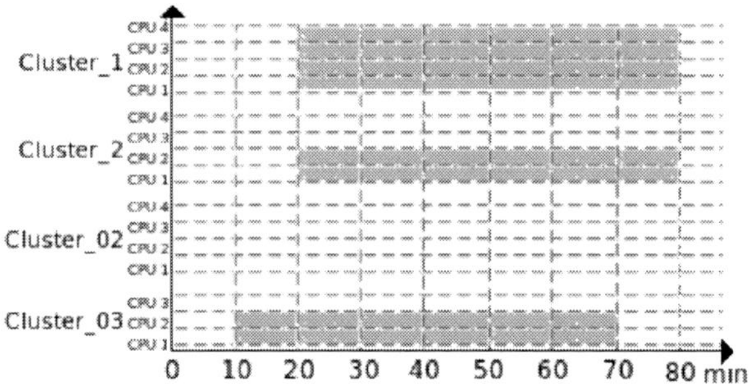


**Figure 5.** Reservations timeline

Figure 5 presents the usage of the cluster processors along the timeline.

*5.4. Conclusion and Evaluation*

In this paper, we presented an approach for advance reservation and co-reservation of cluster resources in a multi-cluster environment system. It adopted a semantic integration of multiple ontologies to provide a more appropriated mechanism to gather information on available grid resources. In addition, it integrated a logic fuzzy algorithm to control distribution of job processes among clusters in co-reservation phase.

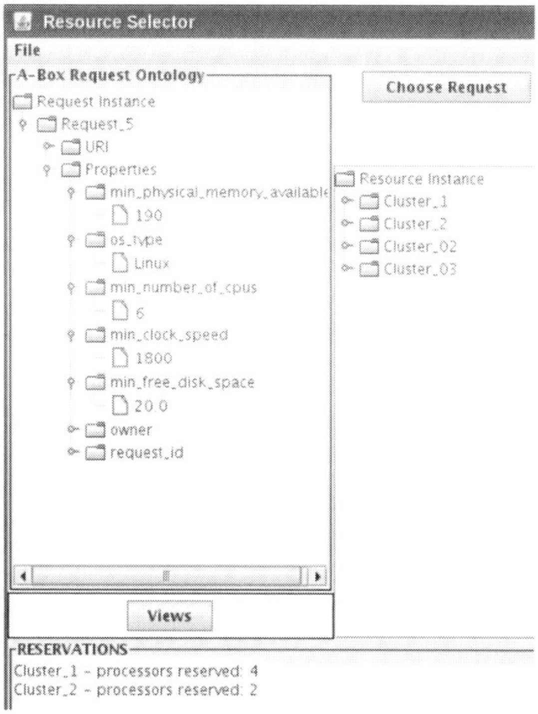


**Figure 6.** Results for user_1

The proposed approach was characterized by the description of the grid resource reservation mechanism and a prototype. This prototype was designed and implemented to make dynamic reservations for large jobs. It also allows the process of matching available resources from different virtual organizations, under different resource descriptions, integrating different ontologies.

The proposed approach also considers the possibility of dividing large jobs that cannot be executed in a single cluster among multiple clusters. These jobs are divided into subtasks among clusters in a controlled way, in order to try to avoid extra fragmentation that increases the cost of communication. It also controls the saturation level of links between clusters at the time the reservations are done. Finally, it avoids conflicts with local reservations, by checking dynamically with local RMSs using a co-reservation approach.

The prototype was tested through simulations done with GridSim, and appropriately achieved the desired goals of selection and the reservation of clusters to attend large jobs in an effective way. The approach employed a human interaction interface as part of the solution, in order to make the interaction more accessible for ordinary users.

As a future research direction, we intend to extend the grid resource reservation system to guarantee different levels of QoS can be met for users. It is also important to experiment with job workflows using the prototype, and to consider ways to optimize performance issues. Additionally, it would be interesting to dynamically show reservations and jobs status during their activities.

## References

[1] Pbs. [online], available: http://www.pbsgridworks.com/
[2] Platform computing inc.: Lsf product suite. [online], available: http://www.platform.com/products
[3] Gridway metascheduler, [online], available: http://www.gridway.org

[4] Takefusa A, Nakada H, Kudoh T, Tanaka Y and Sekiguchi S 2008 *Job Scheduling Strategies for Parallel Processing, (JSSPP'08)* **LNCS 4942** 152–168
[5] Qin J and Bauer M A 2009 *Cluster Computing* **12** 323–340
[6] Barz C, Pilz M and Wichmann A 2008 *8th IEEE International Symposium onc Cluster Computing and the Grid (CCGRID'08), IEEE Computer Society,* 710–715
[7] Siddiqui M, Villazon A, Prodan R and Fahringer T 2005 *9th International Multitopic Conference, IEEE Computer Society* 1–6
[8] Foster I 2002 *Grid Today* **1** 22–25
[9] Silva A P C and Dantas M A R 2007 *19th International Symposium on Computer Architecture and High Performance Computing, (SBAC-PAD'07)* 143–150
[10] Qin J and Bauer M A 2007 *The 20th Annual Symposium on High Performance Computing Systems and Applications, (HPCS'07)* 18–24
[11] Buyya R and Murshed M 2002 *Concurrency and Computation: Practice and Experience (CCPE'02)* **14**(13-15) 1175–1220 URL www.gridbus.org/gridsim/

# Huddersfield University Campus Grid: QGG of OSCAR Clusters

**Dr Violeta Holmes and Ibad Kureshi**

School of Computing and Engineering, The University of Huddersfield, Queensgate, Huddersfield, UK

v.holmes@hud.ac.uk and i.kureshi@hud.ac.uk

**Abstract**. In the last decade Grid Computing Technology, an innovative extension of distributed computing, is becoming an enabler for computing resource sharing among the participants in "Virtual Organisations" (VO) [1]. Although there exist enormous research efforts on grid-based collaboration technologies, most of them are concentrated on large research and business institutions. In this paper we are considering the adoption of Grid Computing Technology in a VO of small to medium Further Education (FE) and Higher-Education (HE) institutions. We will concentrate on the resource sharing among the campuses of The University of Huddersfield in Yorkshire and colleges in Lancashire, UK, enabled by the Grid. Within this context, it is important to focus on standards that support resource and information sharing, toolkits and middleware solutions that would promote Grid adoption among the FE/HE institutions in the Virtual HE organisation.

## 1. Introduction

Grid technology has become mature enough to be deployed not only in the large research institutions, but also in small to medium HE/FE organizations.

The White Rose Grid, a consortium of Yorkshire Universities – Sheffield, Leeds and York – has demonstrated the benefits of sharing distributed resources to support research utilizing Grid computing and distributed high-performance clusters via the Internet [2]. Universities like Oxford and Cambridge have led the way in Campus Grid research [3], [4].

At the other end of the HE/research spectrum, at ELIHE Blackburn College, Lancashire, research and development work in High-Performance Computing (HPC) for teaching parallel computing theory and development of parallel applications has demonstrated that even small to medium HE institutions can deploy computer clusters using open source software and redundant hardware [5], [6].

The National Grid Service (NGS) in the UK has highlighted the need to widen Grid services and include small to medium HE and research institutions under the umbrella of NGS [7].

The University of Huddersfield, classed as a small to medium HE/FE institution, has said that it "is committed to the highest standards of Fundamental and Applied Research and has set a strategy of substantial growth in its research profile and outputs" [8]. Hence, the demand for HPC resources has increased while the funding for such resource is limited.

To respond to this need we have designed the Queensgate Campus Grid (QGG) at the University of Huddersfield, utilizing already available hardware across our schools and departments and open source solutions. We have implemented a unified resource management framework and an intelligent resource brokering system. The University of Huddersfield is spread across 3 campuses: Huddersfield, Oldham, and Barnsley, in Yorkshire UK. The individual campus grids will be integrated into the Huddersfield University Grid.

In this paper we are reporting on our findings based on experience in deploying QGG across different departments and schools in the Huddersfield campus. Our Grid design was lead by the users' needs, in particular Mechanical Engineering and Biochemistry researchers.

## 2. Queensgate Grid – QGG Architecture

To unify the available resources at the University of Huddersfield into a campus grid, Queensgate campus LAN was used as the backbone and the main media for communication. The QGG architectures are heterogeneous, supporting Intel, AMD, and Sun machines, as shown in Figure 1.

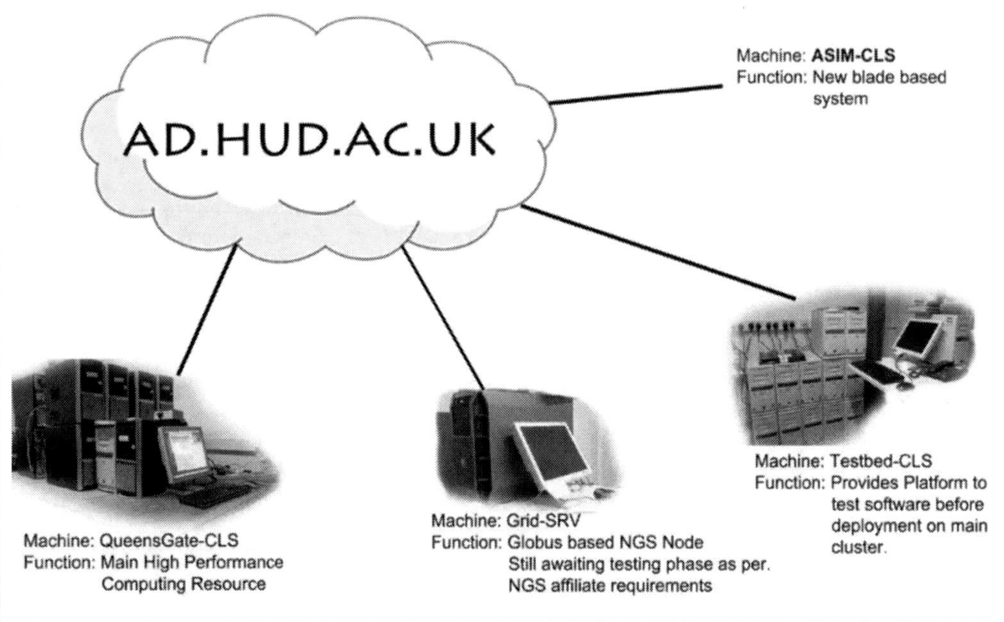


Figure 1. Queensgate Grid System Architecture

We have adopted an incremental approach in designing the university campus Grid, starting by building hardware and software system architecture for departmental/school clusters and applications. Next step involved evaluating cluster performance, speedup and efficiency by benchmarking and timing of simulations. Finally, we unified clusters and grids from various schools within the campus into the QGG.

At the heart of the QGG are three Clusters built using OSCAR cluster middleware. Two of these are shown in Figures 2 and 3.

Initially we acquired 21 Intel Core2Quad machines and used several redundant Intel and Sun machines from the School of Computing and Engineering and Applied Sciences departments to create small departmental clusters based on OSCAR middleware [9], Microsoft HPC Server 2008 R2 Beta 2 [10], and Torque [11] with the MAUI [12] scheduler. The systems currently deployed have increased in size and performance.

### *2.1. The Queensgate Cluster*

The Queensgate Cluster was originally designed to support students and researchers in Mechanical Engineering department, running ANSYS- Fluent for computational fluid dynamics, Abaqus for finite element analysis, and DynamicStudio v3.0 for particle image velocimetry.

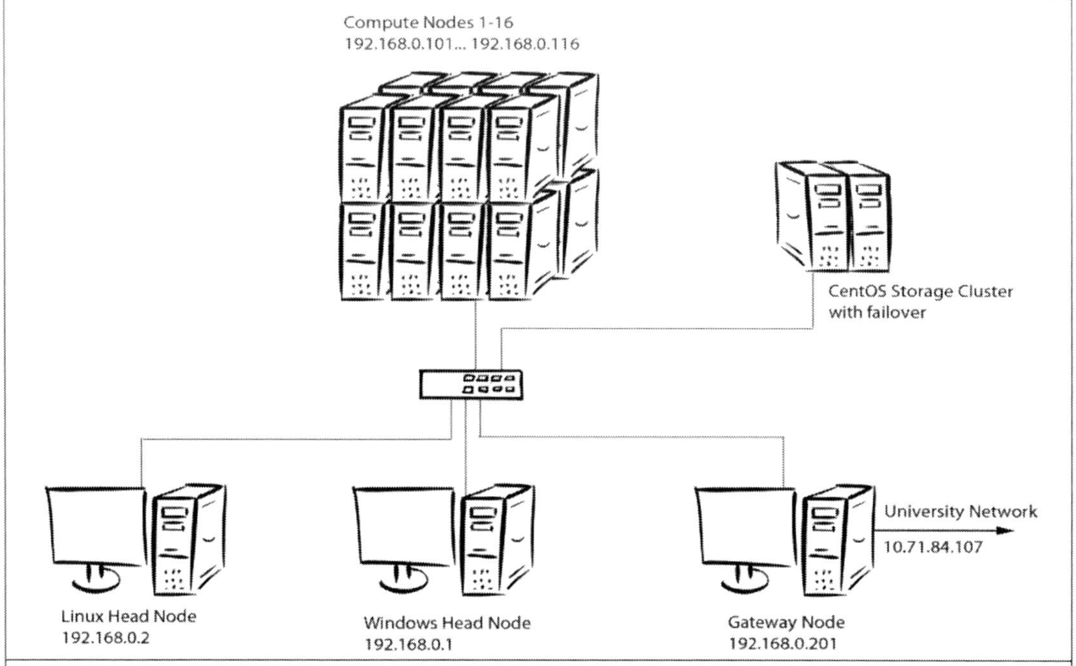


Figure 2. Queensgate Cluster Architecture

Specifications

- 64 Processing Cores
- 168GB of System Memory
- 5.2 TB of Storage
- Gigabit Interconnect
- Single Point of Entry with Failover
- Separate head nodes for each distribution
- 8 Core Storage Cluster for I/O

Software

- Linux CentOS 5.4 ker-2.6 64-bit with Oscar 5.1 b2 Middleware
- Windows Server 2008 Standard 64-bit with HPC 2008 R2 Beta 2 Middleware
- CFD Fluent 32 and 64bit upgrading to Fluent 12
- MPICH environment
- OpenMPI environment

### *2.2. The ASIM Cluster*

The ASIM Cluster was deployed in the School of Applied Sciences, running molecular dynamics software: GAMESS, a Fortran 77 and 90 compiler, DL_POLY, AMBER, LAMMPS, NAMD, METADISE, and NWChem.

### *2.3. The Testbed Cluster*

The Testbed Cluster is used to test different implementations of software and give users a safe environment to test their programs before deploying across the university Grid.

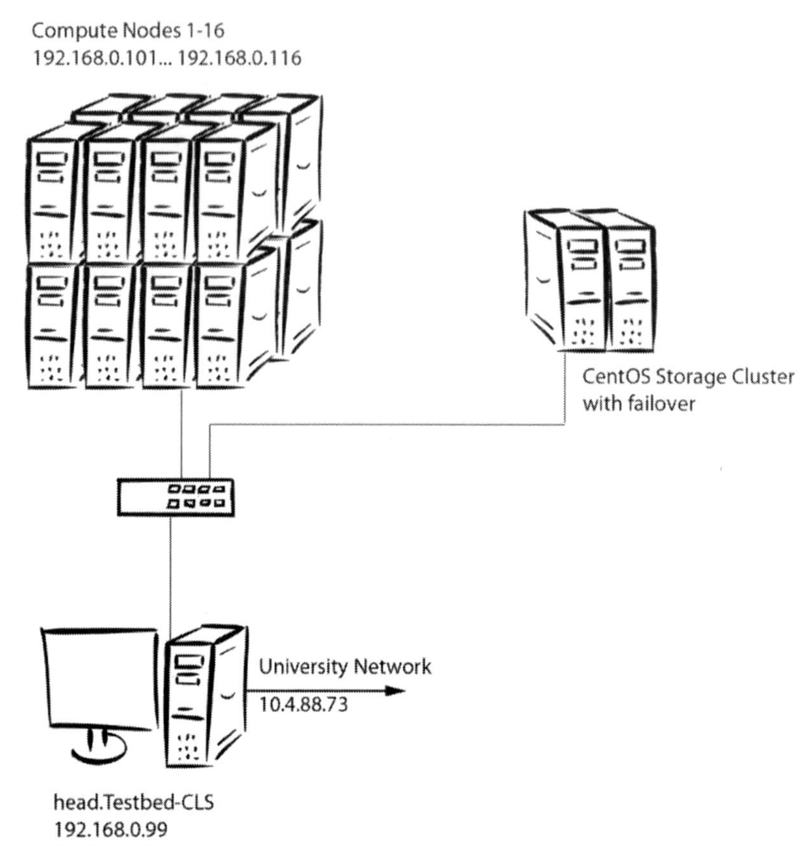


Figure 3. Testebed Cluster System Architecture

Specifications

- 16 Processing Cores
- 42GB of System Memory
- 1 TB of Storage
- Gigabit Interconnect
- Single Point of Entry
- 4 Core Storage Cluster for I/O

Software

- Linux CentOS 5.4 ker-2.6 32-bit with Oscar 5.1 b2 Middleware
- MPICH environment
- OpenMPI environment

These clusters currently run fluid dynamics simulations [13], molecular dynamics simulations, various computation algorithms, and rendering for video and image processing. Figures 4-9 depict the results of certain simulations.

Figure 4.Student Project on Performance of a Range Rover

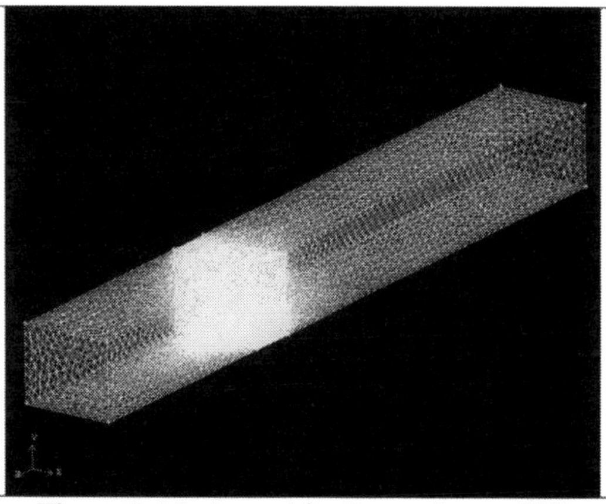

Figure 5. Mesh used for Range Rover CFD Analysis

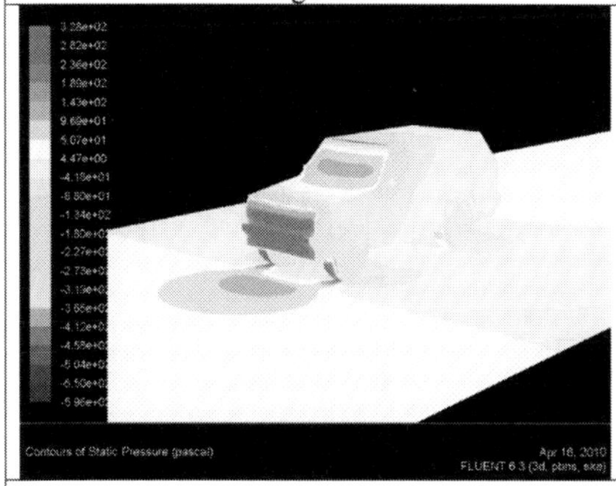


Figure 6. Image of Static Pressure Contours after Simulation

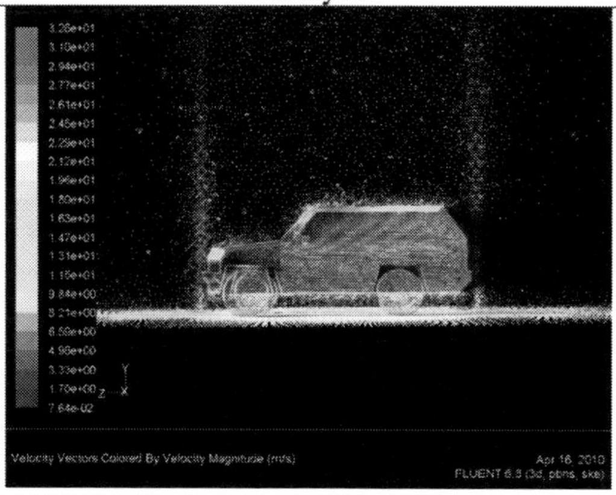


Figure 7. Air Flow Velocity Vector Diagram after Simulation

```
Grid Size

Level     Cells     Faces     Nodes   Partitions
    0   7048243  14449495   1358115            8



       Performance Timer for 7000 iterations on 8 compute nodes
  Average wall-clock time per iteration:                31.046 sec
  Global reductions per iteration:                         146 ops
  Global reductions time per iteration:                  0.000 sec (0.0%)
  Message count per iteration:                             745 messages
  Data transfer per iteration:                          47.325 MB
  LE solves per iteration:                                   6 solves
  LE wall-clock time per iteration:                      7.432 sec (23.9%)
  LE global solves per iteration:                            2 solves
  LE global wall-clock time per iteration:               0.106 sec (0.3%)
  AMG cycles per iteration:                                  8 cycles
  Relaxation sweeps per iteration:                         391 sweeps
  Relaxation exchanges per iteration:                       68 exchanges

  Total wall-clock time:                            217322.798 sec
  Total CPU time:                                  1571160.240 sec
```

Figure 8. Fluent CFD information of Dataset and Simulation CPU Time of 18d-4hr-26min completed in 2d-12hr-22min, 7.2 times faster using eight nodes instead of one

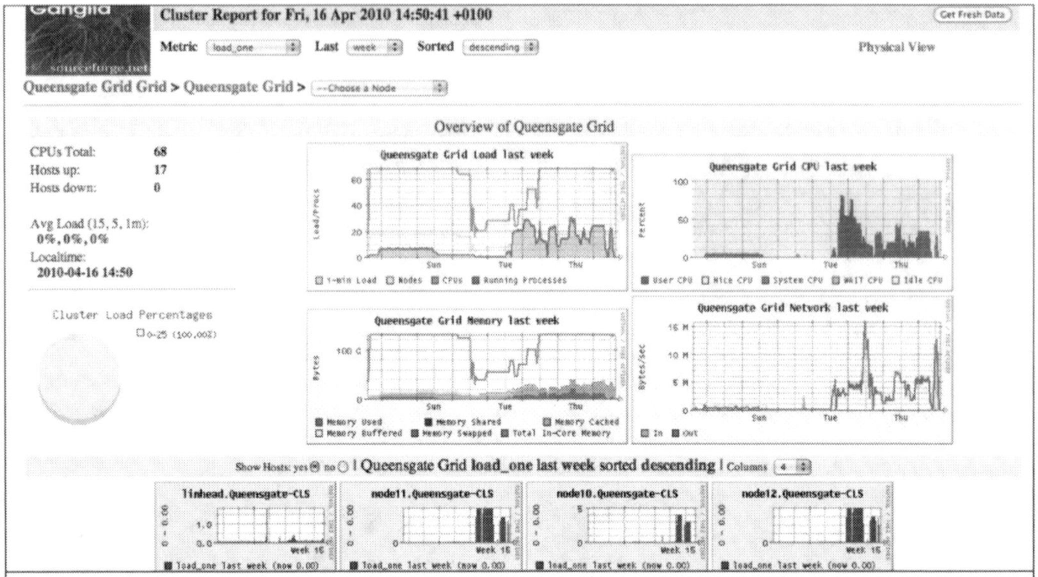


Figure 9. Ganglia Monitor showing activity and changeover in the Queensgate Cluster

### *2.4. The Clusters' performance evaluation*

The Clusters were tested extensively to determine their performance, speed up, efficiency, and power consumption. The benchmarking was done on the basis of LINPACK using ATLAS libraries and is summarized in table 1.

**Table 1. Clusters benchmarking with LINPACK**

| Cluster | Number of Cores | GFlops (max) | Efficiency | Current Max | Power Max | Power/GFlop |
|---|---|---|---|---|---|---|
| Testbed | 16 | 1.5 | ~4% | 7A | 1.6KW | ~1KW |
| Asim | 14 | 28 | 76% | 8A | 1.6KW | 60W |
| Queensgate | 64 | 120 | 84% | 13A | 2.4KW | 20W |

The efficiency of the 2.4GHz P4 machines in Testbead cluster is only 4%, compared to 76% and 84% for Asim (2.3GHz AMD machines) and Queensgate (2.33GHz Cre2Quad machines) clusters respectively. It is clear that the computational power factor has changed due to Hyperthreaded and On-Board/On-chip parallelism.

## 3. QGG Deployment

In order to support grid requirements, there are a number of functions that needed to be provided across the VO – IT resources, servers, storage, and network. Grid administration was required for maximum utilization of the resources and System Integrators to reduce the complexity of the distributed and heterogeneous systems.

In the process of deploying the QGG various innovative solutions were devised.

From the HW/SW appraisal it was evident that a cluster based on Linux systems would not adequately cover the university requirements. Several software packages ran only on the Windows platform (e.g. 3d Studio Max for image rendering, Dynamic Studio v3.0 for particle velocimetry), while others had licenses which were platform locked to Windows (e.g. 3dsMAX). As this project aimed to establish the need of HPC at the University using existing hardware and open-source software, it was neither feasible nor possible to make two Clusters which would run the two platforms.

The solution to this was to implement a hybrid Linux-Windows HPC Cluster that seamlessly and automatically accepts and schedules jobs in both domains and provides dynamic changeover between the two operating systems, based on the jobs' requirements. This was achieved through autonomous agent software and scripts [17].

To support users who prefer to use a graphical user interface rather than a command line input, a Fluent portal was designed and implemented, as seen in Figure 10. The portal provides a web based workflow that allows users to upload their design files (made on their desktop computers), set the FLUENT CFD parameters for the simulation through a GUI interface (created with PHP). The FLUENT specification file is generated in the FLUENT Scheme Language. The user can specify the computational requirement of the jobs. A job file is created using BASH and the PBS JDL, and finally uploaded to the Grid to execute.

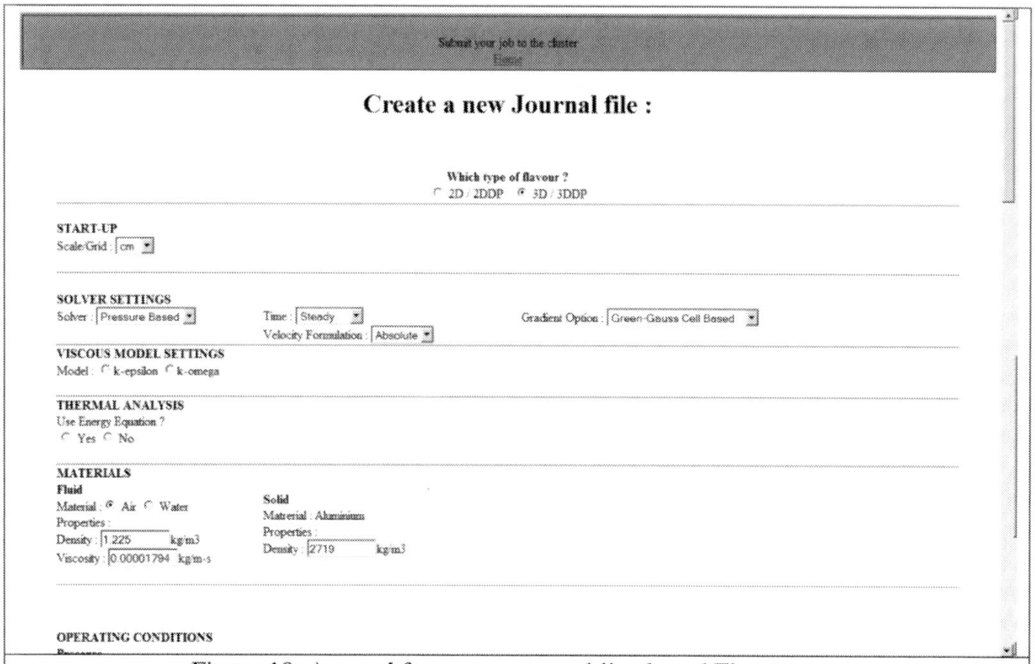


Figure 10. A portal for non-command-line based Fluent users

Currently we are investigating the unification of the computing resources in the University. There are number of possible solutions:

- Condor – to enable the utilization of 5000 PCs that are idle for 60% of the time, especially during the holiday periods
- Globus – a de-facto standard for Grid in the UK, and will enable us to share our resources with the National Grid Service (NGS) community.
- G-Lite – as part of a Grid-Enabled HPC project evaluating a Cloud support for UK e-Science community
- NGS VDT which is based on Globus with GRAM, GSISSH, MyProxy, and is the Grid middleware of choice in the NGS.

## 4. Summary and Future work

In the short term, we have made the QGG a 'cluster of clusters'. The system provides a single point of entry to grid technology [14] [15] for the University's student, staff and research community. Some enterprise work has also been undertaken; QGG provides the precision in simulations required by industry.

We have successfully deployed multiple Clusters using heterogeneous hardware and software, linked via University intranet, and defined the architecture based on the requirements of researchers and students. We have also evaluated a variety of open-source software and hardware platforms and established their suitability for meeting the educational and research needs of students and staff within the Higher Education institutions. Our findings are summarized in Table 2, with CentOS 5.4 being the preferred Linux Operating System.

**Table 2. Linux OS choice**

| **Support** | **Fedora Core 5** | **Fedora Core 9** | **Ubuntu 8.10** | **OpenSUSE 10.1** | **CentOS 5.4** |
|---|---|---|---|---|---|
| New Hardware | | | √ | | √ |
| OSCAR 5 | √ | | | | |
| OSCAR 5.1 (rc1/b1) | | √ | | √ | √ |
| OSCAR 6.0 | | √ | √ | | √ |
| Long Term | | | √ | | √ |
| Application | | | √ | √ | √ |
| Ease of Installation | √ | √ | √ | √ | √ |
| Usability straight out of the Box | | √ | | | √ |
| Non-technical User Friendly | | | √ | | |

A plethora of software has been deployed to meet the needs of our researchers and students.

With over 18000 jobs being completed on the QGG clusters realistic feedback could be gained from the existing users to judge whether the system is effective.

Interviewing the researchers from the Department of Mechanical Engineering has revealed that within the first 2 months of the resource being available, the users changed their project deliverables and started to attempt more complex problems and simulations. The explanation for this change is that the HPC facility has enabled users to increase the quality of their research output.

The Department of Chemistry and Biological Sciences immediately saw the benefits of an HPC system; they already had small HPC clusters and had an access to NGS. However, the staff and researchers immediately felt the benefits of a centralised HPC resource and a Grid connecting the various resources. It made connectivity and usability much easier. Having one point of contact for technical support, resource accessibility and NGS connectivity, simplified the workflow for researchers so they can concentrate on their work instead on problems such as a node in a cluster not being responsive.

Infrastructure wise, software licenses were limiting many users and therefore the full potential of the QGG clusters could not be experienced. The end-to-end latency and interconnects bandwidth also became a bottleneck for an embarrassingly parallel programs as it would not scale well as nodes were increased. For this application users were connecting to the NGS where high-speed Infiniband interconnects are available. Out of the pool of applications deployed on the Queensgate Cluster and the QGG this was the only exception so far, were a decrease in performance was felt on scaling.

In conclusion, we have achieved our main objective, which was to provide a framework for future expansion of HPC resources at the University.

Our future research efforts will be directed towards a Grid/Cloud infrastructure combining available resources from University schools and departments, and from neighbouring FE colleges. We intend to link the geographically-dispersed campuses in Yorkshire and the University Centre at Blackburn College, Lancashire [16]. This will lead to establishing a Virtual Organization comprising a consortium of small to medium colleges and HE institutions in Lancashire and Yorkshire, UK. It will enable resource sharing such as cluster storage, processing power, instrumentation, dedicated software and hardware, and encourage collaboration between our institutions, establishing a framework for Cloud infrastructure for education, industry, business and local government.

**References**

[1] Foster, Ian, and Carl Kesselman.,"The Grid 2: Blueprint for a New Computing Infrastructure" San Francisco, Calif.: Morgan Kaufmann, 2004. Print.

[2] Dew, P. M., Schmidt, J. G., Thompson M and Morris, P., "The White Rose Grid: practice and experience"

[3] Wallom David C. H., Trefethen Anna E., "OxGrid, a campus grid for the University of Oxford" Proceedings of All Hands Meeting 2006

[4] Calleja M., B. Beckles, M. Keegan, M. A. Hayes, A. Parker and M. T. Dove, "CamGrid: Experiences in constructing a university-wide, Condor-based, grid at the University of Cambridge"

[5] Holmes, V and McDonough, T., "The ELIHE High-Performance Clusters for Parallel Computing", Proceedings of IPSI conference IPSI -2006 Marbella, Spain.

[6] Holmes, V. and McDonough, T., "The ELIHE High-Performance Cluster, Proceedings of Cluster 2005 conference (Poster)", Boston, USA.

[7] *NGS | Connecting Infrastructure, Connecting Research*. Web. 16 Apr. 2010. <http://www.ngs.ac.uk/>.

[8] "Research Home." *Research*. University of Huddersfield. Web. 16 Apr. 2010. <http://www2.hud.ac.uk/research/research/>.

[9] *OSCAR Project*. Open Cluster Group. Web. 16 Apr. 2010. <http://oscar.openclustergroup.org/>.

[10] "Windows HPC Server 2008 | Microsoft Supercomputing | Supercomputers." *Microsoft Corporation*. Web. 16 Apr. 2010. <http://www.microsoft.com/HPC/>.

[11] "Torque Scheduler." *Cluster Resources*. Web. 16 Apr. 2010. <http://www.clusterresources.com/>.

[12] "MOAB" *Cluster Resources*. Web. 16 Apr. 2010. <http://www.clusterresources.com/>.

[13] "Fluent CFD." *ANSYS UK*. Web. 16 Apr. 2010. <http://www.fluent.co.uk/>.

[14] "Globus Toolkit." *The Globus Alliance*. Web. 16 Apr. 2010. <http://www.globus.org/>.

[15] "Condor Project Homepage." *UW-Madison Computer Sciences Department Home Page*. Web. 16 Apr. 2010. <http://www.cs.wisc.edu/condor/>.

[16] *Blackburn College Website | Blackburn College Home Page*. Web. 16 Apr. 2010. <http://www.blackburn.ac.uk/>.

[17] Kureshi, I., Holmes, V. and Liang, S., "Hybrid HPC – Establishing a Bi-Stable Dual Boot Cluster for Linux with OSCAR middleware and Windows HPC 2008 R2", Proceedings of All Hands Meeting 2010, Cardiff, UK.

Journal of Physics: Conference Series **256** (2010) 012023 doi:10.1088/1742-6596/256/1/012023

# Backfilling with Fairness and Slack for Parallel Job Scheduling

**Angela C. Sodan and Wei Jin**

University of Windsor, Computer Science, Windsor, Ontario, Canada

Corresponding author's email addess: acsodan@uwindsor.ca

**Abstract**. Parallel job scheduling typically combines a basic policy like FCFS with backfilling, i.e. moving jobs to an earlier than their regular scheduling position if they do not delay the jobs ahead in the queue according to the rules of the backfilling approach applied. Commonly used are conservative and easy backfilling which either have worse response times but better predictability or better response times and poor predictability. The paper proposes a relaxation of conservative backfilling by permitting to shift jobs within certain constraints to backfill more jobs and reduce fragmentation and subsequently obtain better response times. At the same time, deviation from fairness is kept low and predictability remains high. The results of the experimentation evaluation show that the goals are met, with response-time performance lying as expected between conservative and easy backfilling.

## 1. Introduction

Parallel job scheduling requires packing of jobs with different runtimes and different requirements in regards to the numbers of CPU resources. Jobs are typically queued, and different scheduling policies are possible. First-come-first-served (FCFS) is often considered to be a fair scheduling approach but leads to high fragmentation and subsequently high response times if used on its own. Other approaches like priority scheduling (with priorities typically based on job runtimes) or fair-share priority scheduling are often practically applied. Very often the basic scheduling policy is combined with backfilling: 60% of the Top50 machines apply some form of backfilling. Backfilling means that jobs can move to an earlier scheduling position if they do not delay jobs ahead in the queue according to certain criteria. This improves utilization and subsequently average response times.

Two forms of backfilling [10] are commonly used:

- Conservative backfilling: permits moving jobs ahead if they do not delay any waiting job ahead of them. If combined with FCFS, it therefore maintains the fairness of FCFS (except that backfilled jobs receive preferential treatment) and supports prediction of response times. However, conservative backfilling can lead to wasted space and non-optimal utilization and subsequently increased average response time (see Figure 1 a).
- Easy backfilling[1]: permits moving jobs ahead if they do not delay the first waiting job (which prevents starvation). Easy backfilling tends to provide better utilization in high-load

[1] Note that the original scheduler proposing easy backfilling was called EASY [8] and was a combination of FCFS and easy backfilling. In this paper, we consider the backfilling approach as being separate from the basic scheduling policy, i.e., a priority scheduler could also be combined with either conservative or easy backfilling.

phases but also means giving preference to shorter jobs which may push back other jobs (see Figure 1 b). Thus, easy backfilling provides less fairness. Since jobs are often pushed back, predictability is low.

As the contribution of this paper, we present a new slack-based backfilling approach which addresses the above problems and is computationally cheap and easy-to-integrate into the overall scheduling policy. The approach is based on conservative backfilling but relaxes it by permitting constrained delays, called slack. The goal is to increase utilization in high-load phases and subsequently response times by better packing, while keeping the schedule close to FCFS. The latter means that both fairness and predictability are maintained with acceptable deviation. Improved utilization is demonstrated in Figure 1 c. Backfilling Job 6 achieves higher utilization improvement than Job 4 or Job 5. Under the objective of maximizing utilization improvement, therefore Job 6 would be chosen.

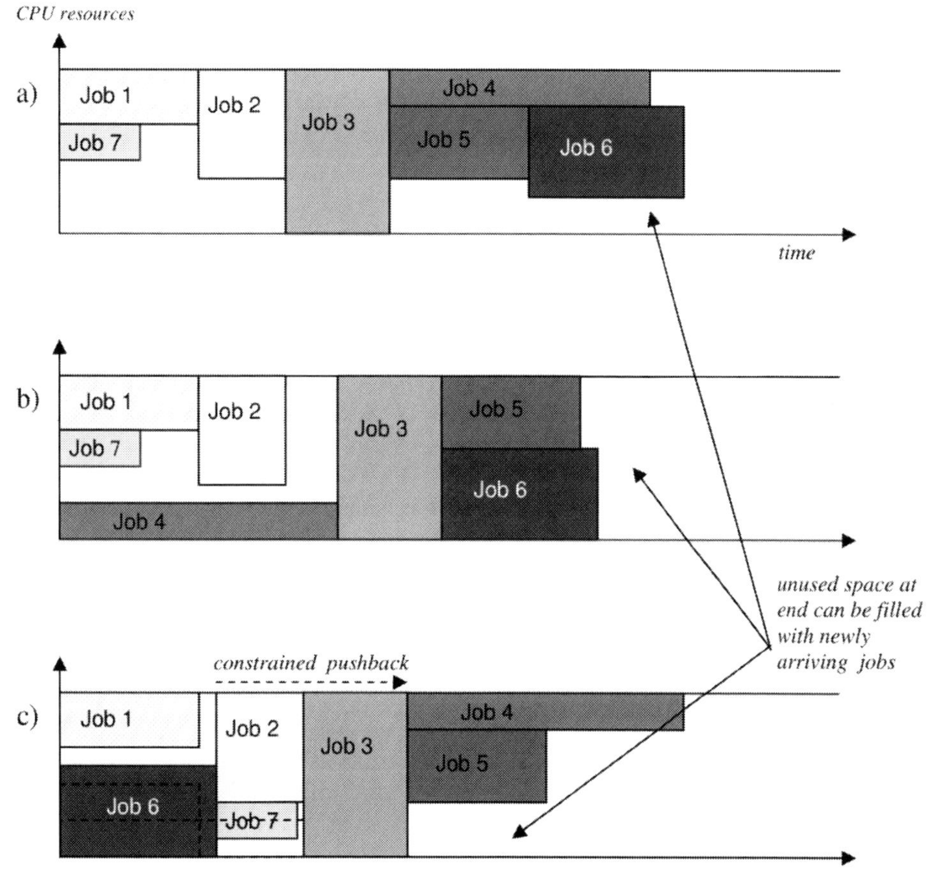


Figure 1. Differences between backfilling approaches: a) FCFS with conservative backfilling: no waiting job may be delayed, b) FCFS with easy backfilling (EASY): only the first waiting job may not be delayed, and c) FCFS with slack backfilling: all waiting jobs may be delayed but only to some extent. With slack backfilling, Job 6 gives better utilization than Job 4 or Job 5 and is therefore chosen (slack constraint is met).

To accomplish these goals, combinations of all possible backfill jobs are considered. This involves trying different combinations with the goal of finding the best combination. To make the search feasible, several heuristics are applied to prune the search tree. In addition, the number of waiting jobs considered for backfilling is limited to a lookahead of a certain number of waiting jobs. This is a feasible restriction, considering that anyway closeness to FCFS matters.

The approach is evaluated via simulation with several workloads. The results demonstrate that slack backfilling maintains high fairness and predictability, while performing better than conservative backfilling.

## 2. Related work

Different definitions of fairness with different corresponding metrics are possible, depending on the criteria: typically seniority, service requirements, or cost. Most common in queuing approaches is to use seniority; with the fairest approach then being first-come-first-served (FCFS). Service-oriented fairness would permit jobs with lower service requirements (shorter runtime) to move ahead of jobs that arrived earlier but run longer. This corresponds to the often applied priority scheduling with higher priority for shorter jobs. However, the metric neglects the difference in arrival times between jobs of different classes. Thus, a combined metric is proposed in [11], which considers both the difference between the service times of two jobs and the difference in their arrival times. The difference between deserved (an equal fraction of the processor time among all jobs) vs. the received service is calculated as positive or negative discrimination. Overall unfairness then is defined as the variance of the individual jobs' discrimination. However, this approach is presented for an M/M/1 queuing system which does not reflect problems in parallel job scheduling like packing varying numbers of jobs into the machine at the same time (space sharing) or applying backfilling.

Some approaches for parallel job scheduling use relative response time (often called slowdown) which is the ratio between response time and runtime. This metric combines wait time (and therefore indirectly differences in arrival times) with runtimes (service times) into a single value. In the context of an M/M/1 system, the argument is raised that mere slowdowns are counter-intuitive since FCFS would be rated as always unfair [1]. In [7], relative expected slowdowns are compared between running and waiting jobs and running jobs preempted if waiting jobs receive significantly worse service defined via the slowdowns. However, slowdown is typically combined with a delay factor. Thus, in [13], $\lambda$-fairness is used which means an approach is fair if no job is delayed by more than a factor of $\lambda$. (response time is not increased by more than a factor $\lambda$.).

$\lambda$-fairness can be considered a special case of slack-based scheduling. In [16], slacks are introduced as delay factors and used to modify the order of the jobs for better utilization. This means that slacks serve as constrained flexibility and reordering is permitted as long as no job is delayed beyond its slack. Slacks are calculated dynamically based upon dynamically adjusted priorities, a static slack factor, and a static average wait time. This idea has been used in various approaches like [4][15][19]. However, in the approach presented in [19], slack factors, applied to wait time, need to be as large as 2 to obtain a benefit.

Using slack as a constraint is supported by the definition of scheduler goodness in [16], which is based on overall happiness defined as the happiness of the unhappiest user.

Another fairness idea is fair share which approximates giving certain users or user groups certain shares of the machine (considering both time and space) over certain time intervals, which then are mapped to priorities [5]. The approach has become widespread but is tuned by many parameters and lacks a theoretical basis.

In [12], two unfairness metrics are proposed and evaluated (without resulting into a preference for one or the other). One metric transfers the seniority/service-time metric of [11] to parallel processing by weighting the resource share according to the size of the job. The second metric determines unfairness as jobs being delayed vs. their relaxed fair start time, which is seniority-oriented. The relaxed fair start time expresses the start time of a job without this job being backfilled (while other jobs ahead in the queue may backfill). In both cases, fairness is calculated after the execution of a workload to rate and compare the fairness of different scheduling policies, as does the original M/M/1-oriented metric of [11]. Our approach is closely related to the second metric as this metric can be modified to apply in dynamic evaluation and works well for non-preemptive scheduling.

In regards to backfilling, typical approaches make backfilling decisions one-job-at-a-time. An exception is the approach in [14] to which our approach is closely related. However, the approach in [14] is applied to both filling and backfilling together (i.e. whenever new jobs are packed into the machine) and is oriented toward easy backfilling: only the first job is guaranteed to be scheduled at the next possible time (resources available). Otherwise, the approach considers the whole combination of jobs (with a certain lookahead) that can potentially be filled/backfilled and selects the combination which provides the highest utilization (defined as the currently used resources vs. the available resources). An optimal solution is found via dynamic programming. However, this approach does not consider fairness. In [15], we have also optimized the next combination of jobs to be scheduled, looking further ahead into the placement of a whole group of jobs and into the resulting schedule and utilization over the corresponding time frame. Jobs are reordered within certain slacks, defined by general slack factors. However, though successfully improving the performance, the approach is expensive.

## 3. Backfilling with Fairness and Slack

### 3.1. Slack-based Backfilling Algorithm

Our proposed slack-based backfilling algorithm consists of two parts which apply at job-submission time and at job-scheduling time. In difference to some other approaches, the job-scheduling time for considering relaxed ordering is exclusively the backfilling time, i.e. at each scheduler invocation, jobs are first filled regularly. However, backfilling then not only relaxes the rules in regards to delays for other waiting jobs but looks at the overall situation and attempts to find the best group of jobs to be backfilled rather than making decisions one-by-one as most other schedulers do. The basic backfilling approach used for estimations for fair start time and as a default (if no combination meeting the criteria can be found) is conservative backfilling. The backfilling approach support two possible objectives:

- Objective 1: closeness to FCFS while reasonably improving utilization and
- Objective 2: high utilization improvement while staying reasonably close to FCFS.

#### *3.1.1. Actions at job-submission time*

Fist, the fair start time of the job is calculated (the term and the idea originate from [12], the exact corresponding term being 'relaxed fair start time'). For the jobs ahead, FCFS with conservative backfilling is applied but the job itself is not backfilled but placed after all jobs submitted before it are packed into the machine. The fair start time is obtained by simulating the schedule ahead. With FCFS and conservative backfilling, this is a feasible approach. Note that when the job is actually scheduled, it may be backfilled. This means that backfilling is not considered a guarantee or potential source of later unfairness if backfilling is no longer possible. Rather backfilling is considered as a potential bonus.

After knowing the job's fair start time, a slack factor $F_{slack}$ is applied to the resulting estimated wait time $T_{wait}$ to define the maximum permitted delay. $F_{slack}$ is calculated in the range $[F_{slack,min}, F_{slack,max}]$ with an exponential function of wait time:

$$F_{slack}(T_{wait}) = min\ (F_{slack,max},\ k_2 * e^{-k1*Twait} + F_{slack,min}) \qquad (1)$$

with $k_2$ to be set such that $k_2 \geq F_{slack,max} - F_{slack,min}$

This means that $F_{slack}$ decreases non-linearly with larger wait times to avoid too much additional delay for jobs that already have to wait a long time. $F_{slack}$ represents the maximum delay possible per job until it is scheduled. However, to avoid using up this slack in scheduling decisions early during the waiting time of the job and to keep some flexibility for later backfilling-decision points, a partial slack $F_{slack,partial}$ is calculated which is the slack that may be used in any particular scheduling decision and is proportional to how long the job already waited ($T_{wait,curr}$). The slack constraint used during backfilling is based on $F_{slack,partial}$ (not on $F_{slack}$). Note that $F_{slack,min} \leq F_{slack,partial} \leq F_{slack}$. The first-time partial slack that may be used in a scheduling decision at any point of time is defined as

$$F_{slack,partial} = F_{slack,min} + (F_{slack} - F_{slack,min})*T_{wait,curr}/T_{wait,est} \quad (2)$$

At later decision points, the slack $F_{slack,curr}$ that was already used should be considered which gives

$$F_{slack,partial} = F_{slack,min} + (F_{slack}/F_{slack,curr} - F_{slack,min})*1/F_{slack,curr}*T_{wait,curr}/T_{wait,est} \quad (3)$$

### *3.1.2. Actions at job-scheduling time*

At backfilling time, the algorithm performs the following steps:

1) Select a limited number $n_{lookahead}$ of backfilling (lookahead) candidates ($S_{cand}$) from the head of the waiting queue.
2) To find a good solution according to the chosen objective:
   - Try backfilling different combinations of jobs from $S_{cand}$.
   - Evaluate each solution according to either Criterion 1 (Objective 1) or Criterion 2 (Objective 2).
   - Evaluate the delays of all non-selected jobs.

   Only solutions passing the corresponding criterion and not violating the slack constraint of any non-selected job will be considered and become part of the possible solutions *solPoss.*
3) Select best solution from *solPoss* according to Objective 1 or Objective 2.

To support the two objectives, the following metrics are calculated which consider those jobs that were delayed vs. their estimated fair start time *FST* and to what extent on average (*OverallUnfairnes,* abbrev. as *OverallUnfair* below) and those jobs that were treated preferentially (*OverallPreferentialTreatment,* abbrev. as *OverallPrefer* below), i.e. started earlier than their *FST.* Note that this fair start time is calculated at job submission time as an estimate, whereas the original posterior evaluation in [12] uses the final scheduling order.

$$OverallUnfairness = \sum_{i \in jobs} \max(AST_i - FST_i, 0)/n \quad (4)$$

$$OverallPreferentialTreatment = \sum_{i \in jobs} \max(FST_i - AST_i, 0)/n \quad (5)$$

with *AST* being the actual start time and *n* being number of jobs.

For evaluation of possible solutions (kept in *solPoss*), either Criterion 1 (Formula 6) or Criterion 2 (Formula 7) are used. Utilization *Util* is measured as the fraction of resources being used.

$$OverallUnfair_k / min\{OverallUnfair_i \mid i \text{ in } solPoss\} \leq OverallUnfairRatio_{max,1} \wedge$$
$$OverallPrefer_k / \{OverallPrefer_i \mid min\{overallUnfair_j \mid j \text{ in } solPoss\}=i\} \leq OverallPreferRatio_{max,1} \wedge$$
$$Util_k / \{Util_i \mid min\{OverallUnfair_j \mid j \text{ in } solPoss\}=i\} \geq UtilRatio_{min,1} \quad (6)$$

This means that the deviation from FCFS should stay below certain thresholds, while utilization is improved beyond a certain threshold.

$$Max\{Util_i \mid i \text{ in } solPoss\} / Util_k \leq UtilRatio_{min,2} \wedge$$
$$\{OverallUnfair_i \mid max\{Util_j \mid j \text{ in } solPoss\}=i\} / OverallUnfair_k \geq OverallUnfairRatio_{max,2} \wedge$$
$$\{OveralPrefer_i \mid max\{Util_j \mid j \text{ in } solPoss\}=i\} / OverallPrefer_k \geq OverallPreferRatio_{max,2} \quad (7)$$

This means that the deviation of the solution from the maximum possible utilization improvement should be below a certain threshold, while the deviation from FCFS stays below certain thresholds.

Note that the thresholds used in Formula (6) and (7) (indexed 1 and 2) may have different values.

Among the possible solutions in *solPoss*, finally the solution with lowest overall unfairness (Objective 1) or highest utilization improvement (Objective 2) is selected. If none of the job combinations meets the corresponding criteria, the job combination with lowest overall unfairness (Objective 1) or highest utilization improvement (Objective 2) is used.

The approach employs branch-and-bound for approaching an optimal solution. Branches in the tree correspond to combinations of jobs, i.e. deeper levels mean adding more jobs to the combination. Several heuristics are used to prune the search tree and make the search feasible. They were mentioned as part of the algorithm description above: branches are pruned if the sum of the job sizes in the combination exceeds the number of available resources, if the slack of any non-selected job is violated, or if the criterion of the corresponding goal is not met. Moreover, the criteria, cf. Formula (6) and (7) express a "good-enough" rather than an optimal solution.

### 3.2. Implementation

The slack backfilling is currently implemented in our preemptive Scojo-PECT scheduler [2] which uses coarse-grain time slices. However, the backfilling approach is usable with any scheduling policy (potentially with even higher improvements due to more backfilling options). Scojo-PECT separates jobs classes based on runtime into short-running, medium-running, and long-running jobs. These classes are scheduled in different time slices, with FCFS per class/slice. The scheduler supports preemption to disk as a default which means that it works well with memory-intensive jobs. Jobs are resumed on the same resources in the following scheduling round in their next slice. Estimation of response times is easily possible if using conservative backfilling since the policy is FCFS.

To minimize fragmentation, unused resources in any slice can be temporarily (until the end of the slice) used by other job classes. This non-type slice backfilling always observes the backfilling rules of the host and of the guest class and only starts new jobs if they can continue running on the same resources in their own slice. Non-type slice backfilling is crucial for good performance and means that actual response times are often shorter than estimated, though preemption overhead can also mean that jobs finish later than expected.

## 4. Experimental Evaluation

### 4.1. Experimental Set-up

Slack backfilling was tested within Scojo-PECT, using simulation and different workloads. Two synthetic workloads (generated by the Lublin-Feitelson Workload Model [9]) and two real traces (from the Feitelson Workload Archive [3]) were used, each with 10,000 jobs. Slack backfilling was compared to both conservative and easy backfilling for each workload. For the characteristics of the workloads, see Table 1.

TABLE 1. WORKLOADS USED WITH MACHINE SIZE $N_{MACHINE}$ AND UTILIZATION *UTIL*

| | Lublin | | CM5 | BLUE |
|---|---|---|---|---|
| | Synthetic 1 | Synthetic 2 | | |
| $N_{machine}$ | 128 | 128 | 1024 | 1152 |
| *Util* in % | 81.5 | 85 | 78 | 74 |
| $RS_M$ | 0.32 | 0.385 | 0.43 | 0.375 |

Scojo-PECT currently makes time-slice decisions in time intervals of 1h: time not needed for an optional short-job slice is split into a slice for medium-running and a slice for long-running jobs. Time shares can be specified as the relative fraction among the slice-lengths between medium-running and long-running jobs ($RS_M$). The experiments used a fraction that corresponds to the results of a common priority scheduler (giving higher priority to short vs. medium and medium vs. long jobs). Jobs were classified as medium jobs (*M*) if running ≥ 10 minutes and < 3 h, and as long jobs if running ≥ 3 h (*L*).

Since Scojo-PECT is a preemptive scheduler, swap cost needed to be modeled. The assumption was that only the amount per specific machine node needed to accommodate the new job is swapped out at job-switch time. Though most clusters are based on Linux and Linux does not support such swapping, small modifications of Linux can make this possible [18]. Each node was assumed to have 8 Gbyte of memory (which is a typical size) with 7 Gbyte being available to applications. Swap cost per Gbyte was assumed to be 14 sec (as per personal discussion with system administrators). A job-memory-requirement distribution was derived from the CM5 trace and from information of system administrators. Basically, most jobs have relative low memory requirements, which means that the jobs running on the same resources in different coarse-grain time slices may all be kept in memory. However, some jobs require more memory and make swap-in/swap-out necessary. *L* jobs tend to have more jobs with high memory requirements than *M* jobs, which was reflected in the job-memory-requirement distribution definition. For backfilling parameters, see Table 2.

TABLE 2. TABLE 1. BACKFILLING PARAMETERS.

| | |
|---|---|
| $n_{lookahead}$ | 20 |
| $[F_{slack,min}, F_{slack,max}]$ | [1.05,1.2] or |
| | [1.2,1.5] |
| $k_1 / k_2$ | 0.000005 / 0.4 ($T_{wait}$ measured in sec) |
| $OverallUnfairRatio_{max,1\,/\,max,2}$ | 1.3 |
| $OverallPreferRatio_{max,1\,/\,max,2}$ | 1.3 |
| $UtilRatio_{min,1} / UtilRatio_{min,2}$ | 1.3 |
| $N_{waitjobs}$ defining high-load phase | 16 |

For the evaluation, average response times *R* and average bounded slowdowns are used. Bounded slowdown is defined as $max(R / max(T_{runtime}, bound), 1)$. The bound prevents too much influence on the overall average from very short-running jobs.

Overall utilization per workload obviously remains the same, even if the packing under backfilling is improved—otherwise, the machine would be overcommitted. However, we can expect that better packing improves the utilization in high-load phases. High-load phases are defined in this work by at least a certain number $N_{waitjobs}$ jobs being in the waiting queue.

### 4.2. Experimental Results

The results are shown in Figure 2. The results shown are for Objective 1 (closeness to FCFS while reasonably improving utilization). The results for Objective 2 surprisingly looked very similar are are therefore not shown. The results for Objective 1 show that the bounded slowdown was improved by up to 27% (Synthetic 1, L). On average, the improvement was 10.6% if utilization was relatively high (Synthetic 1, Synthetic 2, and CM5). There are less improvements for the lower-utilization workload BLUE. The explanation is obvious (and was confirmed by tests that reduced inter-arrival times in the synthetic workloads to make their load lighter): if the load/utilization is low, waiting queues are shorter and there is therefore less opportunity to make improvements in backfilling. Response times are not shown since their relationship was similar.

Most jobs (~ 20%) that scheduled differently than estimated by their fair start time, were treated preferentially (due to backfilling and non-type slice backfilling by the scheduler which tends to decrease response times). Some jobs (~ 1.5%) were delayed within the defined constraints, i.e. were treated slightly unfair as per our definition.

The utilization difference vs. conservative backfilling in high-load phases was only about 1%. Thus, some benefit may also be due to slightly shorter jobs being better served (more easily backfilled) than slightly longer-running jobs.

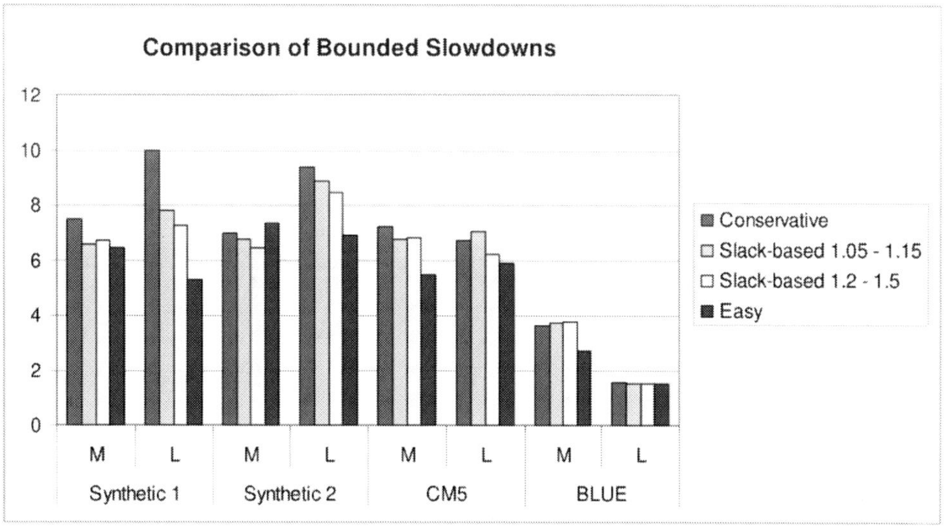


Figure 2. Bounded slowdown for different workloads and different backfilling approaches.

Testing several different $n_{lookahead}$ showed that typically 20 jobs, as used in the presented experiments, typically provide the best results—going much higher does not add much benefit, if any. More detailed evaluations can be found in [6].[2]

Comparing slack-based backfilling to conservative and easy backfilling, clearly shows that slack-based backfilling provides results which lie in between conservative and easy backfilling—which was expected by the design of slack backfilling. Moreover, the slack range has an impact on how much benefit is obtained: for the workloads with higher utilization (Synthetic 1, Synthetic 2, and CM5), the larger slack-factor range [1.2,1.5] obviously tends to provide better performance than the range [1.05,1.2]. However, even a maximum slack of 1.5 is still relatively low, compared to other approaches reported in the literature such as [19].

The search for a good backfilling job combination was very effective and the pruning of the search tree worked very well: the runtime of the simulation was only increased from about 7min to about 10min per workload and parameter setting, i.e. by about 43%.

The predictability remained high: for Synthetic 1, the 95 percentile was 1.12 (*M* jobs) and 1.09 (*L* jobs) if using the slack-factor range [1.05,1.2], and 1.11 (*M* jobs) and 1.19 (*L* jobs) if using the slack-factor range [1.2,1.5]. The maximum prediction errors were 1.32/1.27 and 1.39/1.37. Thus, obviously the larger slack-factor range creates slightly more deviation from the predicted performance, i.e., there exists a trade-off between predictability and performance.

[2] These evaluation do not, however, consider swap cost.

## 5. Summary and Conclusion

As a compromise between the commonly used conservative and easy backfilling, slack backfilling was proposed. Slack backfilling permits limited push backs of jobs and therefore retains much of the fairness of FCFS with conservative backfilling and of its predictability, as the results confirm. As expected, the performance of slack backfilling in regards to average response times and average bounded slowdowns lies between conservative and easy backfilling.

The presented slack-based approach is computationally relatively cheap to compute and easy to integrate into any scheduling policy, since only affecting the backfilling. Therefore, slack backfilling is a promising approach for general usage.

### References

[1] Avi-Itzhak, B., Levy, H., and Raz, D., "Quantifying Fairness in Queuing Systems—Principles, Approaches, and Applicability", Probability in the Engineering and Informational Sciences 22, 2008, pp. 495-517.

[2] Esbaugh, B., and Sodan, A.C., "Coarse-Grain Time Slicing with Resource-Share Control in Parallel-job Scheduling", Proc. High Performance Computing and Communication (HPCC), Houston, LNCS 4782, Springer Verlag, Sept. 2007.

[3] Feitelson Workload Archive, http://www.cs.huji.ac.il/labs/ parallel/workload/logs.html, retrieved May 2007.

[4] Islam, M., Balaji, P., Sadayappan, P., and Panda, D.K., "QoPS: A QoS based Scheme for Parallel Job Scheduling., Proc. JSSPP 2003, LNCS 2862, Springer, 2003.

[5] Jackson, D., Snell, Q., and Clement, M.J., "Core Algorithms of the Maui scheduler", Proc. JSSPP 2001, LNCS 2221, Springer, 2001.

[6] Jin, W., "Backfilling with Fairness and Slack for Parallel Job Scheduling", Master Thesis, University of Windsor, September 2009.

[7] Kettimuthu, R., Subrasani, V., Srinivasan, S., and Gopalasamy, T., "Selective Preemption Strategies for Parallel Job Scheduling", Proc. Int. Conf. on Par. Proc. (ICPP), 2002.

[8] Lifka, D.A., "The ANL/IBM SP Scheduling System", Proc. JSSPP, LNCS 949, Springer, 1995.

[9] Lublin, U., and Feitelson, D.G., "The Workload on Parallel Supercomputers: Modeling the Characteristics of Rigid Jobs", *Journal of Parallel and Distributed Computing*, 2003, pp. 1105-1122.

[10] Mu'alem, A., and Feitelson, D.G., "Utilization, Predictability, Workloads, and User Runtimes Estimates in Scheduling the IBM SP2 with Backfilling", IEEE Trans. on Parallel and Distributed Systems 12(6), June 2001.

[11] Raz, D., Levy, H., and Avi-Itzhak, B., "A Resource-Allocation Queueing Fairness Measure", Proc. SIGMETRICS/Performance, New York, June 2004.

[12] Sabin, G., and Sadayappan, P., "Unfairness Metrics for Space-Sharing Parallel Job Schedulers", Proc. JSSPP Workshop, Cambridge, Springer, LNCS 3834, June 2005.

[13] Schwiegelshohn, U., and Yahyapour, R., "Fairness in Parallel Job Scheduling", Journal of Scheduling 3(5), 2000, pp. 297-320.

[14] Shmueli, E.,and Feitelson, D.G., "Backfilling with Lookahead to Optimize the Packing of Parallel Jobs", J. of Parallel and Distributed Computing 65, 2005, pp. 1090-1107.

[15] Sodan, A.C., Kanavallil, A., and Esbaugh, B., "Group-Based Optimization for Parallel Job Scheduling with Scojo-PECT-O", Proc. HPCS, Quebec City, IEEE, June 2008.

[16] Subramaniam, R., "A Framework for Parallel Job Scheduling", Ph.D. thesis, University of California, Irvine, 1995.

[17] Talby, D., and Feitelson, D., "Supporting Priorities and Improving Utilization of the IBM SP2 Scheduler Using Slack Based Backfilling", *Proc. IPPS/SPDP*, 1999.

[18] B. van Houten, F. Ciesielski, G. Brebner, G. de Jacquelot, M. Riedmann, and H. Strauss, "Advanced Cluster Software for Meteorology", ECMWF workshop on Use of HPC in Meteorology,

Reading/UK, Nov. 2006, available at http://www.ecmwf.int/newsevents/meetings/workshops/2006/high_performance_computing-12th/pdf/Henry_Strauss.pdf.
[19] Ward, W., Mahood, C.L., and West, J.E., "Scheduling Jobs on Parallel Systems Using a Relaxed Backfill Strategy", *Proc. JSSPP*, Springer, LNCS 2537, 2002.

Journal of Physics: Conference Series **256** (2010) 012024 doi:10.1088/1742-6596/256/1/012024

# MPI Enhancements in John the Ripper

**Edward R. Sykes, Michael Lin, Wesley Skoczen**
Sheridan Institute of Technology and Advanced Learning, 1430 Trafalgar Road, Oakville, Ontario, L6H 2L1, Canada

E-mail: {ed.sykes; michael.lin; wesley.skoczen}@sheridanc.on.ca

**Abstract.** John the Ripper (JtR) is an open source software package commonly used by system administrators to enforce password policy. JtR is designed to attack (i.e., crack) passwords encrypted in a wide variety of commonly used formats. While parallel implementations of JtR exist, there are several limitations to them. This research reports on two distinct algorithms that enhance this password cracking tool using the Message Passing Interface. The first algorithm is a novel approach that uses numerous processors to crack one password by using an innovative approach to workload distribution. In this algorithm the candidate password is distributed to all participating processors and the word list is divided based on probability so that each processor has the same likelihood of cracking the password while eliminating overlapping operations. The second algorithm developed in this research involves dividing the passwords within a password file equally amongst available processors while ensuring load-balanced and fault-tolerant behavior. This paper describes John the Ripper, the design of these two algorithms and preliminary results. Given the same amount of time, the original JtR can crack 29 passwords, whereas our algorithms 1 and 2 can crack an additional 35 and 45 passwords respectively.

## 1. Introduction

The password is the most common method of authentication in computing. Used in both low and high security applications, alone or in combination with other authentication methods, the character password remains essential to modern computer security. While the level of security provided by password based authentication depends on a wide variety of factors involving how the password is handled, stored and used, the strength of the password itself depends only on its length and formation.

Password strength is a measure of a password's resistance to guessing. Assuming all other factors of a password based authentication system are immune to compromise, a password may still be guessed. Therefore ensuring that all passwords in a system are of a reasonable strength to resist most guessing attempts is essential, and thus is the purpose of password policy.

John the Ripper (JtR) is an open source software package commonly used by system administrators to enforce password policy. JtR is designed to attack (i.e., crack) passwords encrypted in a wide variety of commonly used formats. It is most commonly used to recover passwords in a computer system and is a useful tool for detecting weak passwords. JtR also has limited application in approximating the abilities of would be attackers whose tools may operate in a similar way.

JtR is a serial program that does not possess native support for parallelization. While cracking short or simple passwords is feasible, stronger passwords are very computationally expensive to crack. Each crack attempt is a reasonably discrete operation that does not directly benefit from previous or future attempts. This makes JtR a prime candidate for parallelization.

Several attempts have been made to parallelize JtR's brute-force attack mode [1], and an attempt to parallelize the wordlist mode has also been made [2]. Furthermore, several script-based attempts to parallelize JtR by running multiple serial JtR processes in parallel have been made but with limited success.

In this research we are interested in the number of passwords that can be cracked in the finite amount of time because this is ultimately what the password-collector ultimately is interested in. We excluded passwords that are impossible to crack in the finite amount of time. In pursuing this goal, two distinct algorithms were designed, implemented and evaluated as enhancements to JtR using the Message Passing Interface. The first algorithm uses multiple processors to crack one password by using an innovative approach to workload distribution. In this algorithm the candidate password is distributed to all participating processors and the wordlist is divided based on probability so that each processor has the same likelihood in cracking the password. The overarching goal of this algorithm is to crack a single password as quickly as possible. The second algorithm is intended to be used by system administrators for systems in which thousands of accounts may be in use. This algorithm divides a potentially large password file equally amongst participating processors. This approach offers efficient load-balancing and fault-tolerant performance.

This paper discusses the single processor version of JtR in Section 2 and related work in Section 3. Section 4 presents the design of our two JtR algorithms. Section 5 presents the preliminary results followed by a discussion and future work.

## 2. John the Ripper

There are many approaches to cracking encrypted passwords. One possible approach might be to reverse engineer the encryption algorithm and recover the password algorithmically, which unfortunately is both exceedingly difficult and highly impractical. JtR cracks passwords by attempting to produce a text string that matches the original password by a process of *guessing*.

Passwords on computer systems are commonly stored together in files, along with other data like usernames and account information. To secure the password files their contents are encrypted, usually with a 1-way hashing algorithm. These algorithms produce an effectively unique summary, or message digest, of the original password. When the password hashes are used for authentication at a later time, the user inputted password is hashed and compared with the original hash stored in the password file. Since hashing algorithms produce effectively unique message digests, if the input was correct, the hashes will match. JtR takes advantage of this system of storage by using the same algorithms used to create the password files to encrypt a series of guesses, then the resulting hashes are compared with the originals.

Producing guesses for use in password cracking can be accomplished in a variety of ways. JtR uses versions of the brute-force and dictionary methods. The brute-force method is the most thorough and is theoretically infallible although extremely time consuming and computationally expensive. It simply exhaustively tries every possible character combination. JtR's implementation of the brute-force method makes use of character frequency tables to try combinations containing more frequently used characters first.

The dictionary method makes use of a user definable list of possible passwords, a dictionary, as a guide for creating guesses. JtR provides a variety of options for applying the text in the dictionary. In addition to trying the dictionary text as is, salts can be added to pad the text and the dictionary text can also be mangled [3].

JtR has four modes of operation; incremental, single, wordlist, and external. The incremental mode simply applies the brute-force method while the wordlist and single modes use the dictionary method. The wordlist mode uses a user-definable text file as the dictionary while single mode generates its own dictionary using data found in the password file, such as the user name and account information. The external mode allows user defined modes of operation to be run using program code created with a subset of the C language.

A wide variety of options can be specified to customize JtR's operation. By default, if JtR is run without additional command line parameters, the single mode will be used first followed by the wordlist mode and finally the incremental mode.

JtR has been extended over the years to work with a wide variety of password file formats. The basic setup, without extensions, supports the following formats:

- Unix crypt(3) hashes:
    - traditional and double-length DES
    - BSDI extended DES
    - FreeBSD MD5
    - OpenBSD Blowfish
- Kerberos AFS
- Windows LM hash

However, JtR has some limitations. JtR's primary functional limitation is its restriction to operation on password files with known formats. It is also non-adaptive and cannot operate on file formats or hashing algorithms that aren't explicitly defined in its source code. Therefore JtR cannot be used directly on an arbitrary password authentication system.

JtR's primary performance limitation is its lack of support for parallelization. It can only make use of a single thread and does not support multi-thread operation natively.

## 3. MPI and John the Ripper

The current state of parallel implementation for JtR using Message Passing Interface (MPI) consists of a basic modification of the incremental mode. This implementation was initially created by Ryan Lim and later refined by multiple authors, notably John Anderson [1-4]. It uses MPI's facilities to divide the workload of the incremental mode by dividing the key space between the available computation processors.

The incremental mode's brute-force attack attempts every possible character combination in search of a hash that matches those listed in the password file. The MPI implementation essentially runs multiple serial incremental mode attacks, but each instance only works with a subset of the overall key space. The primary advantage to this approach is its simplicity and minimal alteration to the original source code. Additionally, there is very minimal communication between processors, each processor calculates its own key space range and performs operations on it.

While this approach effectively parallelizes JtR's incremental mode, it has several limitations. The following list the current areas that are unsolved:

1. Due to minimal communication between processors, if a processor completes its allocated workload it will sit idle until all processors complete.
2. There is no rebalancing of the workload after the initial division.
3. This implementation also doesn't address issues which may arise from one or more processors going offline before execution completes.
4. Finally, this implementation only parallelized the incremental mode.

A separate attempt to parallelize JtR's wordlist mode using MPI has been made by Pippin et al [2]. This implementation is similar in basic concept to Lim's work with JtR's incremental mode in that it takes a minimalistic approach. MPI is used minimally and essentially multiple serial wordlist mode attacks are run, one per available computational processor.

One processor is chosen to be the master which assigns password hashes from the password file to the rest of the processors. Each slave processor then runs the entire dictionary against its assigned hash. Upon completion, the slave processor then asks the master for a new hash. The primary advantages to this method are its simplicity and consistent load balancing.

Unfortunately this method is inefficient with real-world workloads. Each processor must keep a private copy of the entire dictionary which can become very memory expensive, especially with large dictionaries (~0.5 Gbyte). Since each processor must run the entire dictionary against each hash assigned to it, this approach may also become computationally expensive with very large dictionaries or when the dictionary is significantly larger than the password file, as is commonly the case.

## 4. Method

The first phase of this research explored existing MPI implementations of JtR. Several of these implementations were tested and benchmarked for performance. The three primary modes of operation for JtR were analyzed with respect to parallelization. Strategies were also devised for parallelizing each of the three modes, with focus on innovative approaches. As a result, two algorithms were designed to address how to improve the current MPI implementation of JtR.

### *4.1. Algorithm 1: Distributing the Wordlist Contents amongst Participating Processors*

In this algorithm multiple processors are used to crack one password. In this innovative approach the candidate password is distributed to all participating processors and the wordlist is divided based on probability so that each processor has the same likelihood in cracking the password. The overarching goal of this algorithm is to crack a single password as quickly as possible. Figure 1 illustrates the distribution of potential passwords from an original wordlist onto processor 1 to $N$.

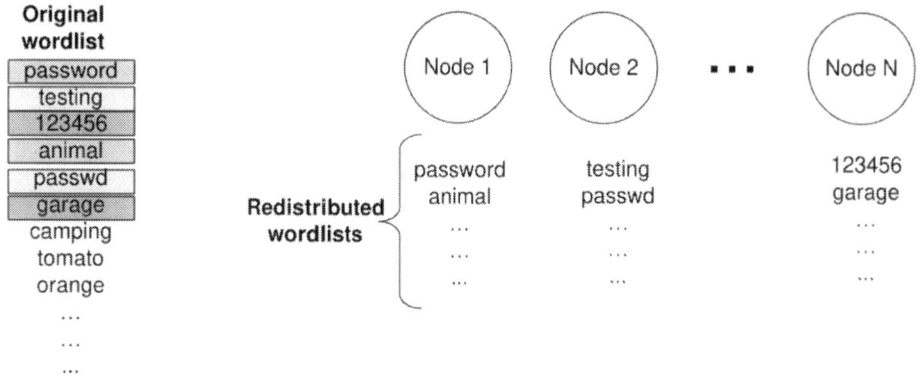


Figure 1. Distributing the wordlist contents fairly amongst participating processors.

The potential benefits of this approach are threefold. One, the most probable passwords are listed at the top of each processor's wordlist thereby increasing the likelihood of finding a match. Two, the wordlist is distributed amongst *N* processors, resulting in gains due to the workload being evenly divided. Three, there is no overlapping of efforts amongst participating processors. In previous versions of JtR, the same passwords are repeatedly cracked by different processors resulting in wasted cpu cycles. Our algorithm ensures that all processors are utilized efficiently collectively attempting to crack the password.

*4.2. Algorithm 2: Password file distribution.*

The second algorithm is intended to be used by system administrators for systems in which thousands of accounts may be in use. This algorithm divides a potentially large password file equally amongst participating processors. This approach offers efficient load-balancing and fault-tolerant performance. Continual monitoring from the root processor tracks the progress of slave processors in terms of how many passwords have been successfully cracked. If a threshold load factor ($\lambda$=50%) is reached, a dynamic and explicit rebalancing occurs in which uncracked passwords are submitted to an underutilized processor. Figure 2 depicts the password distribution used in this algorithm.

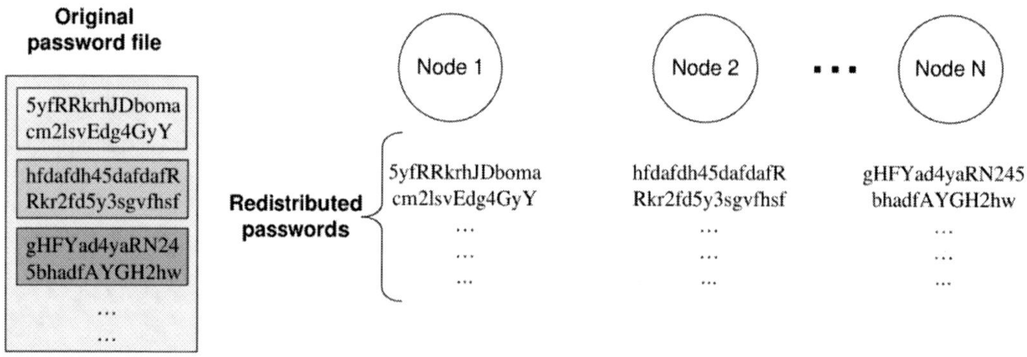


Figure 2. Distributing the password file amongst participating processors while permitting efficient load-balancing.

Benchmarking techniques were used to evaluate the performance of both algorithms. The benchmarks were performed using a known set of passwords encrypted in a chosen format. This password file was subjected to cracking attempts with serial JtR and our MPI implementation, and their results were compared. The dependent variable was the time to crack a given password. The independent variables were the OS, the supercomputer, the supercomputer hardware, the number of processors used, the password hash file and the wordlist. Each benchmark result was averaged over 10 runs to ensure proper benchmarking, repeatability of results, result stability and good result proximity.

## 5. Preliminary Results

In this section we present the preliminary results of our MPI implementations of Algorithm1 and 2 for JtR. Table 1 shows one of the results obtained from an experiment that evaluated the performance of the Distributed wordlist algorithm. This experiment involved 7 passwords, and a wordlist file that contained 3 million words listed in decreasing probability of likelihood. The program was run on 32 processors on the whale SHARCNET supercomputer[1] [1, 5]. Table 1 shows the time to crack each password for the Original John-MPI and our Distributed wordlist algorithm.

Table 1. 7 password entries in the password file, large wordlist (3 millions words), on 32 processors.

| Password | Original John-MPI Time to crack | Distributed worldlist Time to crack |
|---|---|---|
| abc123 | 6 sec | 6 sec |
| guinness | 9 min 43 sec | 9 min 43 sec |
| password | 6 sec | 6 sec |
| erschlangener | 7 days | 12 hours |
| zhora | 9 days | 3 hours |

A second experiment was conducted to evaluate the performance of the original JtR and our distributed password algorithm and to facilitate a comparison between the two. The test was to determine how many passwords could be cracked within a 4 day period using the same supercomputer with 32 processors. A large password file that contained 132 random passwords of variable length up to 14 characters drawn from the full extended ASCII table was used. In this experiment it was found that:

1. the Original JtR cracked 29 of the passwords (21% of total)
2. Algorithm1 (Distributed wordlist), cracked 64 passwords (49% of total)
3. Algorithm 2 (Distributed password) cracked 74 passwords (56% of total)

Table 2 shows these findings.

[1] For details regarding the type of supercomputer used in this research, please see www.sharcnet.ca, specifically: the whale supercomputer: https://www.sharcnet.ca/my/systems/show/27

Table 2. Original JtR, Distributed wordlist algorithm, and Distributed password algorithm experiment findings.

| | Number of Passwords Cracked | % of total cracked | Additional passwords cracked |
|---|---|---|---|
| Original JtR | 29 | 21% | --- |
| Algorithm 1: Distributed wordlist | 64 | 49% | 35 |
| Algorithm 2: Distributed passwords | 74 | 56% | 45 |

## 6. Discussion

The research described in this paper focused on enhancements to the wordlist mode of JtR, which uses the dictionary method for cracking passwords. Based on our preliminary findings we have found significant benefits to parallelizing the dictionary cracking method.

Our first algorithm, using a distributed wordlist method, significantly improves the efficiency of password cracking when a large number of processors are available to crack one or two passwords. The distributed wordlist algorithm can crack over twice as many passwords than the original JtR given the same duration of time in which to work. For several tests, this algorithm was shown to be 72 times faster in cracking passwords than the original JtR. Since this method relies on distributing the wordlist file among processors and wordlist contain thousands of entries, its performance should scale well with the addition of more processors.

One of the primary sources of performance improvement over previous implementations is the greater workload distribution. In previous MPI implementations of JtR's wordlist mode, all processors will attempt all entries in the wordlist on the passwords they are assigned. This results in considerable workload duplication in scenarios where the wordlist file is large while the password file is small. The only workaround was to manually abort and resume individual jobs from time to time. Our approach eliminates this workload duplication, making the first algorithm ideal for these scenarios.

Our second algorithm, using a password file distribution method, has also produced promising preliminary results. This algorithm distributes the workload by dividing the password file. In our implementation, processes are working on different password sets, so workload duplication is avoided.

A previous MPI implementation of JtR's wordlist mode [2] similarly divides the workload along the password file, but does so with the addition of inter-process communication; the processors synchronize with a master processor which assigns passwords to each worker processor. Our approach divides the file at the start of runtime which bypasses the overhead of inter-process communication, since the passwords are distributed among the processors at the start of runtime and synchronization between processors is not necessary.

These are preliminary results, and although promising, more tests are needed to determine how effective our algorithms scale, for instance, by using 128, 256 or more processors. Furthermore, experiments involving larger and more varied entries in the password file are needed to draw conclusive findings.

## 7. Future Research

Future research may focus on combining the two approaches of workload distribution. One possible approach is to use both methods and implement an additional evaluation logic which will determine which method to apply given a specific scenario.

For a scenario where the password list is found to be considerably longer than the number of available processors, the program could use the password distribution technique of algorithm 2. For scenarios where the password list is found to be short, relative to the number of available processors, the second method of workload distribution along the wordlist file implemented in algorithm 1 could be used.

Future work may also explore distributing the password list and the wordlist at the same time. This approach presents a large number of problems involving ensuring proper workload distribution. Any implementation will likely require a high level of inter-process communication and synchronization which will lead to potentially significant overhead.
A new MPI patch was released at near the time of publication of this research {Simpson, 2010 #518}. The patch adds limited workload distribution (auto-splitting in Markov mode), but still does not utilize MPI significantly. MPI is only used to enable each process to know its own relative ID and the total number of processes in a given parallel running group. This implementation should be explored and the inter-process awareness of the distributed wordlist mode can be improved by enabling the processes to notify each other of successful password cracks, therefore avoiding some amount of workload duplication.

Beyond JtR's wordlist mode, future work will include creating an MPI implementation of JtR's incremental mode with a focus on exploring the limitations of current MPI implementations and addressing them. There are no current plans to attempt to parallelize JtR's single crack mode, due to its very short run-time even when run serially. While the single mode can conceivably be parallelized, the benefits of such an endeavour are in doubt.

For incremental mode parallelization, a method of consistent load balancing and processor failure management has yet to be attempted and should be explored. When a processor completes its allocated task, it should be allocated more work taken from the remaining workloads of the other processors. The possible scenario of a processor going offline during execution should also be addressed since execution of incremental mode attacks are very time consuming and graceful processor failure resolution should be explored. This might be implemented by dividing the workload of the failed processor among the remaining processors.

**Acknowledgements**
This work was made possible by the facilities of the Shared Hierarchical Academic Research Computing Network (SHARCNET:www.sharcnet.ca) and the contributions of my student, Michael Lin, and Sheridan Research member, Wesley Skoczen.

## 8. References

1. Lim, R., *Parallelization of John the Ripper (JtR) using MPI.* 2004, Computer Science and Engineering University of Nebraska–Lincoln.
2. Pippin, A., B. Hall, and W. Chen, *Parallelization of John the Ripper Using MPI.* 2006, University of California, Santa Barbara.
3. OpenWall (2010) *John the Ripper's cracking modes.*
4. Anderson, J. (2010) *John the Ripper*
5. OpenWall (2010) *MPI with John the Ripper.* 2010.

# When to Renew Software Licences at HPC Centres? A Mathematical Analysis

**Ge Baolai**[1,1] **and Allan B. MacIsaac**[2]

[1] SHARCNET, The University of Western Ontario, London, Ontario, Canada N6A 5B7
[2] Department of Applied Mathematics, The University of Western Ontario, London, Ontario, Canada N6A 5B7

E-mail: bge@sharcnet.ca

**Abstract.** In this paper we study a common problem faced by many high performance computing (HPC) centres: When and how to renew commercial software licences. Software vendors often sell perpetual licences along with forward update and support contracts at an additional, annual cost. Every year or so, software support personnel and the budget units of HPC centres are required to make the decision of whether or not to renew such support, and usually such decisions are made intuitively. The total cost for a continuing support contract can, however, be costly. One might therefore want a rational answer to the question of whether the option for a renewal should be exercised and when. In an attempt to study this problem within a market framework, we present the mathematical problem derived for the day to day operation of a hypothetical HPC centre that charges for the use of software packages. In the mathematical model, we assume that the uncertainty comes from the demand, number of users using the packages, as well as the price. Further we assume the availability of up to date software versions may also affect the demand. We develop a renewal strategy that aims to maximize the expected profit from the use the software under consideration. The derived problem involves a decision tree, which constitutes a numerical procedure that can be processed in parallel.

## 1. Introduction

In this paper we study a common problem faced by the personnel at many high performance computing (HPC) centres: When and how to renew commercial software licences. Software vendors often sell perpetual licences along with forward update and support contracts at an additional, annual cost, which cover not only technical support, but also the entitlement for future upgrades. The HPC providers have the option to renew or not to renew the update contract for a particular software package, depending on the demand for the software and budget. They may also opt to renew the support contract for multiple years at a discounted price. By paying a fee for the update contract, an HPC provider can protect itself against any incompatibilities that might occur due to the upgrade of their software environment on the deployed systems, such as the operating system, system libraries and compilers. The total cost for the continuing support contract can, however, be costly. One might therefore want a rational answer to the question of whether the option for a renewal should be exercised and when.

In an attempt to study this problem within a market framework, aiming at the self-adjustment of resource supply and demand, we present the analysis of a mathematical problem derived from

[1] Corresponding author.

the day to day operation of a hypothetical HPC centre that charges fees for using resources and software. In the mathematical analysis, we assume that the uncertainty comes from the demand (number of users using a package) as well as the price. Further we assume the availability of the up to date software versions may affect the demand, hence the usage of the resources as well. We develop a software support renewal strategy that aims to maximize the expected profit from the usage of resources.

The derived mathematical problem involves a decision tree, which constitutes a numerical optimization problem that is best solved with parallel processing.

## 2. Related Work

The hardware and software upgrade problem has been studied from a number of different perspectives. In the literature, related, representative work includes the work of Pfening and Telek [1], which aimed to find an optimal maintenance policy for a slowly degrading system, Zhao *et al* [2] which considered the problem of the optimal rejuvenation schedule for an application server, Hartman and Rogers [3] which presented two dynamic programming models for equipment replacement problems and Mehra and Seidmann [4] which considered the problem of finding an optimal timing for software upgrade releases for a monopoly software maker. The nature of the problem under consideration here belongs to a boarder class of problems - revenue management, which has been extensively studied (See for example the classic works [5] and [6]). A problem that is closely related to what we are studying is the application of the real options approach (See for example [7]) to refurbishment of hotels and shopping malls (See for example, [8], [9] and the recent work [10]). In hotel renovation problems, the quality upgrade schedule is to be determined over the life span of a hotel. In our case we use the same methodology applied to a new problem. To our knowledge, the problem we are considering is unique with respect to the objectives we attempt to achieve. We consider an HPC resource provider, in which a charging model is used: for every user, the usage of resources is billed in credits or internal currency. We present a mathematical model based on stochastic analysis and devise a renewal strategy that aims at maximizing the profitability of the HPC provider in terms of either real money or internal currency.

## 3. The Problem Statement

We devise our models based on a number of assumptions; Resources (i.e. time on CPU cores) have values, the resource provider makes a profit by offering available resources to users for the return of currency or credits, and the price of a resource is determined by the demand and may exhibit stochastic motion.

To begin, we introduce the following notation:

- $t$ – Time, with the notation $t+k$ denoting the $k$th renewal time in years;
- $q_k$ – A short notation for time dependent function $q(t)$ at $t = k$;
- $T$ – The end life span, e.g. 5 years.
- $u(t)$ – Demand (number of users);
- $p(t)$ – Price per use;
- $C(t)$ – Cost of renewal;
- $\omega$ – Renewal "path", i.e. a sequence of decisions on whether to renew or not to renew at a number of specified times in year;
- $w_k$ – For a particular renewal path at the $k$th year, an integer that indicates the number of years to renew for.
- $f(t, u, p; \omega)$ – The "cash flow" rate at time $t$, determined by the demand $u(t)$, price $p(t)$, and renewal path $\omega$.

- $V$ – Expected profit.

By making a sequence of decisions at renewal times $t_1$, $t_2, \ldots$, we attempt to solve the problem

$$\max_{\omega} \left\{ \int_0^T f(t, p, u; \omega) dt - \sum_{k=0}^{T} w_k C_k \right\}, \tag{1}$$

over all possible renewal paths $\omega$ subject to certain constraints in $p$ and $u$ and

$$\sum_k w_k \leq T. \tag{2}$$

We shall apply Bellman's principle of optimality [11] to solve this optimization problem backward in time.

In the following sections we consider two models: a simplified, deterministic model that provides an intuitive look at the problem and a stochastic model, for which we explore the solutions within the framework of a Monte Carlo approximation.

## 4. Deterministic Model - A Simplified Case

We first consider a simplified model. To illustrate the method, suppose the time frame to be considered is divided into five intervals which might represent a typical life span of five years. During this time frame, we face a sequence of decisions to make. This can be illustrated by an ensemble of possible upgrade paths, as shown in Figure 1. An example of five years of coverage: $w = \{1, 2, 0, 0\}$ is represented by the line in bold. In this example at the end of year one the contract is renewed for one year. At the end of the year two the contract is renewed for two years. No renewal is needed at the end of year three and at the end of year four the renewal for year five is declined. The first year is assumed to be covered by the initial contract.

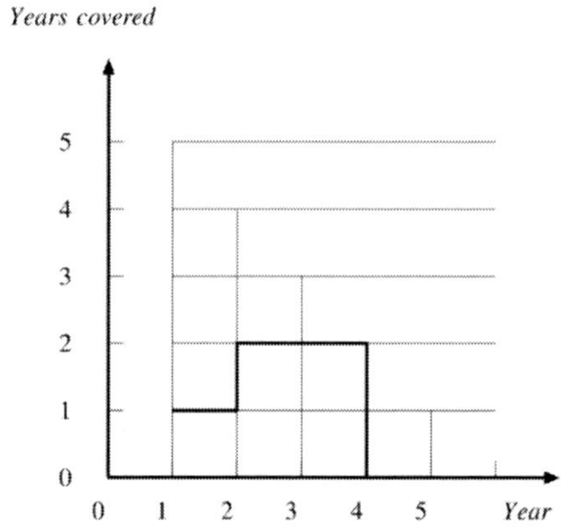


**Figure 1.** Upgrade "paths" in years of coverage. An example of coverage: $w = \{1, 2, 0, 0\}$ is represented by the line in bold. The first year is covered by the initial contract.

An intuitive, tree representation is given in Figure 2, again for a five year period. Each circled number indicates the number of years that would have been covered by a support contract. Each of the triangles on the right resembles the branch immediately left to it at the same level. Our algorithm, which recursively determines the renewal paths, is given below

ALGORITHM 1 *Computing renewal paths. Given the number of years n–the life span– of renewal opportunities, the following procedure determines all the possible renewal paths, stored in two dimensional array* $\omega(:,:)$.

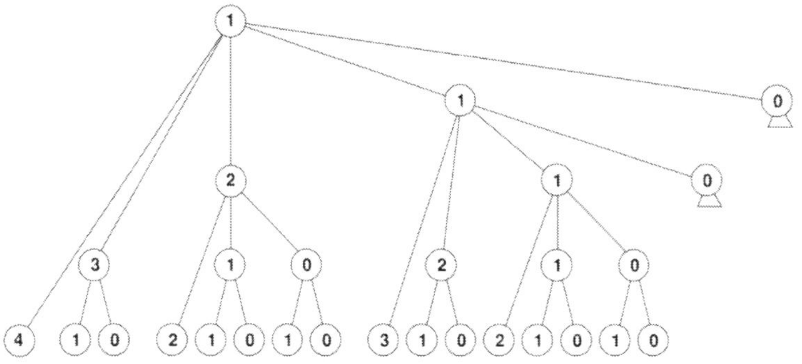


**Figure 2.** Upgrade "paths" in years of coverage – a decision tree for the period of five years, with circled numbers indicating the years of coverage. Each of the two triangles on the two circled zeros represents an identical branch at the same level immediately on the left.

**begin procedure**: path($m$, $t_0$)
    $m$ – Number of years prior to the end $T$;
    $t_0$ – The next year;
    $n$ – Total number of years to consider until $T$.
    $\omega(:,:)$ – Array that stores, for each renewal path $w$, the renewal sequence $w_k$, $k = 1, \ldots, n$;
    $n_p$ – The $n_p$th path, a counter with an initial value 1;
    **if**($t_0 \geq n$), **then**/* Complete the branch, return */
        $n_p \leftarrow n_p + 1$;
        return;
    **end if**

    **for**($y \leftarrow m$; $y \geq 0$; $y \leftarrow y - 1$), **do**/* Create new branch */
        $\omega(n_p, t_0 - 1) \leftarrow y$;
        **for**($t \leftarrow t_0 + 1$; $t \leq t_0 + y - 1$; $t \leftarrow t + 1$), **do**
            $\omega(n_p, t - 1) \leftarrow 0$;
        **end do**
        **if**($0 = y$), **then**
            $t_1 \leftarrow t_0 + 1$;
        **else**
            $t_1 \leftarrow t_0 + y$;
        **end if**
        **call** path($n - t$, $t_1$);
        **for**($t \leftarrow 0$; $t < t_0 + y$; $t \leftarrow t + 1$), **do**
            $\omega(n_p, t) \leftarrow \omega(n_p - 1, t)$;
        **end do**
    **end do**

    return;
**end procedure**

Assume we know the pattern of demand $u(t)$ and that demand drops following a known path when an annual renewal is skipped and hence the upgrades are frozen. When the support contract is renewed, the demand will increase to or remain at the level at the end of previous support contract.

We derive our mathematical formulation by carrying out our analysis backward in time starting at $t = T = 5$ as follows. Let $U_t^k(u,p)$ be the revenue from the use of the software of $k$ years old and $\bar{u}_t^k$ be the average use over interval $[t, t+1]$. Assume, for simplicity, over the entire time period, the price $p$ for use remains constant, as does the cost of renewal, denoted by $c$.

Now at $t = 4$, we denote the expected profit by $V_4^k$, for $k$ year old software, with and without a renewal, by the end of the interval $[4, 5]$, when looking forward at $t$. Note that, it is a common practise that a vendor will only renew the software to the current release and therefore will charge backward the missing upgrades, due to the cost of software development. That is, if the cost for annual renewal is $c$, the cost to renew a $k$ years old software will be $kc$. Therefore we write, denoting $\mathbf{E}(\cdot)$ the expectation operator

$$\begin{aligned} V_4^k &= \max\{\mathbf{E}(U_4^k(u,p)), \mathbf{E}(U_4^0(u,p)) - kc)\} \\ &= \max\{\bar{u}_4^k p, \bar{u}_4^0 p - kc\}. \end{aligned}$$

At $t = 3$, we have these options, not to renew, renew for one year or renew for two years. Therefore, we have[2]

$$V_3^k = \max\{\bar{u}_3^k p + V_4^{k+1}, \bar{u}_3^0 p - kc + V_4^0, 2\bar{u}_3^0 p - (k+1)c\}.$$

We iterate this analysis until $t = 1$, at which we have

$$V_1^0 = \max\{\bar{u}_1^1 p + V_2^1, \bar{u}_1^0 p - c + V_2^0, 2\bar{u}_1^0 p - 2c + V_3^0, 3\bar{u}_1^0 p - 3c + V_4^0, 4\bar{u}_1^0 p - 4c\}.$$

In summary, knowing $V_t^k$, $t = 4, \ldots, 1$, $k = 0, 1, \ldots, t-1$, we can find backward in time the path that leads to the maximum of expected profit, subject to the assumptions we have made on the demand and its response to the upgrade decisions.

## 5. Stochastic Model

We now assume in Eq. (1) the demand $u = u(t)$ follows a stochastic process, as does the price $p(t)$. We use the expectation operator $\mathbf{E}_t$ given the information available at $t$, and rewrite Eq. (1) as

$$\max_{\omega} \left\{ \mathbf{E}_0 \left[ \int_0^T f(\tau, p, u; \omega) d\tau - \sum_{k=0}^{T} w_k C_k \right] \right\}, \tag{3}$$

where $C_k$ is the cost for renewal at the $k$th year and $w_k$ are control parameters

$$w_k = \begin{cases} m & \text{if renewed for } m \text{ years,} \\ 0 & \text{otherwise;} \end{cases} \tag{4}$$

$$\sum_k w_k \leq T, \tag{5}$$

subject to the demand equation

$$du = a(t, u; \omega)dt + b(t, u; \omega)dX, \tag{6}$$

where $a$ and $b$ are known, deterministic functions and $X$ is a stochastic process. To simplify the problem further, we assume the price is constant and consider only the demand $u$.

[2] Note that for a renewal that covers two years, for which the software is kept up to date, we assume the same average uses $\bar{u}_3^0$ and $\bar{u}_4^0$. The same applies to the expression of $V_1^0$ below as well.

Using the notation for renewal times as in the previous section, from (3), define

$$\begin{aligned} V(t, u(t)) &= \max_{\omega} \left\{ \mathbf{E}_t \left[ \int_t^T f(\tau, u; \omega) d\tau - \sum_{k=t}^{T} w_k C_k \right] \right\} \\ &= \max_{\omega} \left\{ \mathbf{E}_t \left[ \int_t^{t+1} f(\tau, u; \omega) d\tau - w_t C_t + \int_{t+1}^{T} f(\tau, u; \omega) d\tau - \sum_{k=t+1}^{T} w_k C_k \right] \right\}. \end{aligned}$$

Denoting

$$\phi(t, u(t)) = \int_t^{t+1} f(\tau, u; \omega) d\tau,$$

$\phi(t, u(t))$ now represents the realization of the cash flow for the period of $[t, t+1]$ when the uncertainty is revealed at $t$. We arrive at the following recursive equation

$$V(t, u(t)) = \phi(t, u(t)) + \max_{\omega} \mathbf{E}_t \left[ V(t+1, u(t+1)) - w_t C_t \right]. \tag{7}$$

The Bellman's principle is interpreted as follows: Upon knowing the demand $u$, which is a function of some uncertainty $X$ revealed at $t$, one is to make strategic decisions to maximize the expected value of $V$ by $t = T$.

To obtain the solution to (7), which involves a series of sub-problems represented by the recursive relation, we use a procedure backward in time. If we know $\mathbf{E}_t \left[ V(t+1, u(t+1)) \right]$, then we can get $V(t, u)$ and subsequently, we can move backwards in time.

Let

$$\psi(t, u(t)) = \mathbf{E}_t \left[ V(t+1, u(t+1)) \right], \tag{8}$$

thus

$$V(t, u(t)) = \phi(t, u(t)) + \max\{\psi(t, u(t)) - w_t C_t\}. \tag{9}$$

In the following section, we will elaborate on a numerical procedure to solve (9) based upon the least squares Monte Carlo (LSM) approach proposed by Longstaff and Schwartz [12].

## 6. Numerical Procedure

In practice, an analytic form of $\psi$ is nearly impossible to derive, hence we seek a numerical solution instead. The idea is to approximate the expected value $\psi(t, u(t))$ when the uncertainty is revealed at $t$. To achieve this goal, we generate $N$ samples of simulated demands $U^i(t)$ and $U^i(t-1)$, $i = 1, 2, \ldots, N$, (also denoted for short $U_t^i$ and $U_{t-1}^i$, respectively) using the demand model $u(t)$ at each $t$. For each $U_t^i$, we have a corresponding realization

$$\phi_t^i = \phi(t, U^i(t))$$

and

$$\psi_t^i = \psi(t, U^i(t) = \mathbf{E}_t \left[ V(t+1, U^i(t+1)) \right]. \tag{10}$$

Given $\psi_t^i$, we obtain from (9) $V(t, U_t^i)$, denoted by $V_t^i$, where

$$V^i = \phi_t^i + \max\{\psi_t^i - w_t C_t\}. \tag{11}$$

We then approximate the expected value $\psi(t, u(t))$ by regressing $V^i$ on $U^i$ at $t-1$. We can then obtain $V(t-1, U_{t-1}^i)$ and so on in the same manner. The maximization of $V^i$ for all simulated sample paths determines the optimal renewal process at time $t$.

We perform this procedure for each renewal path from $t = T$. Note that from (8) $\psi(T, u(T)) = 0$. The following algorithm summarizes the numerical procedure.

Algorithm 2 *Approximating $\psi(t-1, u(t-1))$ given $\psi(t, u(t))$.*

$\psi(T, u(T)) \leftarrow 0$;
$t \leftarrow T$;
**while** $t - 1 > 0$, **do**
  **for** $i = 1, \ldots, N$, **do**
    Generate $U_{t-1}^i$;
    $V^i \leftarrow \phi(t, U_t^i) + \max\{\psi(t, U_t^i) - mC_t\}$;
  **end**
  Regress $V^i$ on $U_{t-1}^i$ to obtain approximated $\psi(t-1, u(t-1))$;
  $t \leftarrow t - 1$;
**end**

To find the approximate solution to the expected value $\psi(t-1, u(t-1))$, we assume a functional form of $\psi(t-1, u(t-1))$ spanned by a linear combination of $d$ selected basis functions on demand $U$

$$\psi = \sum_{k=0}^{d} a_k L_k(U), \tag{12}$$

where $L_k()$ are the basis functions and $a_k$ are coefficients to be determined. With generated samples $U_{t-1}^i$ at $t-1$, and computed values $V^i$, we obtain the following overdetermined linear system in $a_k$

$$V^i = \sum_{k=0}^{d} a_k L_k(U^i(t-1)),\ i = 1, \ldots, N. \tag{13}$$

The coefficients $a_k$ are solved using a standard least squares fitting method.

## 7. Numerical Results

To highlight the utility of this analysis we present some representative sample calculations in the following section. In the first case, we derive the optimal renewal strategy using the deterministic model of Section 4. In the second example, we present two cases with results obtained from numerical simulation using the stochastic method of Section 5.

### 7.1. Deterministic Example

To implement our deterministic model, we first generate yearly demand samples $U_1, U_2, \ldots, U_T$, based on the mean of daily demand of 30 observed (from real logged data). We assign the values $U_1, U_2, \ldots, U_T$ at $t = 1, 2, \ldots, T$, respectively. During the interval between the start of years $t$ and $t+1$, the demand is approximated as a linear function; this interval will be written as $[t, t+1]$. We assume a renewal decision at time $t$ will affect the subsequent demand during the interval $[t, t+1]$. A drop in the demand (in this example by 10%) is assumed at $t$ when a renewal is skipped. By contrast, at $t$, if a renewal is made, the demand will be raised to the same level as that in the previous year. In addition we assume there is no penalty for skipping renewals.[3]

The optimal renewal decision is then relatively straightforward to determine as we calculate the net income along all possible paths. The optimal renewal policy is thus the path of renewals that results in the maximal net income. When the number of years to consider increases, however, the number of renewal options grows exponentially. The work to compute the net

[3] This might not be realistic for some commercial software packages, for which certain penalties often apply if the renewal is not made continuously.

**Table 1.** Deterministic case. Renewal paths and expected revenue (net income) for the price $p = \$0.5$ per use, mean of daily demand 30 and the cost to renew \$1000 with annual increase of 10% for a period of 5 years. The optimal renewal policy is found to be $\{0\ 1\ 0\ 0\ 0\}$.

| No. | Path (years to cover at renewal times) | Expected value of net income (\$) |
|---|---|---|
| 1 | 4 0 0 0 0 | 48175.25 |
| 2 | 3 0 0 1 0 | 49102.85 |
| 3 | 3 0 0 0 0 | 49175.25 |
| 4 | 2 0 2 0 0 | 49077.85 |
| 5 | 2 0 1 1 0 | 49589.65 |
| 6 | 2 0 1 0 0 | 50177.85 |
| 7 | 2 0 0 1 0 | 50102.85 |
| 8 | 2 0 0 0 0 | 50175.25 |
| 9 | 1 3 0 0 0 | 48641.80 |
| 10 | 1 2 0 1 0 | 49619.40 |
| 11 | 1 2 0 0 0 | 49691.80 |
| 12 | 1 1 2 0 0 | 48968.10 |
| 13 | 1 1 1 1 0 | 49479.90 |
| 14 | 1 1 1 0 0 | 50068.10 |
| 15 | 1 1 0 1 0 | 50669.40 |
| 16 | 1 1 0 0 0 | 50741.80 |
| 17 | 1 0 2 0 0 | 50077.85 |
| 18 | 1 0 1 1 0 | 50589.65 |
| 19 | 1 0 1 0 0 | 51177.85 |
| 20 | 1 0 0 1 0 | 51102.85 |
| 21 | 1 0 0 0 0 | 51175.25 |
| 22 | 0 3 0 0 0 | 49537.35 |
| 23 | 0 2 0 1 0 | 50514.95 |
| 24 | 0 2 0 0 0 | 50587.35 |
| 25 | 0 1 2 0 0 | 49863.65 |
| 26 | 0 1 1 1 0 | 50375.45 |
| 27 | 0 1 1 0 0 | 50963.65 |
| 28 | 0 1 0 1 0 | 51564.95 |
| **29** | **0 1 0 0 0** | **51637.35** |
| 30 | 0 0 2 0 0 | 50501.85 |
| 31 | 0 0 1 1 0 | 51013.65 |
| 32 | 0 0 1 0 0 | 51601.85 |
| 33 | 0 0 0 1 0 | 51526.85 |
| 34 | 0 0 0 0 0 | 51599.25 |

income becomes tedious. However one can easily create computer code to generate the complete set of renewal paths by implementing Algorithm 1, and calculating the corresponding net incomes.

Table 1 shows the results for the case where the price per use is \$1, the mean of daily demand is 30 and the cost to renew is \$1000 with annual increase of 10%. The optimal renewal path is found by computing the maximum net income from the revenue generated less the total cost of renewals.

**Table 2.** Results for the non deterministic mode. Renewal paths and expected revenue (net income) for the price $p = \$0.5$ per use, mean of daily demand 30 and the cost to renew \$1000 with annual increase of 10% for a period of 5 years. The optimal renewal policy is found to be $\{1\ 3\ 0\ 0\ 0\}$.

| No. | Path (years to cover at renewal times) | Expected value of net income (\$) |
|---|---|---|
| 1 | 4 0 0 0 0 | 18039.50 |
| 2 | 3 0 0 1 0 | 18366.50 |
| 3 | 3 0 0 0 0 | 13555.00 |
| 4 | 2 0 2 0 0 | 18889.50 |
| 5 | 2 0 1 1 0 | 18805.27 |
| 6 | 2 0 1 0 0 | 13960.50 |
| 7 | 2 0 0 1 0 | 13910.50 |
| 8 | 2 0 0 0 0 | 9123.50 |
| **9** | **1 3 0 0 0** | **19466.00** |
| 10 | 1 2 0 1 0 | 19325.79 |
| 11 | 1 2 0 0 0 | 14553.00 |
| 12 | 1 1 2 0 0 | 19263.14 |
| 13 | 1 1 1 1 0 | 19226.70 |
| 14 | 1 1 1 0 0 | 14414.00 |
| 15 | 1 1 0 1 0 | 14364.00 |
| 16 | 1 1 0 0 0 | 9581.00 |
| 17 | 1 0 2 0 0 | 14453.00 |
| 18 | 1 0 1 1 0 | 14314.00 |
| 19 | 1 0 1 0 0 | 9531.00 |
| 20 | 1 0 0 1 0 | 9481.00 |
| 21 | 1 0 0 0 0 | 4601.00 |
| 22 | 0 3 0 0 0 | 15035.50 |
| 23 | 0 2 0 1 0 | 14929.00 |
| 24 | 0 2 0 0 0 | 10098.00 |
| 25 | 0 1 2 0 0 | 14945.00 |
| 26 | 0 1 1 1 0 | 14841.50 |
| 27 | 0 1 1 0 0 | 10014.00 |
| 28 | 0 1 0 1 0 | 9964.00 |
| 29 | 0 1 0 0 0 | 5082.50 |
| 30 | 0 0 2 0 0 | 9998.00 |
| 31 | 0 0 1 1 0 | 9914.00 |
| 32 | 0 0 1 0 0 | 5032.50 |
| 33 | 0 0 0 1 0 | 4982.50 |
| 34 | 0 0 0 0 0 | 4982.50 |

### *7.2. Monte-Carlo Example*

To illustrate the Monte Carlo method of section 5 we simulate the case of a software package with a daily mean usage of 30 units for a period of five years. Once again we assume there is an annual increase in the cost of renewal of 10%. As in the previous example, we assume there is no penalty when the renewals are skipped. We also assume, exactly as in the previous example, that the demand is affected by the renewals or non-renewals. For each renewal path, at each time of renewal, 10,000 samples of demand are used.

For the least squares Monte Carlo simulation, we have compared the results using a number of

**Table 3.** Results for the non deterministic mode. Renewal paths and expected revenue (net income) for the price $p = \$1$ per use, daily mean of demand 30 and the cost to renew \$1000 with annual increase of 10% for a period of 5 years. The optimal renewal policy is found to be $\{1\ 3\ 0\ 0\ 0\}$.

| No. | Path (years to cover at renewal times) | Expected value of net income (\$) |
|---|---|---|
| 1 | 4 0 0 0 0 | 40079.00 |
| 2 | 3 0 0 1 0 | 40883.00 |
| 3 | 3 0 0 0 0 | 30110.00 |
| 4 | 2 0 2 0 0 | 41979.00 |
| 5 | 2 0 1 1 0 | 41861.00 |
| 6 | 2 0 1 0 0 | 31021.00 |
| 7 | 2 0 0 1 0 | 30971.00 |
| 8 | 2 0 0 0 0 | 20247.00 |
| **9** | **1 3 0 0 0** | **43082.00** |
| 10 | 1 2 0 1 0 | 42902.00 |
| 11 | 1 2 0 0 0 | 32206.00 |
| 12 | 1 1 2 0 0 | 42776.00 |
| 13 | 1 1 1 1 0 | 42753.00 |
| 14 | 1 1 1 0 0 | 31978.00 |
| 15 | 1 1 0 1 0 | 31928.00 |
| 16 | 1 1 0 0 0 | 21212.00 |
| 17 | 1 0 2 0 0 | 32106.00 |
| 18 | 1 0 1 1 0 | 31878.00 |
| 19 | 1 0 1 0 0 | 21162.00 |
| 20 | 1 0 0 1 0 | 21112.00 |
| 21 | 1 0 0 0 0 | 10202.00 |
| 22 | 0 3 0 0 0 | 33221.00 |
| 23 | 0 2 0 1 0 | 33108.00 |
| 24 | 0 2 0 0 0 | 22296.00 |
| 25 | 0 1 2 0 0 | 33140.00 |
| 26 | 0 1 1 1 0 | 32983.00 |
| 27 | 0 1 1 0 0 | 22178.00 |
| 28 | 0 1 0 1 0 | 22128.00 |
| 29 | 0 1 0 0 0 | 11215.00 |
| 30 | 0 0 2 0 0 | 22196.00 |
| 31 | 0 0 1 1 0 | 22078.00 |
| 32 | 0 0 1 0 0 | 11165.00 |
| 33 | 0 0 0 1 0 | 11115.00 |
| 34 | 0 0 0 0 0 | 11115.00 |

different basis functions, including generic polynomials of various degrees[4], Chebyshev, Hermit, Legendre and Laguerre. Our experiments have shown that each of these basis functions gives the same desired accuracy.

The GNU scientific library is used for generating the samples and solving the least squares problem via singular value decomposition.

Table 2 and 3 show the results for two cases: (a) when the price per use is \$0.5 and (b) when the price per user is \$1. The mean of daily demand is set to 30. The renewal strategy shown in

[4] Using more than five terms does not lead to any difference in the results

boldface in each table is the optimal strategy as it maximizes the net income.

## 8. Concluding Remarks

In this paper we have analysed the complex issue of software support renewal policies. This is a problem that requires decision making at multiple stages. The problem is modelled using a mathematical procedure that aims at maximizing the expected value of net income from the revenue generated by the use of the software package minus the total cost of renewals. The mathematical model results in a Bellman equation which involves a series of sub optimization problems and can be solved backward in time. A numerical procedure based on Monte Carlo simulation can be used to obtain the optimal renewal strategy. The optimal renewal strategy is determined by maximizing the expected value of the net income over all possible renewal options.

The computational work required by the Monte Carlo method is generally quite large. One needs to exhaust all the possible paths of renewals and for each path, at each renewal time, to simulate $N$ realizations of the revenue and perform a least squares Monte Carlo simulation by solving an $N \times d$ linear system, with $d$ being the degree of selected basis functions of polynomials, preferably solved by singular-value-decomposition. In the examples shown in this study $N$ was taken at $10,000$. Fortunately the nature of the Monte Carlo approach makes the numerical procedure readily parallelisable. The realizations of revenues for each sample path can be computed independently of each other, and therefore, is in fact perfectly parallelisable.

The purpose of this study is to offer a rational, mathematics-based analysis that aims to provide some insight to software renewal policies. However, the utility and the relevance of the analysis relies on how closely the simulation model can characterize the upgrade schedule and the use patterns of the software under consideration, both of which can be quite diverse and are out of the scope of this study. As well, it is implied in the solution procedure, to which the Bellman's "principle of optimality" is applied, that the uncertainty, that is the demand for the software, is revealed at the time of each renewal decision. This is an ideal assumption, which is rarely true in the real world.

## Acknowledgement

This work was inspired by Friday afternoon discussions with the late Professor Emeritus, Henning Rasmussen of The Department of Applied Mathematics at The University of Western Ontario, who passed away in the summer of 2009. The first author is indebted to Henning for his thoughts, guidance and encouragement to complete this work.

## References

[1] Pfening A and Telek M 1996 Optimal Renewal Policy for Slowly Degrading Systems Working Paper
[2] Zhao T, Qi Y, Shen J, Hou D and Zheng X 2006 *Journal of Communication and Computer* **3** 76–81
[3] Hartman J and Rogers J 2006 *IMA Journal of Management Mathmatics* **17** 143–158
[4] Mehra A and Seidmann A 2008 Optimal Timing of Upgrades over a Software Products Life Cycle Working Paper
[5] Schwartz E S and Trigeorgis L 2001 *Real Option and Investment under Uncertainty* (The MIT Press)
[6] Dixit A K and Pindyck R S 1994 *Investment under Uncertainty* (Princeton University Press)
[7] Anderson C K, Davison M and Rasmussen H 2004 *Naval Research Logistics* **51** 686–703
[8] Wong K C 1993 *The Journal of Real Estate Research* **9** 33–47
[9] Wong K C 1999 *Property Management* **18** 16–24
[10] Rasmussen H 2007 Refurbishment as A Compound Real Option Problem Working Paper
[11] Bellman R 1957 *Dynamic Programming* (Princeton University Press)
[12] Longstaff F A and Schwartz E S 2001 *The Review of Financial Studies* **14** 113–147

# SciNet: Lessons Learned from Building a Power-efficient Top-20 System and Data Centre

**Chris Loken[1], Daniel Gruner[1], Leslie Groer[1], Richard Peltier[1], Neil Bunn[2], Michael Craig[2], Teresa Henriques[1], Jillian Dempsey[1], Ching-Hsing Yu[1], Joseph Chen[1], L. Jonathan Dursi[1], Jason Chong[1], Scott Northrup[1], Jaime Pinto[1], Neil Knecht[1], Ramses van Zon[1]**

[1] SciNet HPC Consortium, University of Toronto, 256 McCaul St, Toronto, ON, M5T 1W5, Canada

[2] IBM Canada Ltd., 3600 Steeles Ave. E, Markham, ON, L3R 9Z7, Canada

E-mail: `info@scinet.utoronto.ca`

**Abstract.** SciNet, one of seven regional HPC consortia operating under the Compute Canada umbrella, runs Canada's first and third fastest computers (as of June 2010) in a state-of-the-art, highly energy-efficient datacentre with a Power Usage Effectiveness (PUE) design-point of 1.16. Power efficiency, computational "bang for the buck" and system capability for a handful of flagship science projects were important criteria in choosing the nature of the computers and the data centre itself. Here we outline some of the lessons learned in putting together the systems and the data centre that hosts Canada's fastest computer to date.

## 1. Introduction — The Current State of Affairs

The key compute systems at SciNet were installed beginning in November 2008 and include two clusters along with a shared 1.5 PB storage system which is visible to all 3,900 compute servers. The 30,000 core General Purpose Cluster (GPC) with a peak theoretical speed of 306 TFLOPS and the 3,300 core Tightly-coupled Capability System (TCS) with a peak theoretical of 60 TFLOPS are all housed in a new green datacentre with a measured average annual PUE of less than 1.2. These two clusters ranked numbers 53 and 16 in the world, respectively, when they first appeared on top500.org (in November 2008 and June 2009).

All systems were fully online and available to the entire Canadian research community as of August 2009. Uptake in terms of system utilization as well as numbers of users and Principal Investigators (PIs) was very quick, as shown in Figure 1. The speed at which the GPC reached 90%+ utilization, particularly since this roughly doubled the number of compute cycles available to Compute Canada researchers nationally, is remarkable. The more specialized TCS system took longer to reach this level of utilization, and because the number of users is smaller, usage is noticeably burstier. A total of more than 200 million CPU-hours were used in the first full year for research projects including cleaner-burning combustion engines, the effect of global warming on the Greenland ice-sheet, regional climate change in Ontario, the role of mantle convection in the formation of supercontinents on Earth, understanding the molecular basis for a substance which appears to eliminate Alzheimer's symptoms in mice, the molecular forces responsible for the remarkable properties of elastin and the body's defence mechanisms, the design of

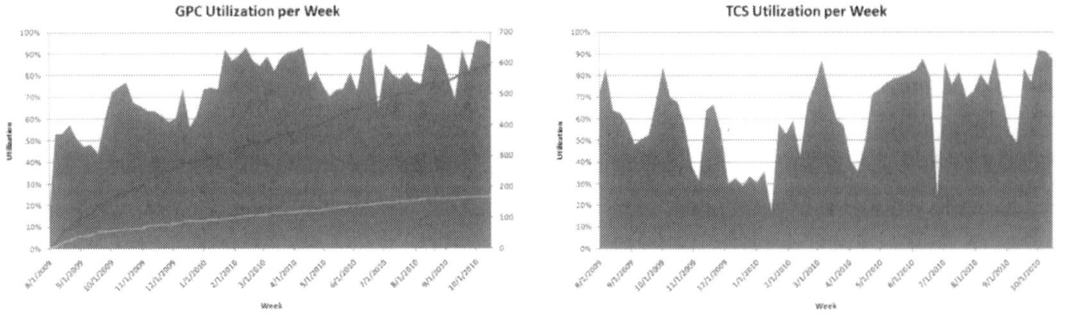


**Figure 1.** Utilization of the General Purpose Cluster (left) shown as shaded area, with number of users and PIs plotted as red and green lines respectively, over the first year of the SciNet systems being publicly available; shown on the right is the utilization data for the TCS.

low-emission aircraft, and the magnetic fields of galaxy clusters. After a year of such work, important SciNet-powered publications in astrophysics [1–17], space science [18], combustion research [19–25], and chemistry [26], are starting to appear in a variety of journals. The ATLAS experiment, that started taking high-energy proton-proton collision data at the LHC in 2009, also has publications that have depended on the analysis facility and crucial simulation data sets that were produced at SciNet[27–30].

SciNet is a consortium of the University of Toronto and its affiliated research hospitals including Baycrest Centre for Geriatric Care, Bloorview MacMillan Children's Centre, Centre for Addiction and Mental Health, Hospital for Sick Children, Mount Sinai Hospital, Ontario Institute for Cancer Research, St. Michael's Hospital, Sunnybrook Health Sciences Centre, University Health Network and Women's College Hospital. Funding comes from the federal government (Canada Foundation for Innovation and the Natural Sciences and Engineering Research Council), provincial government (Ontario Ministry for Research and Innovation and the Ontario Research Fund – Research Excellence) as well as the University of Toronto (the faculties of Arts & Science, Engineering, Medicine and Scarborough).

SciNet is one of the seven regional High Performance Computing (HPC) consortia which make up Compute Canada, an umbrella organization leading the creation of a powerful national HPC platform for research. This national platform integrates HPC resources at seven partner consortia across the country to create a dynamic computational resource. Compute Canada integrates high-performance computers, data resources and tools, and academic research facilities around the country. These integrated resources represent close to a petaFLOPS of computing capability, in addition to substantial online and long term storage with rapid access and retrieval over Canada's national, provincial and territorial high-performance networks.

Working in collaboration, Compute Canada and the university-based regional HPC consortia provide for overall architecture and planning, software integration, operations and management, and coordination of user support for the national HPC platform. As a national organization, Compute Canada coordinates and promotes the use of HPC in Canadian research and works to ensure that Canadian researchers have the computational facilities and expert services necessary to advance scientific knowledge and innovation.

SciNet has a modestly sized support team of 5 system administrators and 4 technical analysts, plus two Chief Technical Officers and a Scientific Director. Nine of the technical staff have science PhDs with a research background in HPC which naturally facilitates interactions with users.

Under the guidance of the CTO, the system administrators are tasked with the challenge of keeping Canada's largest computers running, optimizing system performance, keeping hardware and firmware updated and interacting with vendors to deal with hardware issues.

The technical analyst team, led by the CTO-software, have as their primary responsibility to work with the users to enable their applications to be run at SciNet scales. Another major component of the software team's responsibilities are the tutorials and courses given at SciNet[1]. The analysts also install and maintain software libraries and applications relevant to the users. The size of this group should be constrasted with the computationally very similar HECToR group in the UK which employs 20 full time analysts solely for working with users to develop their codes (*i.e.*, not including "front line support" for compilation issues and the like), or with US laboratories like ORNL where the rule of thumb is "one FTE per three supported projects" (see for instance [31]).

The support team furthermore helps users with a wide range of questions and problems. Apart from login and access problems, these also include questions on how to use SciNet's systems, how to use the systems most efficiently (including how to parallelize and optimize their code), and how to use and/or install libraries. There is no formal ticketing system, but a user can initiate a request for support by emailing a central support list, which all members of the group receive. This informal system remains practical because the group is small and located in the same office, and furthermore allows for a quick response time. The most efficient way of interacting and helping users, however, is through face-to-face meetings, and therefore users are often invited to schedule a meeting with an analyst. Unfortunately, this is not always an option for remote users in which case video links have been successfully used.

## 2. History

The development of SciNet at the University of Toronto stretches back to 1999 with the timing of an award of $7.4M from the Canada Foundation for Innovation (CFI) and the Province of Ontario for the founding of PSciNet, an acronym which then stood for the Physical Sciences computing NETwork. This funding was received in response to an application prepared by a group consisting of astrophysicists from the Canadian Institute for Theoretical Astrophysics, chemical physicists from the Department of Chemistry and planetary physicists from the Department of Physics. The funds were invested to acquire three distinct computer systems, each one designed to serve the special needs of each of these collaborating groups and to be operated by them as separate systems.

A second proposal to CFI for further PSciNet development was funded in 2003 for $11.2M, this time to a group consisting of high energy experimental particle physicists from the Department of Physics, planetary physicists from the Department of Physics and a third group consisting of aerospace and mechanical engineers from the University of Toronto Institute for Aerospace Studies (UTIAS) and from the Department of Mechanical and Industrial Engineering. This funding was employed to acquire two new cluster systems as well as an upgrade to the vector system employed by the planetary physics group.

The SciNet consortium, with the "P" for "Physical" dropped from the acronym, was established in 2005 through the continuing collaboration of all five of these previously involved groups, joined by their colleagues from the areas of computational biology, genomics and bioinformatics at both the University of Toronto and the ten research hospitals affiliated with it. SciNet participated in developing the Compute Canada response to the 2006 National Platform Fund (NPF) call for proposal from CFI and was allocated $15M of the total amount awarded to Compute Canada in December 2006. The CFI award was matched by the Ontario provincial government and supplemented by the University of Toronto in order to provide SciNet with a total capital budget of $32.8M.

[1] Upcoming courses are listed at `https://support.scinet.utoronto.ca/courses/`, and course material is generally posted on the support wiki, `https://support.scinet.utoronto.ca/wiki/index.php/SciNet_User_Support_Library`

Complications in administering the NPF award across 7 consortia and more than 15 institutions delayed the issue of the SciNet Request for Proposals (RFP) until Jan 2008. After reviewing all proposals a final contract with IBM was signed in July 2008 for the construction of the datacentre (in an existing building) and the installation of two clusters and storage. Renovations began in late Aug 2008, the first cluster (the TCS) and storage system were installed in November and opened to friendly users in December, the datacentre was fully completed in February 2009 and the installation of the largest cluster (the GPC) began in March with the arrival of the first IBM iDataPlex servers based on the brand-new Intel Nehalem CPU architecture. Friendly user period for the GPC began in May and both systems were fully opened to researchers from across Canada at the beginning of August 2009.

## 3. Planning and Acquisition

Beginning even before the NPF Call for Proposal, the SciNet Technical Advisory Committee (STAC) actively engaged in planning and preparation for a future HPC system(s) by conducting user surveys, studying existing HPC facilities at University of Toronto and other sites in North America and consulting with industry concerning datacentre design as well as HPC equipment futures and trends. By the time the NPF proposal was submitted in mid-2006, it was clear that local researchers required significant capacity and capability computing and that the wide-range of research needs (from high-resolution global climate modeling to high-throughput data analysis for ATLAS and various Cosmic Microwave Background experiments) would likely require two systems with different compute node characteristics and network architectures. The requirement of some workflows to utilize both compute systems and a desire to use storage space as efficiently as possible drove the consensus towards a common, shared storage system.

During the planning process, it also became clear that the total power, cooling and space requirements being contemplated were far in excess of what was available in existing datacentres on campus. Extensive searches for appropriate space on and off campus and consultations with various co-location facilities led to the conclusion that SciNet would have to build or renovate a new datacentre (though solutions based on shipping containers would be considered). An off-campus, single storey, slab-on-grade "brownfield" site with adequate space and power was identified in 2007 and rented in 2008 ready to be converted into a datacentre.

After much deliberation, the STAC concluded that the SciNet requirements for two compute systems, storage and a datacentre coupled with the constraints of fixed capital and operating budgets gave rise to an optimization problem for system size and datacentre cooling design that was likely best solved by vendors. For example, the capability system was likely to be more expensive per Flop and to generate more heat per Flop than the capacity system though the dollar differentials depended on exact system type and configuration. In addition, possible compute system candidates required different cooling system designs (e.g. air or water-cooled systems) with some of the air-cooled versions requiring significantly greater airflows than normal in standard raised-floor datacentres. It was therefore decided to issue a single RFP requesting four components: datacentre, storage and two compute systems. Vendors were given a total capital budget and, in addition, were required to provide a 5-year operating budget (for power and all hardware and infrastructure maintenance) that fit within the SciNet operating budget. SciNet reserved the right to "cherry-pick" the various components from different responses.

The *de novo* nature of putting together the SciNet datacentre, two compute systems, and storage, allowed vendors a great deal of flexibility; but there were strong restrictions on the resulting solution – there were fixed capital and 5-year operating budgets dominated by power costs, but within that envelope SciNet obviously wanted the best possible systems for its mix of problems which included both throughput computing and tightly coupled, massively parallel jobs. The SciNet RFP, rather uniquely, gave the vendors all four components in the bid and maximized their flexibility in designing and pricing the entire package.

**Table 1.** GPC RFP tentative minimum specifications

| | serial | parallel | shared-memory |
|---|---|---|---|
| Peak theoretical Flops | > 300 Tflops | > 100 Tflops | n.a. |
| RAM | $\geq$ 2 GB per core | $\geq$ 2 GB per core | 128 and 256 GB per node |
| Total RAM | 60 TB | 20 TB | n.a. |
| Largest non-blocking single MPI job | 256 cores | $\geq$ 5,000 cores | 8-16 cores |
| Network connection | $\geq$ 100 MB/s/core | $\geq$ 1 Gb/s/core | 100 MB/s/core |
| Network latency (node to-node) | $< 60\ \mu s$ | $< 5\ \mu s$ | $< 5\ \mu s$ |
| I/O to disk (sequential, large blocks) | $\geq$ 1 MB/s/core | $\geq$ 2 MB/s/core | $\geq$ 2 MB/s/core |
| number of simultaneous I/O streams | 10,000 | 5,000 | 100 |

SciNet presented participating vendors with a challenging problem of optimization. The overall solution was constrained by a single point-in-time funding model which provided both upfront capital and some minimal long-term operational funding. No long term sustainable source of funding could be assumed and thus the problem was one of balancing the capital acquisition costs of hardware (and in essence performance) against the facility implementation cost and the long-term operational costs including both power and personnel. Vendors were presented with the details of the physical space that had been acquired (empty slab space) and local power availability (~4MW) along with a fixed budget for five years of both operating and capital funding.

The SciNet RFP was comprised of the three main computing infrastructure components along with requirements for construction of the data centre. In order to maximize the flexibility and allow for innovation at the vendor level, the barest minimum specifications for the systems were given. Initially a two-phase deployment was planned, but discussions with vendors led to a single phase installation with a moderate budget withheld for future storage upgrades.

The minimum specifications for the GPC, which was to have the largest user base and would support the most varied science projects, were structured to be capable of simultaneously running three key types of applications: serial, MPI and large-shared memory jobs. Serial jobs were expected to use up to 75% of all the available cycles on the GPC. The ATLAS component (discussed further in Section 8) needed to be able to run up to 4,000 serial jobs on an x86 Scientific Linux 4.0 or equivalent platform. Parallel MPI jobs were expected to be able to use up to at least 10,000 cores for a single job and up to 25% of available cycles. Large, shared-memory, serial or OpenMP jobs were expected to use up to 16 cores and 128-256 GB of RAM.

No particular network infrastructure was specified and the possibility of an inhomogeneous cluster was allowed. A Linux-like environment, provisioning on demand and an integrated scheduling system were preferred. The tentative minimum specifications are outlined in Table 1

The TCS was envisaged to serve a more restricted community of users who required a system able to efficiently integrate models in which the coupling between processes was extremely tight. These included coupled climate system simulations as well as other problems in which hydrodynamic processes played a central role. The installed system was expected to be able to deliver at least 20 Tflops if the architecture was of a modern parallel-vector design in which multiple CPU nodes are interconnected at high-bandwidth or a higher peak processing capacity if some other architecture was proposed that could deliver an equivalent capability in terms of work load. Two benchmarks were provided for vendors that were expected to be performed on their proposed solutions with the performance results playing a key role in selecting the winning bid.

Numerous criteria were specified for the storage subsystem but the overarching preferred guideline was to have a high-performance single subsystem accessible by all nodes in the GPC and TCS. Five main storage areas were envisioned: user `/home`, fast parallel `/scratch`, longer term storage on `/project`, a small space for backups and finally dCache space for the ATLAS project. The approximate delivered sizes for these areas in Phase I was expected to be at least 20, 600, 200, 100, 500 TB respectively (total of 1.5 PB) and in Phase II, 50 TB, 1.5 PB, 2 PB, 100 TB and 1.3 PB (total of 5 PB).

As a large amount of storage was being requested, allowance was made in the RFP for lower cost, high-density enterprise SATA-style drive technologies or equivalent to be used, but with a RAID6 or equivalent technology. Given the diverse populations to be served, specifying the exact performance criteria necessary was deemed too difficult and user case scenarios were presented instead give some approximation of the expected access patterns and performance required for various scientific projects. A rule of thumb was that all processing cores in the GPC should be able to simultaneously access the storage system with at least 1 MB/sec read/write.

The SciNet GPC was to provide the ATLAS Tier-2 Analysis Centre and, as such, had certain requirements that were specified in the RFP and that are summarized in Section 8.1. Vendors were asked to describe their technique and hardware and software requirements for implementation of dCache and how to integrate and interface this subsystem with the large SciNet disk storage and the component of the GPC capable of running the Atlas software.

The final component of the storage request was for tape backup solutions with a minimum of four LTO-4 or equivalent tape drives and 200 slots.

To ensure that the consortium received the best possible value for money it was left to vendors to complete this optimization by either working together in partnerships or building a consortium of their own to address the various diverse computational, data storage, networking and facilities requirements. This resulted in several unique and unusual pairings of vendors, but in the end IBM presented the most complete solution that achieved the goals of SciNet.

To complete the optimization of the solution IBM brought on board several partners which included:

- Ellis Don – General Contracting
- WZMH Architects – Architectural
- Hidi Rae Engineering – Mechanical Engineering
- Lapas Engineering – Electrical Engineering
- Modern Niagara – Mechanical Contractor
- Guild Electric – Electrical Contractor
- Johnson Controls – Chiller & Automation

This team completed several iterative design runs and cost projections for various datacentre configurations, efficiencies and long-term power costs to design the optimum physical facility, each time working along with the IBM HPC Engineering team to re-calculate the additional computational capacity that could be added after each round of optimization and cost savings, this in turn typically increased the electrical and mechanical load until finally a balance was reached.

It was clear that presenting the opportunity to optimize the complete facility and solution as a single business case allowed all vendors to provide the maximum possible return on our investment, with all vendors capable of negotiating with their individual partners in a more powerful way than SciNet could have one at a time. This forced close integration also resulted in much more tightly integrated solutions being presented to SciNet, with little wastage, bloat

**Figure 2.** Partial view of the computer room in the SciNet data centre. In front, the GPC iDataPlex racks; at the very back, the TCS and disk.

or unnecessary equipment or elements in any of the proposals. While each vendor approached the problem with a different strategy overall, several unique and compelling solutions were presented, but IBM was selected in part due to a slightly higher level of integration between the components, particularly at the support level.

Throughout the process all vendors were challenged to demonstrate that a very high degree of due diligence had been undertaken in forming their estimates, particularly around estimates of power consumption, efficiency and future power costs. Of these, only future power costs have been left to market forces, with IBM committing to both consumption and overall facility efficiency targets as part of the contractual obligations. The price of power, however, is beyond the scope of what most large IT vendors will accept in terms of operating risk and SciNet is responsible for this portion of funding.

## 4. Resulting System Architecture

The resulting system designed and delivered by IBM consisted of two main compute elements, a single shared storage element, all connected with a primary 10G Ethernet network, and sub GigE and Infiniband networks. The Tightly-coupled Capability System (TCS) consists of 104 IBM p575 nodes each with 16 Dual-Core Power6 CPUs at 4.7 GHz and 128 GB of RAM connected with DDR Infiniband, totaling 3,328 cores running AIX 5.3L as the operating system. The General Purpose Cluster (GPC) consists of 45 racks of 84 IBM iDataplex dx360 M2 nodes each with 2 Quad-Core Intel Xeon E5540 CPUs at 2.53 GHz and 16 GB RAM, totaling 3,780 nodes and 30,240 cores. The entire GPC is connected with Gigabit Ethernet, and 864 nodes (6,912 cores) are also connected via DDR Infiniband. The operating system is Linux CentOS 5.3.

The primary disk storage is provided by two DataDirect Networks (DDN) DCS9000 couplets with 1790 1 TB SATA hard drives, for a total usable space of ≈1.4 PB. The primary file system mounted by all compute nodes is IBM's General Parallel File System (GPFS, [32]) with three primary partitions; `/home` 14 TB, `/project` 365 TB, and `/scratch` 465 TB. The remainder of the disk space is used for the ATLAS project and under dCache. The systems as currently set up are shown in Figure 2.

The job scheduling software used is Adpative Computing's Moab Workload Manager [33] with the open-source TORQUE as its resource manager on the GPC and IBM's LoadLeveler on the

TCS. The Extreme Cluster Administration Tookit known as xCAT [34] is used for provisioning the operating systems of both the TCS and GPC nodes as well as the various infrastructure nodes.

## 5. Data Centre and Power Efficiency

The cost – and time required to build – power infrastructure downtown, combined with the lack of available quality space for a data centre (requiring lots of square footage, high ceilings and flexibility for mechanical systems) drove SciNet to actually place an advertisement in a national newspaper in its search for space. Any plausible rental cost, even going out 10 years in the future, would have been less than the one-off price of bringing a power feed of the size required into downtown Toronto. Thus, the data centre and the main SciNet offices are not co-located.

In early 2008, SciNet rented 12,000 square feet of space in an industrial/business warehouse complex roughly 30km north of the main University of Toronto campus (St. George campus). The current rental agreement is for five years with two options to renew for a further five years. The space is self-contained, has its own double-wide loading dock and two exterior walls with an under-utilized high-voltage power line running immediately along one of the exterior walls and a dedicated, separately metered 600V, 4,000 A power feed. The building construction is single storey, slab-on-grade with 15 feet from slab to beams with a metal slat roof above.

The rented space originally had no interior walls and the landlord was amenable to modifications of the exterior walls and roof which gave great flexibility in the the design of the datacentre and the location of the power and cooling infrastructure. The electrical (270 sq feet)and newly-constructed mechanical (1,900 sq feet) rooms are located along an exterior wall in such a way that the entire chiller assembly (shipped on a 40'x15' skid) had to be moved only a short disatnce directly from the loading bay into its final location. The cooling tower sits on a reinforced section of the roof immediately above the chiller. Two offices provide adequate room (360 sq feet)for staff who are usually on-site only to perform maintenance functions in this lights-out datacentre.

The computer room itself is a 3,100 square foot (290 $m^2$) rectangular room surrounded by corridors on all sides and has an 18" raised floor used only for the water piping, power and network cables. Transfomers (600V to 480V and 600V to 208V) are located on the exterior side of the machine room walls (in the hallways) with the corresponding distribution panels on the interior side of the wall. Three 35-ton air handlers provide air-cooling (on top of the raised floor) to the one row of racks which is not completely water-cooled (disk storage, switches and the Power 6 systems which are 25% air-cooled).

A significant research area being addressed with the SciNet machines is that of climate change and global warming, which is why creating one of the greenest datacentres in the world was of key importance in this project. This concern drove the design towards using water-cooling and a cooling tower. A traditional modern datacenter generally uses at least 33% of the energy going into its centre for cooling and other non-computing power consumption; however, SciNet and IBM have successfully created a centre that uses less than 20% towards these areas.

### *5.1. Water Cooling*

One major design consideration for the facility was the appropriate heat extraction system. As the overall solution was approaching 2MW in power consumption the efficiency and floorspace dedicated to the cooling system was significant. IBM considered several cooling options, including traditional Computer Room Air Handlers (CRAH), compact In-Row Air Handlers, Above Rack Air Handlers, and a variety of coolants including phase-change materials and refrigerant based solutions. The use of water to the equipment was a direct result of the ease of implementation, relative safety of the fluid, and excellent heat transfer properties when correctly maintained.

**Figure 3.** The 735T centrifugal chiller used in the SciNet datacentre, with some of the associated plumbing.

IBM chose to use direct-to-rack water cooling for the vast majority of the equipment in the facility. The TCS was based on the IBM POWER6 p575 and required direct water cooling both to the rack via a rear-door heat-exchanger and to the processors themselves via a direct contact heat plate and appropriate tubing. For the GPC IBM proposed the use of the innovative iDataplex rack form-factor which had the capability to support a very large rear-door heat-exchanger capable of removing up to 30kW per rack when properly configured.

The decision to use water as the primary coolant was cemented by the relative ease by which SciNet was able to implement a "free-cooling" solution to further augment the overall efficiency of the facility as discussed later.

This decision was not without challenges however. The primary chiller for the facility is a 735T centrifugal chiller shown in Figure 3 with variable speed drive and an optimum output temperature between 4°C–7°C. To ensure the chiller could operate most efficiently the highest temperature that could be used would be 7C, which itself was too low to be used directly in a rear-door heat-exchanger (see Fig 4. The IBM rear-door heat-exchanger design is a remarkably simple one, using basic copper tubing and fins to create a very large, passive radiator with no moving parts, and no components other than the inlet and outlet Parker quick-connects. This passive design results in excellent reliability and very low cost, but also raises the potential for condensation should the inlet temperature drop below the facility's dew-point [35].

The same challenge is not present on the POWER6 hardware, which has the ability to self regulate with redundant built-in active heat-exchangers at the bottom of each POWER6 rack. As a result the two major computational elements of the SciNet complex are fed by two different water loops – the POWER6 equipment is directly fed the 7°C water from the chiller, and the iDataplex equipment is fed from a secondary loop moderated via a large heat-exchanger in the mechanical room to a stable 12-13°C. Further control of the dew-point in the room is completed by using active humidity management in one of the three 50T air handlers that provide air cooling to the small amount of equipment that was incompatible or impractical with rear-door heat-exchangers (45U non-standard storage rack, network switching with large amounts of cabling, etc.).

The result of the dual-loop system, with a very small amount of humidity control is that all heat from the 30,000+ Intel Cores in the GPC, and about 80% of the heat from the 3,300 POWER6 cores is directly transferred into the facility water supply and later returned to operating temperature via the cooling towers and 735T chiller. The flow-rate of the water through the system is carefully controlled to optimize the heat-removal against the operating

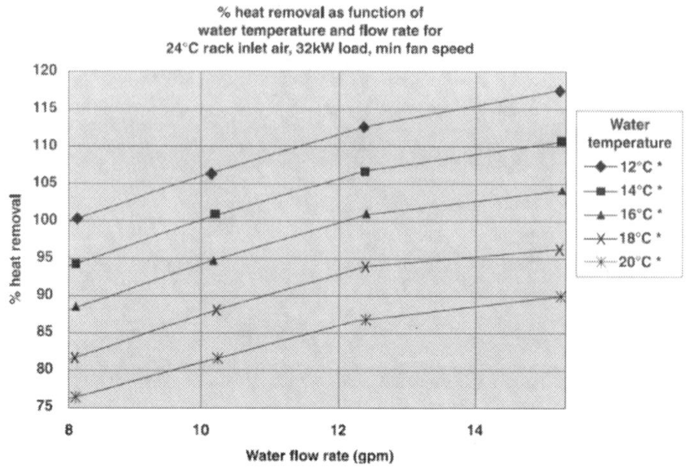


**Figure 4.** Typical performance of the heat exchanger, 32 kW heat load; from [35]

**Figure 5.** Cooling pipes being laid out below what will be the machine room floor during the construction of the data centre.

temperature of the equipment, with known capabilities for the rear-door heat-exchangers.

All cooling infrastructure was located below the floor, allowing plenty of space for the large schedule 40 piping required to handle the water flow as shown in Figure 5.

It is this aggressive use of water cooling that permits such a high efficiency for the overall data centre. Without the water cooling, the air temperature in the datacentre can reach 50°C in 15 minutes; thus any failure (or maintenance) of the cooling requires a rapid shutdown of the entire data centre.

### *5.2. Free Cooling*

One of the major design elements of the solution was the overall solution efficiency. To attain a target PUE of 1.16 it is necessary to exercise all available natural benefits. The SciNet facility is located at approximately 43°50N latitude and lends itself to the potential for significant periods of time with temperatures capable of supporting a variety of "free" cooling methods. During the design of the facility the potential for augmented natural cooling or any form of economization was quickly identified with several options evaluated including:

- Air Economizers (Air exchange cooling)
- Water Side Economizers
- Heat Transfer to Local Business (operational cost offset)
- Phase-change Solutions (Stored Cooling)

To determine the viability of free cooling solutions the first analysis was to determine the climate patterns in the vicinity of the facility. While part of the SciNet research group includes significant expertise in climate science it was extremely easy to obtain local temperature profiles through general HVAC suppliers. Local temperature data is available at a fairly granular level in most of North America from ASHRAE certified HVAC suppliers as it is commonly used in the "Bin Method" for calculating local heating and cooling costs, where hours spent in temperature bins over the course of a year allow you to estimate the amount of HVAC required; tabular data for Toronto is given in Table A1 in the Appendix.

This data helps guide the possible use of ambient cooling, in this case the data shows that of the 8760 total yearly hours, on average 2811 hours are spent at or below 0°C/32°F and as many as 5529 hours are spent at or below the maximum water temperature of 12C/53°F. Put differently this indicates that it is reasonable to expect that approximately 32% of the year is spent below 0°C/32°F and upwards of 63% of the year is spent with external temperatures near our highest water feed temperature. The opportunity to benefit from the use of this ambient temperature profile is enormous.

As the decision to utilize water cooling to the racks had already been made, in part for density reasons, the use of air-side economizers and external air-exchange systems was deemed too costly, and complicated for this environment. For many environments this would make perfect sense and is actively being researched and developed by many vendors including IBM and Intel Corp. One particular area of continuing research is the long-term effect of air impurities, be they chemical or particulate in nature and whether this has any impact on the lifespan or reliability of the equipment. In essence a better understanding of the effect of these impurities would help guide the implementation cost of these solutions as extensive filtering would increase the cost. We did not pursue this option.

The extensive use of water suggested we primarily focus on water-side economizers, heat-transfer or sale to local business (as a cost mitigation strategy) or phase-change solutions. As the facility is in an industrialized neighbourhood a search was initiated to determine the viability of heat-sale. Unfortunately distance issues and the relatively low $\Delta T$ between inlet and outlet temperatures also eliminated this option. Evaluation of Phase-change solutions was also completed with IBM's expertise resulting from the implementation of a large phase-change solution at IBM's manufacturing facility in Bromont, Quebec [36]. However, while this design holds significant potential for the type of variable commercial workloads, the magnitude and consistency in the cooling requirements at a large HPC facility would have required significant phase-change infrastructure and a longer term ROI horizon than the 3-5 years necessary for this type of facility.

Upon evaluation of all other options it was clear that water-side economization was the most appropriate path forward. Put simply, a water-side economizer uses just the pre-existing pump hardware of the mechanical chiller solution to ensure proper fluid flow, but bypasses the chiller itself, relying instead on just the external indirect cooling tower for heat removal. This bypass can be configured in a variety of modes, variably adjusted to provide partial, or full free-cooling depending on the external ambient temperature. While it is possible to begin to leverage this ambient cooling any time the external temperature is at or below the 7°C temperature of our primary cooling loop the mechanical bypass systems inherently increase the complexity of the solution and introduce additional potential failure modes. The potential for issues with the transition from total mechanical (chiller) cooling to partial and then full free-cooling increase proportionally to the number of transition events. As a result in practical application use of the water-side economizer bypass tends to be limited to sections of the year where average temperatures approach 3°C or below on a routine basis, and after transition to full free-cooling minimal cycling back to mechanical cooling is experienced. This decision has worked well, providing an excellent balance between efficiency and the risk of failure associated with the many automated mechanical valve changes that must occur to completely bypass the main chiller.

### *5.3. Power Metering*

In order to assess and monitor cooling efficiency in a datacentre it is critical to meter and record as much data as possible. A total of 22 circuits are metered in the SciNet datacentre in order to separately monitor the power used by the water pumps, chiller, cooling tower, lights, UPS, the TCS (in groups of 2 racks) and the GPC (in groups of 3 and 4 racks). This granularity is fine enough to measure the PUE.

A sample weekly plot is shown in Figure 6; during this period one can see that the GPC typically draws ∼750kW, the TCS draws 400kW and the PUe varies between 1.12 (corresponding to a roughly 6 hr period overnight Thurs-Fri during which the outside wet-bulb temperature was low enough for the system to use free instead of mechanical cooling) and 1.25. Data over an entire year of operation is currently being analyzed and appears consistent with an average PUE less than 1.2.

The SciNet datacentre qualifies for hourly spot-market electrical rates as a commercial customer using more then 50kW but less than 4MW of power. Rates are set by the Independent Electrical System Operator (www.ieso.ca) and the average weighted price YTD (Oct 2010) is 4.0 cents/kw-hr. Unfortunately, this basic consumption charge averages out to just 45% of the actual power bill which includes various charges for debt retirement, transmission and, most significantly, to account for the difference between the spot market price and the rates paid to regulated and contracted generators. Total datacentre power consumption for the first full year of operations (1 Aug 2009 to 1 Aug 2010) was 11.8M kW-hrs.

## 6. Network Overview and Design

The SciNet computing facility is comprised of multiple networks that are integral to the overall functions of the cluster, storage and data centre itself.

The management functions of the clusters and storage as well as infrastructure control and monitoring elements of the data center are handled over multiple subnets. These are mostly Gigabit Ethernet networks that run over Blade Networking Technologies (BNT) G8000 class switches and converge in a Force10 C300 Director class switch.

A high-performance low-latency network connects nodes of the TCS POWER6 based p575 cluster together. The network is a quad-plane DDR Infiniband network utilizing QLogic 9240 Director class switches. The network is configured for full bisectional bandwidth and currently provides an 80 Gbps connection to each node but has the capability of being upgraded to 160 Gbps per node. This network serves as both the message passing interface for parallel codes as well as providing data access to the GPFS filesystem for the TCS Nodes.

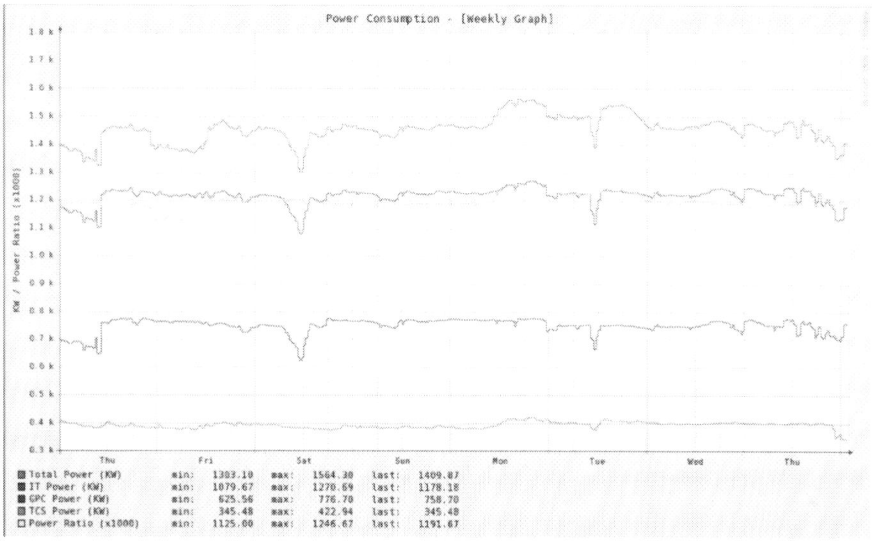


**Figure 6.** Power consumption (in kW) over a one week period by, from top to bottom; entire datacentre, IT equipment, GPC and TCS. The light green line is the PUE (ratio of total power to IT power) which varies from 1.125 to 1.25 over this period

Analysis of all the potential user codes that would run on the GPC at SciNet showed that a large portion of the expected workload would be actually be suitable to a lower bandwidth, higher latency network. However, there was still a requirement for some codes to have access to a high-bandwidth low-latency network. Based on this analysis it was decided that the optimal network design for the GPC would actually be a hybrid of two networks. As costs of high-bandwidth low-latency networks can approach up to 30% of the overall cost of a cluster, this tailoring of the networks to meet the workload requirements allowed SciNet to procure a much larger cluster than if it was decided to build a uniform high-bandwidth network to all the nodes. The network design for the GPC connected all of the nodes to a Gigabit Ethernet Network and almost 25% of the nodes to a full-bisection DDR Infiniband network configured in a CLOS Fat-Tree topology. This 2-tier DDR Infiniband network utilized Mellanox Connect-X Host Channel Adapters (HCA) connected to QLogic 9024 Leaf Switches which then uplinked to QLogic 9240 Director Class Core Switches.

By far the most interesting network in the SciNet facility is the Gigabit Ethernet network that connects almost four-thousand devices, including all of the compute nodes in both the GPC and TCS clusters. While each node is only connected by a Gigabit connection, the sheer number of devices leads to challenges in designing a network that can sustain a reasonable throughput for all of these nodes concurrently. Aggregation can be accomplished by utilizing 10-Gigabit Ethernet between switch layers but very few director class Ethernet switches can handle full bandwidth at this scale. At the time of the acquisition of the GPC, the Myricom Myri-10G line of switches were the only 10-Gigabit Ethernet switches on the market to support full-bisectional bandwidth across all of their ports. However, this extreme level of performance carried with it some trade-offs. The most important of which was that they only supported flat layer 2 networks. No higher level OSI function such as routing or even VLAN support was present. This limitation is what fundamentally led to the SciNet GPC compute network being constructed as potentially the largest flat layer 2 network in the world with approximately eight-thousand MACs (four-thousand nodes, each with a host and IMM MAC). The challenges and lessons learned from implementing, maintaining and supporting this network have provided deep insight into scalability of the Ethernet protocol at its limits.

The GPC Gigabit Ethernet compute network is configured as a 2-tier network with forty-two nodes connected by Gigabit Ethernet to a BNT G8000 Series switch. The G8000 is then uplinked to a core Myri-10G switch over a single fibre optic 10-Gigabit link. This is configuration is repeated ninety-two times for a total of 3,864 GPC compute nodes. Additional infrastructure nodes and all 104 TCS nodes are also connected to the GPC Network in this same fashion. The TCS nodes were added to this network in order to extend the same GPFS filesystems across both the GPC and TCS clusters. A common network is required for this in order to support token passing, locking and other miscellaneous functions. The GPC network is not a full-bisection network - blocking of approximately 4:1 is encountered at the BNT G8000 edge switches as the Gigabit links are aggregated into a single 10-Gigabit Uplink. Even with this moderate level of blocking the GPC compute network provides a massive amount of bisectional bandwidth for an Ethernet network. The design of this network and it's high throughput and modest blocking factor was a significant contributing factor in the achievement of this cluster's 16th place ranking in the June 2009 Top500 list and it's distinction as the only Gigabit Ethernet cluster in the top 80 list entries.

### *6.1. Static ARP Tables*

One of the considerations in building a large flat Ethernet network is handling broadcast traffic. On smaller networks this is usually not a concern as broadcasts are typically a relatively small percentage of the overall traffic. However, as the number of endpoints in a network grows, the amount of time a Network Interface Controller (NIC) spends processing broadcast traffic from all the other endpoints grows linearly. As any given NIC has a fixed amount of bandwidth and processing capability the amount of time spent processing broadcasts grows as an N problem. To compound this issue, the internal architecture of the Myri-10G switch which provides the high-bandwidth full-bisection Ethernet network, does not perform as well with excessive broadcast traffic.

On the SciNet GPC network the most significant contributor to broadcast traffic was the Address Resolution Protocol (ARP). This protocol resolves IP addresses to MAC addresses for unicast communication in a local subnet. It was found that as the the number of endpoints on the network crested beyond 1500 nodes or 3000 MACs, ARP traffic started to consume a significant amount of each NIC's processing resources and led to issues such as flow-control lock-ups and the NC-SI hangs that are described later in this document. In order to reduce the overall amount of broadcast traffic on the network, it was decided that we would implement and maintain a Static ARP table that contained mappings for every node in the network. While this is not practical for many networks, it is perfectly feasible for a tightly controlled internal network where endpoint devices and not leaving and joining on a constant basis. With the Static ARP table inserted into the ARP cache of every node as it booted the broadcast traffic on the network due to the ARP protocol was essentially reduced to zero. In order to handle the general housekeeping of node and NIC hardware failures and replacements, a few small program were written in order to maintain the static ARP table and ARP caches across all the nodes.

- `./replacenode` - A script written to clean out the xCAT database and Static ARP table of a particular node that was being removed from the cluster. This is mostly used for hardware service replacements that would effect the MAC of the node (i.e. System board and/or NIC).
- `./updatearp` A script that updates the static ARP table and dynamically updates the ARP Caches of all running nodes with the ARP entry of a newly discovered node. This is used after HW replacement of a system board or NIC.

The implementation of the Static ARP table was probably the single most important configuration change in stabilizing the GPC network for production use. It is possible that in the future, with some of the planned upgrades, the static ARP system may no longer be a requirement. However, currently it is still in full production use.

### *6.2. Other Network Considerations, STP and Flow-Control*

Another lesson were learned during the deployment of the GPC network was that a slight misconfiguration in one of the standard Ethernet protocols, that may not affect a smaller network, could have a dramatic effect at this very large scale.

By default all switches in the SciNet GPC network were enabled with Spanning Tree Protocol (STP). This protocol searches for and eliminates 'loops' in the network. We found that even with RSTP (rapid spanning tree protocol) the amount of time it took the network to converge could cause interruptions to jobs and filesystem I/O in the network. This problem was exasperated by a misconfiguration on a few ports that had nodes attached but were configured as switch to switch links (i.e. not 'edge' ports) This had the effect that anytime this link was reset for any reason it forced the entire network to run STP and re-converge. One of these ports happened to be connected to a p520 GPFS NSD Server which was having issues with momentary link resets (described in section 7.3). Since this port was not configured as an edge port, the network would end up in regular STP loops trying to re-converge before the next link reset. Once these configuration issues were found and corrected the instances of STP having to re-converge the network were significantly reduced. However, in the end it was decided to disable STP completely across the GPC network. The reason for this decision was that the stalls in network traffic caused by STP outweighed its potential benefit of finding and correcting a loop in the network. As the physical topology of the GPC network was well understood and is of a relatively 'simple' repeating design, the likelyhood of introducing a network loop was extremely low. Local security at the facility also makes it unlikely that anyone will connect a rogue switch into the network and create a loop.

Another challenge encountered scaling up the GPC network for production use was the implementation and control of the 802.3x flow-control protocol. Flow-control is an extremely valuable low-level protocol that allows network devices to limit in-bound traffic rates by issuing pause frames to stop from becoming overloaded. This is much less costly than a network device dropping packets and relying on higher levels of the network stack, like TCP/IP, to request re-transmission of the lost data. The potential pitfalls of flow-control is that rogue or misbehaving devices have the potential to lockup the entire network by continuously sending pause frames. On small networks this is unlikely. However, at the scale of the GPC network, with its massive amounts of aggregate throughput and sheer number of endpoints it becomes statistically more likely that a network device can become overloaded or even enter an unstable state. This was seen on multiple occasions during the implementation of the GPC network. The first observed occurrence of this was when a compute node encountered a rare hang condition in the Network Controller Sideband Interface (NC-SI) between its host NIC and IMM baseboard management processor (see Figure 7. When this NC-SI channel locked up and the host NIC could no longer pass packets up to the IMM processor, the buffers on the host NIC filled and then it started flooding the network with pause frames essentially locking up the entire network.

After an in-depth investigation it was found that excessive broadcast traffic was leading to the hang condition of the NC-SI channel. This discovery was one of the original motivators for the development and implementation of the Static ARP table in the GPC cluster. IBM later released updated uEFI and IMM firmware revisions that stabilized the NC-SI channel and made the likely hood of these types of lockups much less likely.

As the potential for excessive flow-control pause frames from rouge or misbehaving devices always exists on the GPC Network additional steps were taken to ensure that a single device

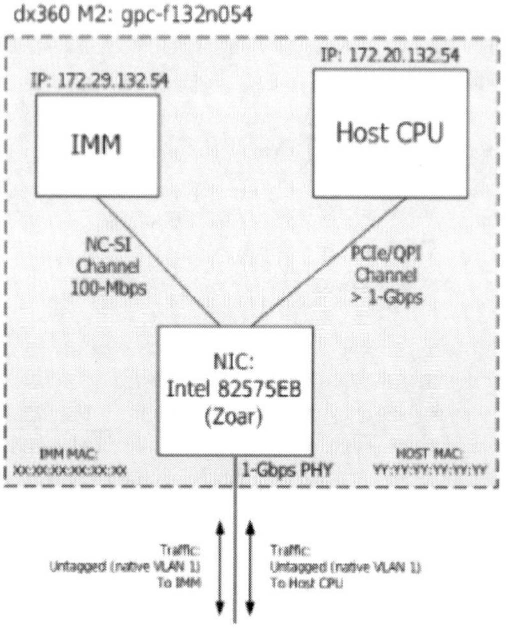


**Figure 7.** Schematic showing the Network Controller Sideband Interface (NC-SI) between host NIC and IMM baseboard management processor.

could never lockup the entire cluster indefinitely. In working with the development teams from Myricom an additional function was added to the Myri-10G switch which allows it to monitor the percentage of pause frames that it receives on any given port and disable flow-control on that port if it exceeds a configurable limit. With flow-control disabled the port would start dropping packets as opposed to forwarding pause frames. This would allow the rest of the network return to normal operation while only devices on the given port would be effectively blocked out until the flow-control situation was resolved, at which point the switch would re-enable flow-control on the port and return it to the network. This configuration parameter, now available on all Myri-10G switches, is:

```
PausedPercentLimit = 63
```

where the setting is in one-hundredths of a second and represented in hexadecimal; 0x63 is 0.99 or roughly 1 second.

### *6.3. 802.3ad, Myricom, and System P*

In order to increase bandwidth through to the GPFS filesystem from the GPC network, multiple 10-Gigabit Ethernet links from each GPFS NSD Server were 'bonded' together using the 802.3ad Link Aggregation Control Protocol (LACP) and connected directly to the Myri-10G switch, as opposed to going through a BNT distribution layer. Early performance testing showed that downstream traffic from the GPFS Servers to the compute nodes in the GPC network scaled with the multiple links. However, the opposite was not true. Upstream traffic from the compute nodes to the GPFS Servers never exceeded the capabilities of a single 10-Gigabit link. After much investigation it was discovered that the MAC addresses of all of the compute nodes were even in the 6th (last) hexadecimal octet. This was the result of the MAC burn-in process used by Intel for their dual-port 82575EB controller which was the on-board NIC of every compute node.

The first five octets of both ports were identical and the last octet always varied by one with the first NIC port always being even and the second NIC port always being odd. Since all compute nodes were connected to the first NIC port this meant that all the MAC addresses were even. The Myricom protocol for LACP balancing did a simple modulo 2 on the source MAC, and since the result was always 0, the traffic was always balanced to a single 10-Gigabit port on the GPFS Servers. Once this was identified, Myricom modified their LACP balancing algorithm to hash the entire MAC address before determining the modulo 2 result. This resulted in good balance of upstream traffic and increased bandwidth to the GPFS Servers. This firmware updated again made it into the production stream and now is part of every Myri-10G switch sold.

By far the most unique issue identified during the deployment of the GPC network was a small but troublesome link reset that occurred between the 10-Gigabit Ethernet NICs on the IBM System p520 GPFS NSD Servers and the Myricom Myri-10G switch. At first, with the STP Edge port misconfiguration, these link resets would cause STP to constantly re-converge the network. However, once the STP configuration was corrected and then ultimately disabled. The link resets became an oddity that didn't seem to have a major impact on the overall network deployment. It wasn't until the GPFS IO testing ramped up that it was determined that the link resets were having a negative effect on the overall performance of the filesystem. Engineers from both IBM and Myricom engaged to identify and resolve the issue. Testing was done with the p520 10-Gigabit adapters and other vendor's switches as well as other 10-Gigabit Ethernet adapters and the Myricom switch. Issues were only ever seen with the combination of the p520 10-Gigabit adapter and the Myri-10G switch. After using an inline network sniffer/analyzer on a similar setup at the Myricom labs it was found that the p520 10-Gigabit adapter was releasing a packet that should have been filtered out and that the Myri-10G switch was miss-interpreting this packet as a link error and cycling the port. Firmware was developed and provided for both the Myri-10G switch to correct the packet miss-identification as well as for the p520 10-Gigabit adapter to filter out the packet before it ever hit the wire. Applying either of these updates would resolve the link resets but both were applied to the SciNet system and both fixes are now part of IBM's and Myricom's standard firmware releases.

### *6.4. Future of the GPC Network: Projects and Investigations*

Two projects are currently under investigation for the future of the GPC Ethernet Network. The first project is looking at doubling the number of 10-Gigabit uplinks from the ninety-two BNT G8000 switches. This would effectively half the overall blocking factor of the network and bring it close to 2:1. Enough ports are available on the core Myri-10G switch to support this upgrade and the BNT can support up to four 10-Gigabit uplinks. It is planned that smaller scale testing will be completed on a portion of the GPC cluster to determine what effects this increased bandwidth will have on a variety of workloads. As the BNT G8000 can support the additional 10-Gigabit uplinks in a variety of configurations it is expected that several of these will be explored during the small scale tests:

Two 10-Gigabit uplinks from the same dual-port XGE uplink module, which connects to a single Broadcom ASIC inside the switch. Uplinks will be configured in an 802.3ad LACP bond to the Myri-10G

Two 10-Gigabit uplinks, each from distinct dual-port XGE uplink modules, which connect to separate Broadcom ASICs inside the switch. Uplinks will be configured in an 802.3ad LACP bond to the Myri-10G.

Two 10-Gigabit uplinks, each from distinct dual-port XGE uplink modules, which connect to separate Broadcom ASICs inside the switch. The G8000 will be configured with an internal VLAN, so that 1-Gigabit ports on each Broadcom ASIC will route to the respective uplink module on that ASIC. This configuration essentially carves the switch into two smaller switches, each with their own dedicated ASIC and uplink module.

The second GPC Network project currently being investigated at SciNet is the potential re-configuration of the IMM baseboard management controllers in the compute nodes to support VLAN tagging. Currently the GPC Network supports close to 8000 MACs. Approximately half of those MACs are the IMM modules in the compute nodes. The addition of a VLAN tag to traffic originating, or destined to, the IMM would provide the opportunity to segregate and divert this traffic at the level of the BNT G8000 switches before it reaches the core Myri-10G switch. This would reduce the number of MACs that the Myricom switch would have to support from 8000 to 4000. A schematic of the proposed change is shown in Figure 8.

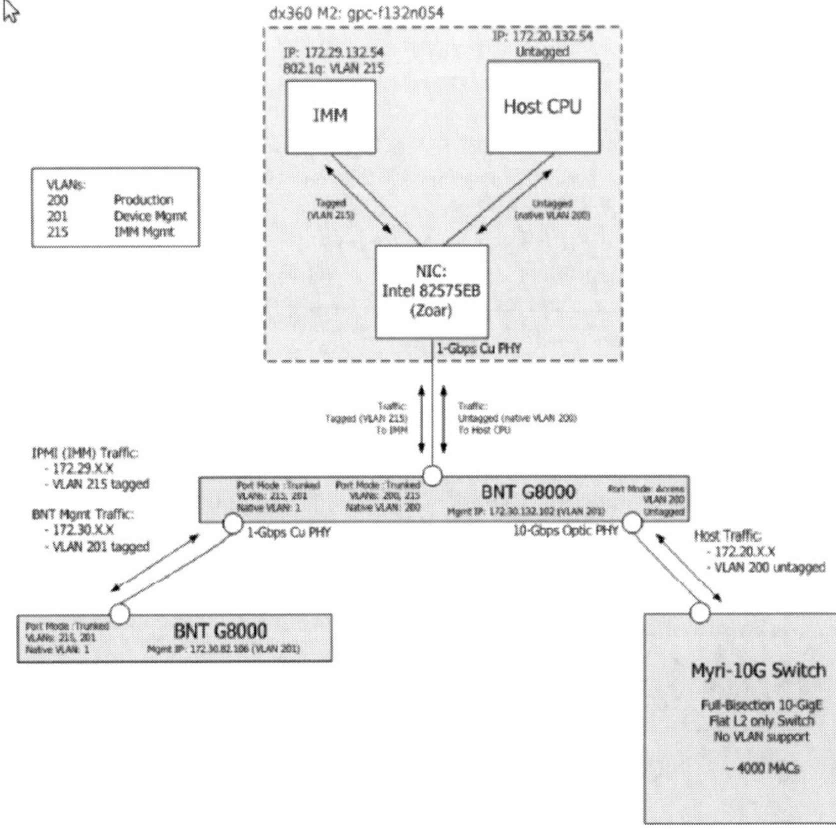


**Figure 8.** Schematic showing proposed VLAN tagging to segregate and divert IMM traffic at the level of the BNT G8000 switches before it reaches the core Myri-10G switch.

This reconfiguration would require minimal additional hardware to support the new IMM VLANs and could have an impact on further reducing occurrences of broadcast storms and excessive flow-control on the Myri-10G. This would ultimately benefit both the end user and filesystem traffic for the GPC Ethernet Network.

There are currently some minor technical challenges with the Linux device drivers for both Intel and Broadcom NICs in this environment. Default drivers bundled with current Enterprise Linux Distributions filter VLAN tags once they are loaded. This effectively cuts off the out-of-band communication with the IMM once the Operating System is loaded. In lab testing it has been shown that the current versions of the Broadcom drivers do not exhibit this same behaviour, and the Intel driver can be set to bypass its VLAN tag filter in one of three ways. These OS considerations are not expected to cause any setbacks should SciNet decide to move forward with this project.

At this point in the project, plans have been constructed to deploy this new architecture to the SciNet GPC Network and integrate it with the current xCAT management and infrastructure environment, as shown in Figure 9.

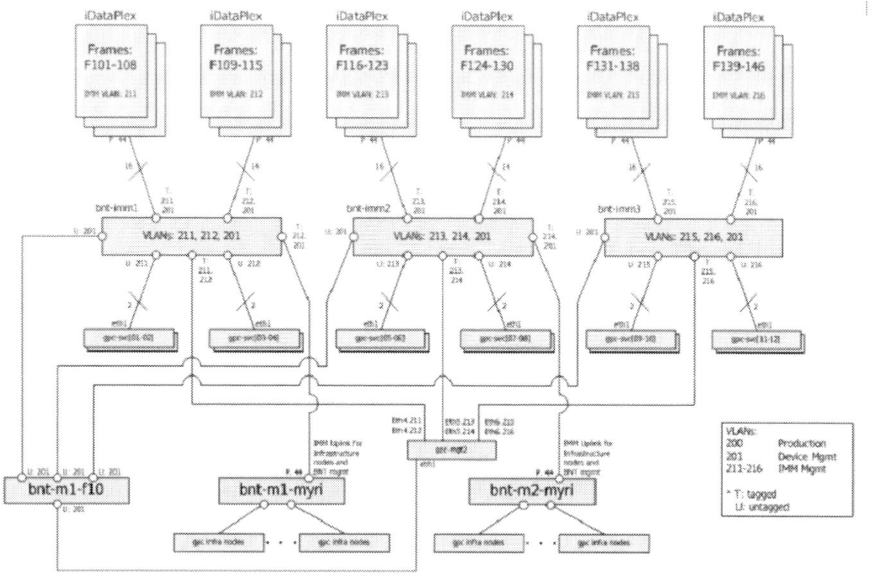


**Figure 9.** Schematic showing proposed VLAN tagging to segregate and divert IMM traffic at the level of the BNT G8000 switches before it reaches the core Myri-10G switch.

## 7. File System and Storage Design

### *7.1. Considerations for a Stateless, Diskless environment*

The enormous scale of the computing resources at the SciNet facility demanded that alternative methods be considered for the deployment and management of software in the clusters. With just under four-thousand independent computing devices in the facility, a traditional deployment methodology of installing an operating system software stack to a hard drive on each individual node would present several administrative challenges. One of the most common and problematic issues for clusters of this size is software drift. This is the variability, over time, of the software and configuration files contained within each node. This most often occurs as a result of "silent" failures in bulk push updates of software patches and/or configuration files which are not accepted by just a few nodes out of the entire cluster. As these kind of updates are repeated over time the software variability across the cluster increases. This can eventually lead to symptoms such as job failures or "jitter" in the cluster.

Stateless computing resolves many of the administrative challenges of extremely large scale clusters by removing the stateful implementation of the operating system software from the hard drive on each compute node. Instead a common operating system image is developed on a management node and is sent out to many nodes as they boot from the network. The image is expanded and run from RAM in the node and the "personality" of each node can be determined by hardware unique identifiers such as the MAC address. Stateless computing can resolve challenges such as software drift and can also be a valuable tool to test and deploy software patches or even different customized operating system images. It is important to note that stateless computing does not automatically imply that nodes are diskless. Hard drives can still be installed in nodes and used for other purposes, such as swap or scratch space. However, in extremely large scale deployments such as the GPC system at SciNet, Mean Time Between Failure (MTBF) and overall power consumption of the system can become a concern. One of the devices most prone to failure in a compute node is the hard disk, and the more drives you have the more likely you are to have a failure. As you approach four-thousand hard drives in a clustered system the MTBF can be measured in hours or days. In addition, four-thousand hard drives consume a considerable amount of power, their elimination from the cluster nodes contributed considerably to reducing the overall power draw of the cluster.

Implementing the stateless computing model and eliminating the local hard drives from the cluster compute nodes has not been without its challenges. The first challenge encountered was to deal with software components that require nodes to maintain state across reboots. One such package is GPFS, which requires that each node maintain an up to date copy of the GPFS cluster configuration files. It is not possible to include these files into the stateless operating system image as they are always changing. Instead a server was configured in the environment to provide the up to date GPFS configuration files to any node that requested them in real time. The stateless compute nodes were then configured to query this server for these files on boot before starting the GPFS daemons. This allowed the stateless node to have an up to date copy of the GPFS configuration files and successfully join the GPFS cluster on boot. It should be noted that other technologies are now available to handle stateful software requirements in a stateless environment. The xCAT Management tool that provides the stateless framework at SciNet now supports a new paradigm called "Statelite". This operating model combines all the benefits of stateless while allowing administrators to configure specific stateful files in an operating system image. Each node boots from the common stateless image but refers to its own copy of the specified stateful files. This implementation is being investigated by SciNet but is not currently in use.

Another challenge specifically related to the absence of hard drives in the GPC nodes is the handling of memory allocations without the benefit of a swap space. While it is fairly obvious that running high performance computing codes outside of the bounds of a node's physical memory capabilities is not very "high performance", not having the benefit of this buffer space changes a code from running slowly to outright failing. Interesting effects were seen on the GPC as user codes attempted to malloc() more memory than was physically available on the node. The Linux Out-of-Memory Killer (OOM) would hunt down processes to terminate in order to free memory for the requesting user process. Ironically, the process typically targeted was the GPFS daemon, which consumes moderate space as it allocates the GPFS pagepool (i.e. Filesystem cache) space in with the daemon. Killing GPFS would potentially allow the user application to complete its malloc() but it would be left without a filesystem to work with and would typically fail because of this. This issue has been somewhat mitigated by adjusting the OOM killer configuration to not target the GPFS daemon process. However, this is not a perfect solution as there will always be user codes that attempt to allocate more memory than is physically available on the system.

### *7.2. GPFS and "Green" Provisioning*

The stateless capabilities of the xCAT Management tool when combined with Adaptive Computing's Moab Intelligent Cluster Scheduler allow the SciNet GPC to operate in a dynamic provisioning "green" mode. In this mode of operation, the Moab scheduler is able to instruct xCAT to provision certain nodes with specific operating system images based on either policies or criteria within the job. If nodes complete a job and remain idle with no new workload to backfill them, then Moab can instruct xCAT to power-off these nodes in order to reduce the overall power consumption of the cluster. If new workload arrives in the Moab queue and all powered on nodes are busy then it can instruct xCAT to turn-on nodes that are powered off and provision them with the correct Operating System for the workload in the queue.

Several challenges were encountered in operating the dynamically provisioned environment at the large scale of the GPC System. Some of these issues were related to the Ethernet network and others were related to how certain software reacted to the dynamic environment at this scale.

Both computing clusters in the SciNet facility have common access through an extremely large Ethernet network. With approximately eight-thousand MAC addresses it is potentially one of the largest flat layer 2 Ethernet networks in the world. There were many considerations and technical challenges in building this network that are discussed in Section 6. As it related to dynamically provisioning the cluster, uncertainties in some network operations such as DHCP, TFTP, etc, could potentially lead to nodes not completing their boot and/or provision after being instructed to do so by Moab through xCAT. The Moab Service Manager (MSM), which is the interface between Moab and xCAT, uses a polling mechanism to query the state of nodes through xCAT. If a node did not completely boot then Moab would wait for the polling interval for provisioning to complete and then when it detected that not all nodes completed the provisioning it would start another cycle. This could lead to extremely long lead times before jobs began actual execution on the cluster and even could cause thrashing due to the occassional node boot failures when sufficiently large numbers of nodes were being provisioned.

Moab's MSM also instructed xCAT to probe all the Integrated Management Modules (IMM) on the compute nodes every 5 minutes. It used this information for multiple purposes including the "amber light avoidance" feature which alerted Moab to potential hardware issues on specific nodes so that they could be avoided for job scheduling. Due to the nature of the IMM and the network this IMM probing would lead to broadcast storms that would potentially disrupt all network traffic. The challenge of broadcast storms on the network has been one of intense focus throughout the bring-up and ongoing operation of the cluster. While attempting to avoid or minimize broadcast storms, the MSM polling intervals were gradually increased from 5 minutes to 15 minutes, and then 30 minutes. Finally MSM polling was disabled when the "green" mode was turned off. This has resulted in a much more stable network.

The final challenge with the dynamic provisioning environment was encountered in the shutdown procedure for the nodes. Once idle nodes had been inactive for a certain time period Moab, through MSM, would instruct xCAT to turn off these nodes through the "rpower" command which would be synonymous to pulling the power cord on a PC. Since the software stacks on these nodes were stateless it was originally thought that this would have no adverse effects. What was found through observation though was that the GPFS filesystem would need to run through a recovery process if a node in the cluster suddenly disappeared without properly exiting the cluster. This pseudo log rollback is done to ensure consistency of the filesystem. If many nodes were powered off within a close time span then this could potentially lock-up the GPFS cluster for other users while the recovery operations took place. This could cause jobs on other nodes to fail as they exceeded their own timeouts for I/O completion. While it was never fully tested, since "green" mode was disabled, modifying the Moab/MSM/xCAT interface to instruct a node to properly exit the GPFS cluster before executing the rpower command should provide significantly more stable operation.

As described above, there have been many challenges with the dynamic provisioning at this ultra scale. The increase in utilization of the SciNet computing systems has been so rapid that there has actually been very little opportunity for power saving gains from "green" mode operation, as almost all nodes are busy all the time. That being said, several projects are currently under investigation that should further improve the performance of the Ethernet Network, and this in conjunction with further optimizations of the Moab/MSM/xCAT interface should allow the "green" mode to be re-enabled in the future.

### *7.3. Managing GPFS*

GPFS is a high performance clustered file system, and its managers (both for token and for file system traffic) play a critical role in the overall performance and stability of the SciNet cluster. Upon installation of the TCS (104 compute nodes), SciNet's eight NSD servers were used as the GPFS managers. This setup worked adequately without issues. However, file system hangs and inaccessibility started to occur after the addition of the GPC, which resulted in a 40-fold increase in the size of the GPFS cluster. It was only then that the GPFS support team identified that NSD servers cannot simultaneously function as filesystem managers for a large cluster. The first attempt to remedy this situation was to use KVM virtual machines on x3550's as the GPFS managers, which do not have very stringent hardware requirements. This approach, however, failed completely. In the end, the GPFS managers were moved to dedicated machines, which greatly improved performance. SciNet is currently using eight GPFS managers attached to two management switches.

There are many factors that affect filesystem performance. One of the most important is network congestion. GPFS relies entirely on the underlying network for all I/O and token communication. As described above, the network of the SciNet machines is implemented as one big flat network, a common practice for a GPFS cluster. However, due to the size of the cluster, which was at the time the largest single-domain GPFS cluster ever built, this design posed a huge challenge to SciNet. Network congestion, especially broadcast storms, brought down our GPFS cluster many times in the early days. IBM Canada's HPC team identified this issue and implemented static ARP tables in our system. This approach drastically reduced the network congestion, and GPFS stability was greatly improved.

In an effort to further reduce network traffic, SciNet reduced the use of Moab Service Manager (MSM) queries, which acquire information such as status and power utilization from each GPC node. This traffic goes through the management network, which shares the same physical wire as GPFS, and can therefore inadvertently affect the data network greatly. As mentioned above, disabling "green" mode helped reduce the load on the Ethernet network, thus increasing the stability of GPFS. As proposed by IBM Canada's HPC team, SciNet is currently investigating the possibility of improving the network, and hence GPFS performance, by separating management traffic from the data network through virtual LAN tagging.

Other factors that affect GPFS performance encountered at SciNet were:

- Time synchronization across the entire cluster.
- Critical GPFS parameter settings such as page-pool size and file system block size.
- Making sure that filesystems span a sufficiently large number of disk spindles in order to maximize both bandwidth and IOPs.
- Ensuring that all network equipment, such as NICs and switches, have up to date firmware that results in more stable operation.
- Aside from system manageability factors, GPFS is not immune to careless or misinformed users, who do not use the system efficiently. Educating users about the shared nature of the SciNet resources, particularly the filesystems, can weed out inefficient I/O practices, such as incurring in huge numbers of IOPs, reading and writing files too frequently and in small blocks, and creating millions of small files. At SciNet we are constantly monitoring the systems' CPU, memory, and I/O usage. We thereby identify issues and problem codes, and help users optimize their codes and I/O access patterns to improve their jobs' as well as the overall system's performance.

### *7.4. GPFS, Small File I/O, and Users*

GPFS is designed and tuned to be a high-performance parallel HPC filesystem; that is, it makes possible very high-bandwidth parallel access to disk. As with other parallel HPC file systems, it is designed for large block file I/O, and is not especially suited to sustaining large numbers of I/O Operations (IOP) that perform small reads and writes. This fairly straightforward and relatively unavoidable fact of computer systems engineering has caused the largest number of individual issues; not so much technical, but rather of an educational nature.

Some users, particularly those new to HPC, saw our systems as fairly familiar linux systems and thus believed they could do what they had been doing on their desktop — which, after all, worked[2] — and simply multiply that by 1000 jobs. We have seen users use $P^2$ files for communications between processes (writing "messages" from process $i$ to process $j$ in file $ij$, and using file permissions as a sort of semaphore system); we have seen users reading and writing individual floating point numbers to files from deep within their computational kernels. And of course, we have seen more mundane cases where each task writes a few hundred small (few kB) files, and then by multiplying this by 1000 or so jobs results in million-file directories or directory trees. In addition, users frequently write out real data as ASCII text, which is both typically written out in many small pieces, requires more disk space, and is significantly slower due to the cost of conversion to string.

Of course, if these practices only hurt the user's own performance it would not be a huge issue; however, unlike per-node memory or CPU usage, the filesystem is a shared resource, and the performance of the overall filesystem is extremely vulnerable to even a single user doing a lot of IOPs on even just a handful of nodes. Making this even more urgent in the first year of operation was that the large amount of excess network traffic from the few sources mentioned previously in this document, combined with heavy IOPs usage (on the GPC, all administrative network traffic goes over the same network as I/O – which, on the Ethernet GPC nodes, is also the MPI network) resulted in a filesystem which was far less stable than we would have liked with the older GPFS 3.2 series. The newer 3.3 series is significantly more stable, but the ability for a single user to starve out others for IOPs even from just a few nodes remains.

On a more mundane level, since everything on the two clusters shares a very small number of file systems, a few users with very many small files (and until the most recent series of user outreach campaigns, users with tens or hundreds of millions of files were not unusual) can enormously slow down important system processes which must traverse the file system file by file, such as backup, quota enforcement, and scratch purging. To help address that issue we recenty introduced strict quotas per user and per group on the amount of data and the number of inodes that are allowed on /scratch.

The above considerations have made very active and detailed system monitoring, so that particular users can be selected for focused and mandatory educational opportunities, crucial to ongoing operations. In some cases where little memory is used, user codes can be trivially modified to use ramdisk rather than local disk for high-IOP stages of computations, with final results being tar'ed and sent to GPFS in one operation. In other cases, code has to be significantly rewritten to perform well on shared filesystems (but this almost invariably results in improved operation even on desktop machines). However, one of our biggest users—the ATLAS collaboration—has code and workflow defined elsewhere and can not easily make such changes; for them, finding solutions was much more difficult.

### *7.5. Hierarchical Storage Management*

Implementing a hierarchical storage management scheme (HSM) is a pilot project started in July/2010 with a select group of users, and is still in progress.

What we would like to offer users is a way to offload/archive data from the most active file systems (scratch and project) without necessarily having to deal directly with the tape library or "tape commands".

[2] or so they believed

In the initial stages we explored how far we could go by just extending the functionality of our current backup system, a Tivoli Storage Manager installation composed of a TS3310 tape library with 396 slots and 4x LTO4 drives.

IBM manuals describe a fully integrated TSM/HSM solution, with "near online" access capabilities of data on tapes, and apparently perfectly suitable to our facility. What we quickly realized was that in practice it could take anywhere from a few minutes to many hours to either migrate or recall HSM files, depending on their size and their numbers. Most importantly, HSM recall operations run at the highest priority by design on the TSM system, and therefore we observed contention for access of the tape drives while regular backups were taking place. That made the "near online" idea of a "tape based" file system extension impractical from our perspective, at least in the sense that users should not expect to access HSM files from running jobs directly.

Nevertheless we didn't give up, and decided to further push the "archive" part of the idea, and we came up with a reasonable compromise: we devised a 2-step migration system and deployed a 15 TB transient file system called `/repository`, accessible from the datamover servers. On step 1, users may relocate data as required from the active file systems to `/repository` in a number of ways, such as copy, move, tar or rsync. "Transient" refers to the fact that `/repository` works like a "Black Hole": on step 2, `/repository` is constantly being emptied in the background (by the HSM daemons), even while users relocate data in from other file systems. What is left behind is the directory tree with the HSM stub files and the metadata associated with them. In this scenario, on step 2 users also have the option to manually migrate or recall files between `/repository` and the "tape system" with simple commands such as 'dsmmigrate' or 'dsmrecall'.

We also found that performance of this 2-step process was as much a function of the number of files as of the amount of data. Until we had worked diligently with our users to substantially reduce the number of small files on the active file systems we would not be able to proceed. Ideally we would like to see average file sizes greater than 100MB in `/repository`, and recommended that users stage their data ahead of time in large tar-balls.

At this point we still see the contention for tape drives, since we only have 4 units, and are considering a new tape library just for HSM purposes.

## 8. Large Hadron Collider Computing Grid and the ATLAS Experiment

The SciNet infrastructure provides the largest of the five ATLAS Tier-2 Analysis Centres in Canada (see reference [37] for a more complete description of the Worldwide Large Hadron Collider Computing Grid (WLCG) and the ATLAS-Canada Computing Model). The resources provided at SciNet equal those provided by WestGrid, which has its Tier-2 centres distributed across the University of Victoria, Simon Frasier University, and the University of Alberta. The sheer size of the SciNet Tier-2 centre and the design of the SciNet systems provide some unique challenges to deploying the WLCG middleware and the ATLAS software stack and operating the analysis centre successfully. There is a continuing effort to improve the efficiency of the site in addition to increasing its size to meet the commitments to the experiment. The support is provided by two dedicated FTE's.

The original ATLAS request for computational hours only represented a few percent of the total compute capacity of the ethernet-portion of the GPC. Therefore the grid middleware and ATLAS software could not completely dictate the design for the SciNet infrastructure. However, there was already extensive experience at the University of Toronto with running a Tier-2 centre on a dedicated high-energy physics compute cluster (with 220 dual single-core nodes and 30 TB of storage) which informed some of the RFP requirements and original design choices. We describe some of the original approaches attempted and various mitigation scenarios that were developed as experience was gained at the much larger scale of SciNet. The requested storage infrastructure for ATLAS always represented a significant fraction of the planned SciNet deployment and the

IBM response utilizing their DCS9000 series storage, as described in Section 4, provided a common solution for ATLAS and for the rest of the SciNet users.

### *8.1. Storage Resource Management and dCache*

The components making up the storage element (SE) and storage resource management (SRM) layers were dictated by the previous experience with the dCache filesystem [38] and the fact this was the infrastructure supported by the Canadian ATLAS Tier-1 centre based at TRIUMF [39]. As TRIUMF had already successfully deployed and was running a DDN-based solution provided by IBM, the SciNet ATLAS storage infrastructure offered in the RFP response from IBM was essentially modelled on this solution[3]. Deploying dCache on top of GPFS was not considered at the time as this presented additional layers of complexity which were unnecessary. Given the initial problems with stabilizing the GPFS operation with thousands of nodes accessing the storage via Ethernet, as described in Section 7, it would likely have proven disastrous for the first few months of the SciNet ATLAS Analysis centre. Some thought was given during the RFP process to using the STORM [40] system on top of GPFS (or parallel NFS), but this was always considered as a phase II approach rather than for initial deployment. Again, in hindsight, the data storage solution has been the most successful and stable component of the ATLAS components. The DDN platform with RAID6 deployed over the disk drawers in a silo and verify-on-read and write has proved very stable and almost maintenance free from the dCache perspective with no detected data loss or corruption in the first year of operation. There are not many single points of failure in the chain, although there have been enforced downtimes associated with single DDN controller failures to ensure that there was no chance of data corruption.

The current system provides about 530 TB for ATLAS and is scalable to $\sim$1 PB. Figure 10 shows the growth in the data stored at the SciNet ATLAS Tier-2 data centre since going online. About 5-7 TB of data can be moved a day through the 1 Gbps lightpath to TRIUMF (see Section 8.5).

### *8.2. WLCG Middleware and ATLAS Software*

In the initial deployment of the Tier-2 centre in 2009, the logical choice was to install the grid middleware and ATLAS software on the GPFS filesystem designated `/project`. After much testing in the late fall and early winter it was proven that the performance on GPFS, particularly for the ATLAS software area, was abysmal. This was essentially due to the fact that an ATLAS software release (of which there are dozens) consists of a few GB of data but 100k-1M files, many of which are very small (few kB). Software installation jobs, that are initiated centrally from an ATLAS centralized server at INFN (Italy), would take 2-3 times longer to run at SciNet than at other Canadian Tier-2 sites on similar nodes. Running large numbers of ATLAS analysis jobs also demonstrated that read performance would not scale as many files would be kept open during processing and small amounts of reads would occur throughout the data processing. Increasing the GPFS page pool size, as discussed later, did improve matters, but not dramatically.

A major drawback of the centralized ATLAS software deployment system is that installation of the complete set of releases can take several weeks to accomplish, especially during the initial phase of operation where not all services are robust or hardened. ATLAS software is dependent on `pacman` [41] which is extraordinarily sensitive to the absolute pathname. This installation

[3] The dCache storage is provided by four x3650 dual-cpu 8 GB RAM 'pool' nodes that connect on the backend via a fibre-channel switch to the DDN controllers and connect to clients via the SciNet core Myricom Myri-10G switch. The dCache administration level is provided by four x3550 single-cpu nodes that provide the databases for the filesystem and other services plus an additional four x3550 single-cpu nodes that provide the 'door' services for external transfers via protocols such as GridFTP and GSIDCAP

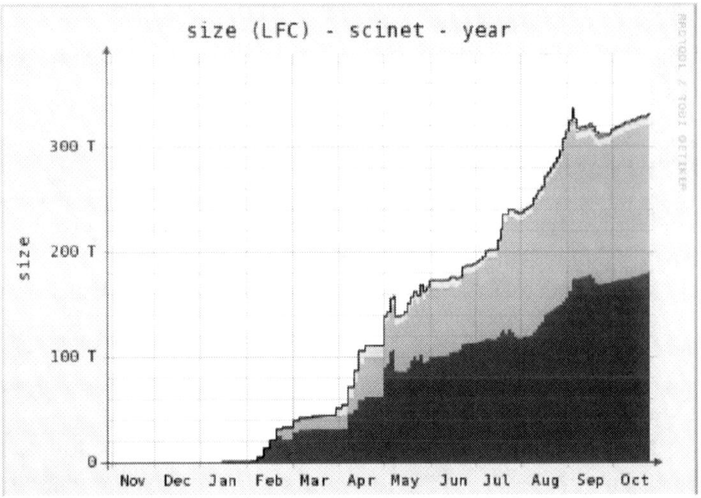


**Figure 10.** Data storage growth at the SciNet ATLAS Tier-2 Data Centre. Tan represents the event summary data and blue represents the reduced analysis data formats. Older reprocessed datasets are deleted when newer data becomes available.

process was repeated a couple of times in the first few months of operations, once as the paths for the software were changed after deployment had begun and disk configurations were changed and when a second compute element was put online to load balance the job management from the grid.

In early 2010, an NFS server was installed on one of the dCache pool nodes, utilizing one 16 TB tier from the DDN storage as the backend storage. This server proved to scale extremely well, up to 1000's of simultaneous Monte-Carlo simulation jobs during a large production run during the summer. Since then, the DDN tier has been replaced with a RAID0 array constructed from four 146 GB SAS drives. Testing is underway as well on a RAID0 with four 250 GB SATA drives. Very good performance is seen as the NFS caching on the server side scales very well for the hundreds of clients reading essentially the same files as compared to the caching that needed to be done on the client side on individual nodes when GPFS is used.

As discussed in Section 7.4, even backing up the ATLAS software on GPFS was an issue due to the millions of files which created long traversal times for the backup software. Instead, backups are made on NFS servers and tarballs are copied to GPFS areas for storage to tape for long term archiving.

### *8.3. ATLAS HEP Compute Nodes and GPFS*

As discussed in Section 7.1, the GPC compute nodes are diskless and are managed by xCAT and Moab and boot via PXE. As there is only 16 GB RAM per node, keeping the OS and software layers in the deployed image to a bare minimum was deemed essential to efficient use of the systems. The original image deployed under CentOS 5.3 was about 275 MB and has stayed roughly at this size, with modules provided from the GPFS `/project` space adding the necessary software packages and compilers for most SciNet user codes and applications.

It was always understood that the HEP applications, in particular the grid and ATLAS codes, would require substantially more in their compute image than the "standard" vanilla compute image based on CentOS 5.x. In particular many standard rpms usually found on workstations are expected by the ATLAS pilot and Athena code distribution [42], in addition to some legacy rpms to support the 32-bit builds that were still being used in early 2010. A HEP image was designed based on the standard SciNet compute image, with additional packages originally adding about 300 MB to the footprint. The hope was that these images might converge in time, but a number of factors make this unlikely. The number of required packages for successful operation of grid jobs, though finite, has caused the image to grow to about 1 GB. Another major difference in the requirements is that the HEP jobs need external network connectivity to operate with the Grid and with ATLAS job and data management layers. This connectivity operates through a NAT node which also serves as the main data mover node for SciNet (via 10 Gbps Ethernet). To date, no other user group outside of HEP has needed or requested external connectivity for the compute nodes.

An additional requirement from ATLAS that has yet to be implemented fully at SciNet is that each running job should have access to 2 GB of physical RAM and 3 GB of virtual memory, usually provided by swap. This requirement for ATLAS sites was only requested by the experiment in the last year, so was not in the original design or RFP. Various schemes are being tested to see if providing swap space to the kernel over either GPFS or on an NFS server is practical. The concern with any swap space provided over a networked file systems is that the node performance could be severely compromised, so this needs to be studied before going into large scale deployment. To date, very few jobs have been lost due to lack of memory on the node, but this has occurred. The jobs that tend to have these large footprints are user analysis jobs, rather than the more stable and better controlled simulation production jobs. Some of these failures have been critical as the node would run out of memory and start killing essential services e.g. the GPFS daemon which has a large footprint. Various mitigation techniques have been tried with limited success in telling the OS and Out-Of-Memory killer to sacrifice user applications first, but this does not always help.

One technique in the short term that could help if the available memory becomes a major issue is to limit the number of concurrent jobs on a node. Moab can manage the available memory as a resource to allow nodes to be underfilled. This is not desirable long term or for all jobs as this would be an inefficient use of the available cores on a node, so some tailoring based on which jobs have the large memory footprint would be useful. We have torque-submit helper scripts that can request the necessary resources on the fly, but there is no current mechanism to predict which jobs from the ATLAS job schedulers have these requirements, so this is a manual process at the moment.

Another requirement from the experiment that has doubled in the last year is that ATLAS now requests the availability of 20 GB of scratch disk space per running job. This assumes the potential for up to 14 GB for input files, 5 GB for output and 1 GB for logs etc. The SciNet GPFS `/scratch` space has been used to provide the pseudo-user account home and scratch directories, but not without occasional hiccups. As discussed in Section 7.4, too many files or subdirectories in a directory can prove very inefficient; even simple operations such as listing the files in a directory can sometimes hang for many minutes. After a few months of operation, the `/scratch` space was split off from the `/home` areas for the pseudo-user accounts under which the grid jobs run. Each node now manages its own subdirectory for `/scratch`, the intent being the GPFS file token management for a particular area is limited to a single node. This approach seems to be effective in limiting some of the large wait times that were occasionally seen.

Another approach being investigated is to provide the local scratch for a node with a loopback device, i.e. a large block file is created on GPFS and then mounted on the node with an ext-2 or ext-3 file system built on top of the loopback device. The hope here is to present GPFS with a single file and space-token to manage and all other file operations would be kept local to the node. Given the very variable job load and file access patterns that continually run at SciNet, making definitive statements about improvements or lack thereof from configuration changes takes time to investigate. Some of these approaches were attempted in the early days of the ATLAS deployment on the cluster, but providing space via iSCSI targets or loopback devices did not prove helpful to throughput performance. However, as the SciNet systems' usage has grown, other issues have emerged with the large number of clients in the GPFS cluster, so minimizing the ATLAS contribution to the pounding of the file systems in a desirable goal.

One major improvement that was discovered in the early days of running while trying to scale the ATLAS athena code to hundreds of concurrent processes was the GPFS pagepool size. To limit the use of physical RAM on a node, this parameter, which controls the caching available to the local GPFS daemon, was originally set to 256 MB in the compute image during the cluster deployment. Extensive testing and detailed measurements of up to 8 analysis jobs running on a node demonstrated that long run-times were being generated by occasional pauses in the GPFS file system feeding data to running processes. These pauses could last seconds to even minutes occasionally. Increasing the pagepool size steadily from 256 MB to 8096 MB demonstrated that the "sweet" spot for ATLAS applications (at least in this test) was around 1 GB and the gains were marginal for anything larger. This large efficiency gain was shown to hold with many nodes running concurrently, though there still is the expected performance impact to having hundreds of jobs simultaneously accessing the same GPFS file system. For a while, the HEP image was deployed with 1 GB and the generic SciNet image was run with 512 MB, but as updating the pagepool size dynamically meant transmitting this information to all other 4,000 nodes, this proved inefficient for node startup times and general GPFS stability, monitoring and performance. All images now use the 1 GB pagepool size.

### *8.4. ATLAS Job Scheduling*

From the outset, SciNet planned to schedule both the GPC and TCS via node and not by core. Grid jobs however need to be scheduled by core, so special routing queues were set up in Moab and Torque to accept serial jobs scheduled through the grid globus job managers. There were various issues that needed to be resolved to ensure these were used efficiently, including some debugging of the Moab software that could cause occasional overprovisioning due to a race condition.

ATLAS requests 48 hours of walltime on modern CPUs for job completion. This synchronizes well with the SciNet scheme as no user job gets more than 48 hours of wallclock time. Since green provisioning has been turned off for the last few months, scheduling these is not a major issue, as these serial jobs only go to the serial queue which can only run on the nodes provisioned with the HEP image. However, we anticipate that green running may be desirable in the future, and tailoring the request time of the grid jobs to be more commensurate with the actual running time would benefit them, as Moab can back-fill schedule node usage much better during the transition times when nodes are being switched from one flavour of image to another. The semantics for providing these job times do not exist in the current WLCG setup; the only way to implement something on this scale currently would be to advertise different length queues to the grid, but no job-matching in the work management schedulers makes use of this information to the authors' knowledge.

We cap the total number of concurrent jobs to protect the storage servers and to meet our commitments to the ATLAS collaboration. These caps have been extended on occasion to encompass special processing requests. In particular, a very large production of simulated data[4] was run over a few weeks during the spring for the jet energy scale calibrations and systematic error studies that were vital in getting some of the first data published from the LHC which set new limits for the production of new particles [27–29]. During this production we ran up to 4,000 ATLAS jobs simultaneously.

To accommodate the large number of jobs, we quickly realized that a single compute element was insufficient to handle the job load from the grid without introducing a bottleneck, especially

[4] Over 3 million events were generated in 60,000+ jobs consuming about 360,000 cpu-hours.

as there tended to be slightly bizarre interactions between the local schedulers and the factories producing the pilot jobs that were sent to the site[5]. We deployed a second compute element which took a little time to configure as both compute elements were scheduling jobs to the same pseudo-user accounts and to the same compute nodes and had to share the same "view" of the universe. Most other large sites that have dedicated resources tend to split their clusters and queues amongst their compute elements eliminating some of the possible confusions and configuration issues. This was not an option at SciNet as we needed to be able to share and steal as many free slots as we could to accomplish the production quickly to meet the deadline of a workshop. This dual compute element configuration has proved useful long term, giving us a level of redundancy. To date, the SciNet Tier-2 has run over 700,000 ATLAS jobs utilizing about 3.5 million cpu-hours.

As mentioned earlier, green provisioning proved to generate a load on GPFS that was deemed unacceptable. In addition, transitioning between the HEP and standard image via dynamic provisioning was contributing to some of the issues and has since been disabled. The length of the job queues from the ATLAS work load management systems for both the simulation and user analysis schedulers does vary frequently albeit with fairly long periodicity. Without dynamic provisioning, daily manual checks are required to ensure that a suitable number of HEP images are deployed and are not sitting idle. Measuring the local idle job queue is not sufficient as the pilot factories are designed to always keep a certain number of jobs queued even if they are not destined to receive a payload.

### *8.5. Optical lightpath to TRIUMF*

It was decided during the initial RFP phases and deployment of the data centre that the simplest wide-area network configuration was to utilize the University of Toronto core network. SciNet appears as just another department in the `utoronto.ca` domain. To accommodate a 1 Gbps point-to-point lightpath connection from the datacentre to TRIUMF, VLAN tagging needed to be turned on to allow routing of traffic on the internal Layer-2 network from the LCG nodes to a separate VLAN which connected to the CANARIE[6]-provided fibre that is hosted on the SciNet edge switch before it hits the main campus router. This was fairly simple to accomplish. Upgrading this link to a full 10 Gbps in the future may require lighting up further wavelengths and providing additional ports on the main Force 10 edge switch at the data centre.

## 9. Issues of Scale

A variety of other issues arose over SciNet's first year of operation that were caused by system software simply not scaling up to the size of the GPC.

Very few pieces of software, even HPC-specific tools, come "out of the box" ready and tuned to be run on almost 4000 nodes. One of our first experiences with this was our scheduler, where compiled-in limits for maximum number of jobs were far too small, 32-bit integers overflowed, and communications between the resource manager and the nodes was happening far too frequently on a 4000 node system, flooding the network. However, we had an excellent working relationship with our scheduler vendor (Cluster Resources, now Adaptive Computing) and patches were made for us nearly immediately, and fed back into the shipping product.

Similarly, MPI libraries—even commercial MPI libraries—were completely unable to launch several-thousand-task jobs (a) reliably, and (b) in a reasonable amount of time, sometimes taking literally hours before mpirun and/or `MPI_Init()` would complete. OpenMPI, which had excellent Torque/PBS integration, did somewhat better in this respect using the `tm` interface

[5] ATLAS jobs are run via pilot jobs that arrive at the site and ensure a healthy environment before pulling a payload with the real job from centralized PanDA databases [43]. If no payload is found, the pilots self-terminate after a couple of minutes.

[6] Canada's Advanced Research and Innovation Network [44]

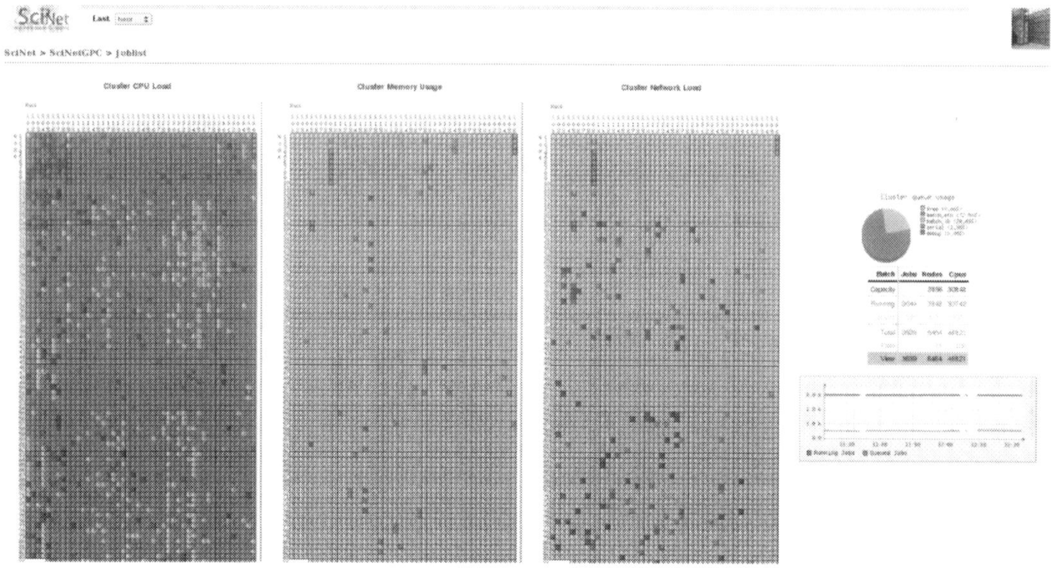


**Figure 11.** An experimental GPC-wide dashboard using Ganglia and Job-Monarch.

to start jobs, but it was still unusable for thousands of tasks. It was only with the advent of the hydra process manager in MPICH2, which is now being introduced into the newest version of IntelMPI, that we can routinely start jobs of this scale. MPI issues remain in some cases – running large, all-to-all dominated jobs (e.g., FFTs) on Infiniband, something like memory pinning requires too much memory and jobs can fail; running the newest OFED stack ameliorates this but does not eliminate the problem.

On any system, monitoring is crucial, and with the shared file systems monitoring the behaviour of user jobs is particularly urgent. But no dashboard-type monitoring systems run easily out of the box on our large system, as the monitoring has to be very light-weight and have a very small network footprint. Our sysadmin team has developed an experimental GPC dashboard (Fig 11) based on logging infrastructure developed in house, which is already enormously helpful in giving us an 'at a glance' look at the behaviour of the entire system and the jobs on it, allowing us to find issues before they become problems.

I/O continues to come up as an issue in sometimes surprising ways. Beyond continuously working to improve users' jobs, both for their sake and for those of other users, smaller surprises emerge; for instance, our original file system (`/scinet`) containing frequently used libraries (fft, mpi, etc) and applications, did not require a great deal of space, so we originally had it on a very small filesystem containing only a few disks; however, because it was being repeatedly accessed by almost all jobs on almost all nodes, it had to be placed on a much larger filesystem so that it could be striped over many spindles to give us the bandwidth and IOPs we needed, or else it would cause a bottleneck for jobs starting up.

## 10. Conclusion

To borrow a line from complex systems research, "more is different". Building, operating, and running code on a system of this scale is very different than dealing with smaller systems. By maximizing flexibility, and having a small agile team of large-scale computing specialists working closely with vendors, the SciNet experience has been a success and continues to be a challenge and a learning opportunity.

The SciNet systems have enabled computational research at a scale unprecedented within Canada. From January 2010 to the end of October, researchers have used approximately 200 million CPU hours on SciNet systems to perform simulations and data analysis, pushing forward the fields of biomedicine, climate research, astrophysics, aerospace, chemistry, high energy physics, and many other research areas. We hope the lessons contained within this document help the designers and maintainers of future systems reach this level of productivity even faster.

**Acknowledgments**
SciNet is funded by: the Canada Foundation for Innovation under the auspices of Compute Canada; the Government of Ontario; Ontario Research Fund - Research Excellence; and the University of Toronto.

## Appendix A. Temperature data for Toronto

Table A1 gives temperature data for Toronto.

**Table A1.** ASHRAE Toronto weather bin data.

| Average bin temp (°F) | OAW (gr/lb) | OAH (BTU/lb) | Bins per time period 12am-8am | 8am-4pm | 4pm-12am | Total Bin Hours |
|---|---|---|---|---|---|---|
| -32 | | | 0 | 0 | 0 | 0 |
| -27 | | | 0 | 0 | 0 | 0 |
| -22 | | | 0 | 0 | 0 | 0 |
| -18 | | | 0 | 0 | 0 | 0 |
| -13 | | | 0 | 0 | 0 | 0 |
| -8 | -9.0 | -3.3 | 1 | 0 | 0 | 1 |
| -3 | -7.0 | -1.8 | 6 | 0 | 0 | 6 |
| 2 | 1.0 | 0.6 | 39 | 13 | 15 | 67 |
| 7 | 7.0 | 2.7 | 57 | 29 | 52 | 138 |
| 12 | 11.0 | 4.6 | 110 | 83 | 81 | 274 |
| 17 | 16.0 | 6.5 | 151 | 84 | 131 | 366 |
| 22 | 21.0 | 8.5 | 177 | 151 | 135 | 463 |
| 27 | 26.0 | 10.5 | 257 | 224 | 226 | 707 |
| 32 | 30.0 | 12.3 | 330 | 213 | 246 | 789 |
| 37 | 35.0 | 14.3 | 247 | 234 | 260 | 741 |
| 42 | 39.0 | 16.1 | 202 | 185 | 242 | 629 |
| 47 | 43.0 | 17.9 | 226 | 212 | 186 | 624 |
| 52 | 49.0 | 20.1 | 286 | 202 | 236 | 724 |
| 57 | 53.0 | 21.9 | 310 | 180 | 259 | 749 |
| 62 | 57.0 | 23.7 | 219 | 201 | 249 | 669 |
| 67 | 61.0 | 25.6 | 184 | 217 | 217 | 618 |
| 72 | 65.0 | 27.4 | 98 | 277 | 231 | 606 |
| 77 | 67.0 | 29.0 | 19 | 238 | 95 | 352 |
| 82 | 70.0 | 30.7 | 1 | 122 | 45 | 168 |
| 87 | 73.0 | 32.3 | 0 | 45 | 12 | 57 |
| 92 | 77.0 | 34.2 | 0 | 10 | 2 | 12 |
| Total | | | | | | 8760 |

Journal of Physics: Conference Series **256** (2010) 012026 doi:10.1088/1742-6596/256/1/012026

## Appendix B. SciNet by the Numbers

Table B1 gives some interesting (to us!) statistics about our systems.

**Table B1.** SciNet by the numbers.

| | |
|---|---|
| Number of jobs run so far | 3,407,060 |
| File systems (usable size) | |
| home | 14 TB |
| project | 364 TB |
| scratch | 466 TB |
| dCache | 536 TB |
| Peak number of files | |
| `/scratch` | 320 million |
| `/project` | 110 million |
| Current number of files | |
| `/scratch` | 40 million |
| `/project` | 20 million |
| `/scratch` space used | |
| Peak | 461 TB |
| Typical | 410 TB $\pm$ 23 |
| `/scratch` Purging cycle | not accessed in 90 days |
| average # of files/directory | 100 |
| current file type distribution | |
| binary | 90% |
| ascii | 9% |
| generic | 1% |
| Support emails received | 15,640 |
| RAM DIMMs on GPC | $\sim$ 30,000 |
| DIMM failures | 150 in the last six months |
| Slots in tape library | 396 |
| Backed up data | 294 TB over 310 LTO4 tapes |

## References

[1] Lovelace G, Scheel M and Szilagyi B 2010 *Preprint arXiv:1010.2777*

[2] Alvarez M and Abel T 2010 *Preprint arXiv:1003.6132*

[3] Pfrommer C and Dursi L 2010 *Nature Physics*

[4] Duez M, Foucart F, Kidder L, Ott C and Teukolsky S 2010 *Classical and Quantum Gravity* **27** 114106

[5] Pang B, Pen U, Perrone M, Treibig J, Wellein G, Hager G, Altisen K, Liu Y, Moy M, Wang Y *et al. Architecture* **163** 643

[6] Chluba J, Vasil G and Dursi L *Monthly Notices of the Royal Astronomical Society*

[7] Fowler J, Acquaviva V, Ade P, Aguirre P, Amiri M, Appel J, Barrientos L, Battistelli E, Bond J, Brown B *et al.* 2010 *The Astrophysical Journal* **722** 1148

[8] Foucart F, Duez M, Kidder L and Teukolsky S 2010 *Preprint arXiv:1007.4203*

[9] Battaglia N, Bond J R, Pfrommer C, Sievers J L and Sijacki D 2010 *Preprint arXiv:1003.4256*

[10] Marriage T A *et al.* 2010 *Preprint arXiv:1010.1065*

[11] Sehgal N *et al.* 2010 *Preprint arXiv:1010.1025*

[12] Dunkley J *et al.* 2010 *Preprint arXiv:1009.0866*

[13] Das S *et al.* 2010 *Preprint arXiv:1009.0847*

[14] Hajian A *et al.* 2010 *Preprint arXiv:1009.0777*

[15] Marriage T A *et al.* 2010 *Preprint arXiv:1007.5256*

[16] Menanteau F *et al.* 2010 *Preprint arXiv:1006.5126*

[17] Chluba J and Thomas R 2010 *Preprint arXiv:1010.3631*

[18] Ivan L, Northrup S A, Groth C P T and De Sterck H 2010 *Proceedings of the 18th Annual Conference of the CFD Society of Canada*

[19] Saffaripour M, Zabeti P, Dworkin S, Zhang Q, Guo H, Liu F, Smallwood G and Thomson M 2010 *Proceedings of the Combustion Institute*

[20] Northrup S A and Groth C P T 2010 *Proceedings of the 18th Annual Conference of the CFD Society of Canada*

[21] Jha P K and Groth C P T 2010 *Proceedings of the Combustion Institute/Canadian Section Spring Technical Meeting*

[22] Groth C P T, Lin W, Hernández-Pérez F E and Gülder Ö L 2010 *Proc. of the Combustion Inst./Canadian Section Spring Technical Meeting*

[23] Jha P K and Groth C P T 2010 *Proc. of the 18th Annual Conf. of the CFD Society of Canada*

[24] Hernández-Pérez F E, Yuen F T C, Groth C P T and Gülder Ö L 2010 *Proc. of the Combustion Inst.* vol 33

[25] Charest M R J, Joo H I, Gülder Ö L and Groth C P T 2010 *Proc. of the Combustion Inst.* vol 33

[26] Zhou Y and Segal D 2010 *The Journal of Chemical Physics* **133** 094101

[27] Aad G *et al.* (ATLAS) 2010 *Phys. Rev. Lett.* **105** 161801 (*Preprint* 1008.2461)

[28] Aad G *et al.* (ATLAS) 2010 *Preprint arXiv:1009.5069*

[29] Aad G *et al.* (ATLAS) 2010 *Preprint arXiv:1009.5908*

[30] Aad G *et al.* (ATLAS) 2010 *Preprint arXiv:1010.2130*

[31] Dursi L J, Bohlender D, Wadsley J and Kavelaars J J 2010 *White Papers for the 2010 Canadian Astronomy Long Range Plan* URL http://www.casca.ca/lrp2010/Docs/LRPReports/CDandN_WP.pdf

[32] URL http://www-03.ibm.com/systems/software/gpfs/

[33] URL http://www.clusterresources.com/products/moab-cluster-suite/workload-manager.php

[34] URL http://xcat.sourceforge.net/

[35] IBM 2009 IBM Rear Door Heat eXchanger for the iDataPlex Rack Installation and Maintenance Guide URL http://publib.boulder.ibm.com/infocenter/idataplx/documentation/topic/com.ibm.idataplex.doc/dg1dimst.pdf

[36] IBM 2007 IBM Canada's innovative water cooling project recognized for innovation and energy efficiency URL http://www.ibm.com/ibm/environment/news/enerstat_2007.shtml

[37] ATLAS Canada Collaboration 2006 URL http://atlas-canada.web.cern.ch/atlas-canada/documents/AtCanComp-PublicV1.pdf

[38] URL http://www.dcache.org

[39] URL http://atlas.triumf.ca

[40] URL http://storm.forge.cnaf.infn.it

[41] URL http://www.archlinux.org/pacman

[42] 2005 *The Athena Control framework in production, new developments and lessons learned*

[43] Nilsson P *et al.* (ATLAS) 2008 *PoS* **ACAT08** 027

[44] URL http://www.canarie.ca

# AUTHOR INDEX

**Institute of Physics**
Dirac House, Temple Back
Bristol BS1 6BE UK

CURRAN ASSOCIATES INC
proceedings.com